DIFFUSION

FORMALISM AND APPLICATIONS

DIFFUSION

FORMALISM AND APPLICATIONS

SUSHANTA DATTAGUPTA

CRC Press
Taylor & Francis Group
Boca Raton London New York

CRC Press is an imprint of the
Taylor & Francis Group, an **informa** business
A CHAPMAN & HALL BOOK

CRC Press
Taylor & Francis Group
6000 Broken Sound Parkway NW, Suite 300
Boca Raton, FL 33487-2742

First issued in paperback 2019

ISBN-13: 978-1-4398-9557-3 (hbk)
ISBN-13: 978-0-367-37924-7 (pbk)

Visit the Taylor & Francis Web site at
http://www.taylorandfrancis.com

and the CRC Press Web site at
http://www.crcpress.com

Contents

Preface

This book covers more than two centuries of the history of diffusion. What started from the French school of mathematical physicists, primarily Fourier and Laplace, who introduced us to two distinct paradigms of diffusion—physical and stochastic—diffusion is a topic that still occupies our space with its various ramifications. It is a subject that intertwines mathematics with observed phenomena, illustrated by Brown's findings of the random zigzag motions of pollen particles in a tube of water kept at a fixed temperature. Thus almost a century after Fourier and Laplace talked about physical and stochastic diffusion, Einstein unified the two disparate concepts through his path-breaking work on Brownian motion. Almost simultaneously, the subject gave birth to an independent branch of mathematics—probability theory—through the work on Wiener processes, Polya's random walk ideas, Gauss–Ornstein–Uhlenbeck processes, dynamical semi-group of the Markovian type, and so on. The subject has also given rise to another branch of mathematics called stochastic calculus, as one transits back and forth from Langevin to the Fokker–Planck approaches to diffusion. Diffusion in a structurally disordered medium, which leads to what is called anomalous diffusion, is another important concept that has gripped the attention of physicists and mathematicians alike.

It is remarkable that what may appear to be a mathematically abstruse subject has had so much impact on materials science. Thus defect-diffusion, whether of point defects such as interstitials or vacancies, or extended defects such as grain boundaries and dislocations, has had enormous influence on topics related to general concepts and radiation damage, spinodal decomposition, and other metallurgy-related topics, in particular.

Spectroscopy, or what light through its interaction with matter can reveal about the properties of the latter, is another domain in which diffusion plays a ubiquitous role. Whether it is Moessbauer or neutron spectroscopy of nuclei or Raman and infrared spectroscopy of molecules, diffusion, via its avatars of translational or rotational diffusion, leaves a deep imprint on experimental data.

Diffusion and cooperative interactions are another area of contemporary interest in the chemistry and physics of materials and biology. Thus, reaction–diffusion equations are at the core of pattern formation, morphogenesis, and fractal colony growth of biological species. When it comes to biological systems, diffusion is the central issue in molecular motors and intracellular transport.

Finally, the question boils down to the subatomic arena where one asks what happens to an elementary particle such as an electron or a proton that undergoes diffusive motion. Their dynamic behavior, in interaction with their surroundings, leads to dissipative motion, which invariably necessitates the investigation of diffusion in the quantum domain. Therefore, that study directly impacts the contemporary subjects of dissipative tunneling, Landau diamagnetism, coherence-to-decoherence transition, quantum information processes, electron localization, and so on.

This book is an attempt to cover all these subjects in, what I hope, is a coherent weaving together of a variety of topics that have occupied the attention of some of the great minds of physics, mathematics, and chemistry. The objective is to familiarize the reader with the necessary formalism and to expose her (or him) to a glimpse of possible applications to practical issues. I hope that the suggested applications will be representative enough to allow the reader to venture into newer terrains not covered in this book.

Sushanta Dattagupta
Visva-Bharati, Santiniketan

Acknowledgments

This book is dedicated in honor of the 80th birthday of my guru Martin Blume, who introduced me to the fascinating world of diffusion, and in particular, to its application to spectroscopy. As a young graduate student coming to America in the 1970s from an Indian hierarchical system, Blume's unobtrusive style of doctoral supervision was a revelation. He gave me adequate space to become an independent physicist, yet was always available to suggest new references to the subject and to provide friendly support in times of need.

Working as a research student at the Brookhaven National Laboratory also opened my eyes to the importance of talking to experimental physicists about spectroscopy and scattering methods although the topic of my work was theoretical. That training held me in good stead for collaborative work with both experimentalists and theorists over the years. In addition to Marty Blume, I therefore owe a great deal of gratitude to the following colleagues with whom I have worked on different topics of diffusion: Girish Agarwal, Amnon Aharony, V. Balakrishnan, Ora Entin-Wohlmann, Gautam Gangopadhyay, Rupamanjari Ghosh, Shmuel Gurevitz, Sanjay Puri, M. Sanjay Kumar, Subir Sarkar, Herbert Schober, Kurt Schroeder, Subhasis Sinha, Ajay Sood, P. A Sreeram, Lukasz Turski, and Gero Vogl. Papers written with them figure in this book, as do published work with my doctoral and master's students: Malay Bandyopadhyay, Ebad Kamil, Jishad Kumar, Manas Roy, Jagmeet Singh, and Tabish Qureshi. I am deeply thankful to all of them.

The J. C. Bose Fellowship of the Department of Science and Technology in the Government of India was a great help in supporting this project. I would like to thank Vivekananda Maji for inputting the manuscript and my Kala Bhavan colleague of Visva Bharati, Ashok Bhowmik, for his help in designing the cover.

About the Author

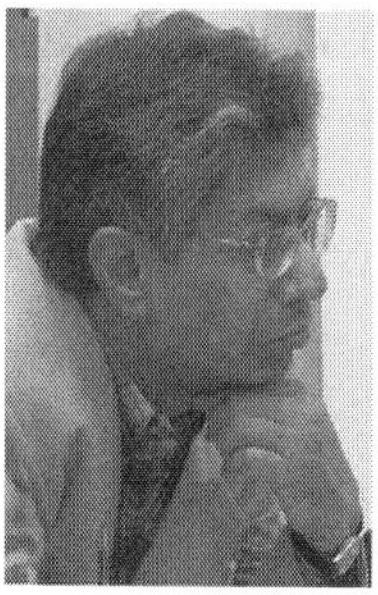

Sushanta Dattagupta is currently the vice chancellor of Visva Bharati, Santiniketan, a university founded in 1921 by Asia's first Nobel Laureate in Literature, Rabindranath Tagore. Following university education in Calcutta and a year's stint as a lecturer in physics at Presidency College, Dr. Dattagupta carried out doctoral and postdoctoral work at Brookhaven National Laboratory and Carnegie Mellon University, respectively. He returned to India in 1976 to begin a career as a scientist at the Reactor Research Center in Kalpakkam (1976–1981), and went on to serve the Central University of Hyderabad (1981–1986), Jawaharlal Nehru University (JNU) in New Delhi (1986–1999), S. N. Bose National Center for Basic Sciences in Kolkata (1999–2005), and the Indian Institute of Science Education and Research (IISER) in Kolkata (2006–2011), before moving to Viswa Bharati in September 2011.

Aside from publishing nearly 150 papers in international physics journals as well as three books, *Relaxation Phenomena in Condensed Matter Physics* (Academic Press, 1987), *Dissipative Phenomena in Condensed Matter* (coauthored with Sanjay Puri; Springer-Verlag, 2004), and *A Paradigm Called Magnetism* (World Scientific, 2008), and serving both research centers and universities, Dr. Dattagupta has been involved in institution building, as dean of the newly founded School of Physical Sciences in JNU, director of the Bose Center and founder–director of IISER. He is now deeply committed to nurturing holistic and boundaryless education.

Dr. Dattagupta is internationally acclaimed and has held visiting positions in Germany, Italy, the United Kingdom, the United States, and Israel. He has been a fellow and vice president of all three of the major science academies of India, and is also a fellow of the World Academy of Science, located in Trieste.

Section I

Classical Diffusion

1

Introduction to Brownian Motion

1.1 Introductory Remarks

Diffusion pervades all walks of life. It impacts every aspect of basic sciences—biology, chemistry, geology, and physics, not to mention mathematics. Inevitably, technological applications to blood circulation, aerosols, powders, pollutants, colloidal suspensions, oil exploration and myriad other areas have also greatly benefited from studies of diffusion.

Starting from Fourier's analysis of what we may call physical diffusion by observing heat propagation in a rod (Fourier 1807) more than two centuries ago (1807–1811), diffusion has undergone a dichotomous evolution into physical and stochastic diffusion (Narasimhan 2009, 2010). The stochastic theme was pioneered by another French mathematician, Laplace (1809), around the time Fourier's treatise was published. Laplace's work led to the celebrated random walk and central limit theorems. The following sections will review historical developments, culminated by Einstein's contributions (1905, 1906) which, in a sense, unified physical and stochastic diffusion, albeit in the context of Brownian motion (Brown 1828, 1829).

1.2 Fourier Equation

Fourier was interested in determining how the temperature T of a one-dimensional rod changes with position x and time t. He viewed a solid as a continuous medium in which heat flows by the route of conduction. Assuming that radiation loss can be ignored and that the rate of heat transfer between two points on the rod is proportional to their temperature difference and inversely proportional to their distance, he wrote the following equation:

$$\rho c \frac{\partial T}{\partial t}(x,t) = \mathcal{K} \frac{\partial^2 T}{\partial x^2}(x,t). \tag{1.1}$$

In the equation, ρ, c and $\mathcal{K}$ are physical parameters designating mass density, specific heat, and thermal conductivity of the solid, respectively. Fourier consciously ignored radiation leak and the weak temperature dependence of c and $\mathcal{K}$ in order to reduce the equation to a tractable, linear form. By combining the physical coefficients in the form:

$$D = \frac{\mathcal{K}}{\rho c}, \tag{1.2}$$

where D is the thermal diffusivity, Equation (1.1) can be cast into the familiar form of a diffusion equation:

$$\frac{\partial T}{\partial t}(x,t) = D\frac{\partial^2 T}{\partial x^2}(x,t). \tag{1.3}$$

In his 1807 monograph, Fourier also introduced the three-dimensional Cartesian version of Equation (1.3):

$$\frac{\partial T}{\partial t}(x,y,z;t) = D\nabla^2 T(x,y,z;t), \tag{1.4}$$

where ∇ was subsequently called the Laplacian. Equation (1.4) has the structure of a parabolic partial differential equation. While it presupposes isotropic diffusion, later years have seen generalization to anisotropic diffusion wherein the diffusivity is a tensor.

The solution of Equation (1.3) [or Equation (1.4)] is totally dependent on the nature of the boundary condition. Assuming initially that at $t = 0$, the temperature is localized at the arbitrary point x of the one-dimensional rod (that can be guaranteed by a focused heat source at $x = 0$), we have

$$T(x,t=0) = \bar{T}_0\delta(x - x_0), \tag{1.5}$$

where $\bar{T}_0$ is a constant with a dimension of temperature times length and δ is the Dirac delta function. Further assuming that the temperature falls to zero at either side of an infinitely long rod, the boundary condition reads

$$T(x = \pm\infty, t) = 0. \tag{1.6}$$

Accordingly, the solution of Equation (1.3) can be written as

$$T(x,t) = \frac{\bar{T}_0}{(4\pi Dt)^{1/2}}\exp\left[\frac{-(x-x_0)^2}{4Dx}\right]. \tag{1.7}$$

On the other hand, if one considers a finite rod of length L and adopts the reasonable boundary condition that the temperature does not leak out at the boundary, i.e., its gradient is zero, we have

$$\frac{\delta}{\delta x}T(x,t)\Big|_{x=0,L}=0. \tag{1.8}$$

In that case, and with the same initial condition as in Equation (1.5), the solution of Equation (1.3) reads

$$T(x,t)=\bar{T}_0\frac{1}{L}\sum_{n=-\infty}^{\infty}\exp\left(-\frac{\pi^2 n^2 Dt}{L^2}\right)\cos\left(\frac{n\pi x}{L}\right)\cos\left(\frac{n\pi x_0}{L}\right). \tag{1.9}$$

In his earlier work, Fourier was interested in heat propagation in bounded domains only and therefore devised solutions that were trigonometric functions, as in Equation (1.9). Because the convergence of geometric series was not well established then, Fourier's monograph submitted to the French Academy of Science was rejected by the referees, two of whom were Lagrange and Laplace. Interestingly, even the notion of function was not well founded in the early 19th century.

It is worth noting that the temperature at a point (now on a three dimensional rod) is related to the quantity of heat ∇H stored in an elementary volume ∇V around the point, and is given by

$$T=\frac{1}{\rho c}\frac{\nabla H}{\nabla V}. \tag{1.10}$$

The temperature T is thus an intensive quantity given by the ratio of an extensive quantity: the heat content to the total heat capacity (= $\rho c \nabla V$) of a volume ∇V. Further, an equation of continuity (a consequence of a conservation condition in this case) of the heat over a volume is the concept behind Equation (1.3). Indeed a conservation principle is at the heart of other examples of physical diffusion such as molecular diffusion in materials, flow of galvanic currents, pressure-driven fluid flow in resistive media, and other phenomena.

1.3 Random Walk

Shortly after Fourier submitted his monograph on the theory of propagation of heat in solids, Laplace (1809) proposed what may be called stochastic diffusion. The idea was given a more solid basis almost a hundred years later

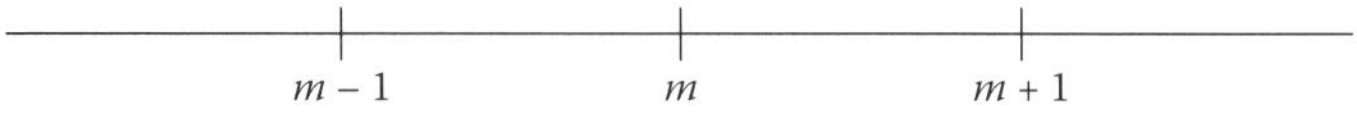

FIGURE 1.1
Random walk on a line.

(Pearson 1905) through a formulation of the random walk problem. We shall illustrate Laplace's ideas through the simpler analysis of a one-dimensional random walk. Assume that a walker takes successive steps to the right or to the left at random. After a total of N steps, she finds herself at a site m: $-N \le m \le N$ (Figure 1.1). The question is what is $P_N(m)$, i.e., what is the probability of finding the walker at m after N steps? Let n_1 be the number of steps to the right and n_2, the number of steps to the left. Then

$$\begin{aligned} N &= n_1 + n_2, \\ m &= n_1 - n_2, \\ m &= 2n_1 - N. \end{aligned} \tag{1.11}$$

Assume that successive steps are delinked from past history and are statistically independent. The probability of any one given sequence of n_1 steps to the right and n_2 steps to the left is

$$W(p,q) = (p)^{n_1} \cdot (q)^{n_2}, \tag{1.12}$$

where $p(q)$ denotes the elementary probability of a step to the right (left). Therefore,

$$P_N(m) = \frac{N!}{n_1!n_2!} W(p,q) \,. \tag{1.13}$$

Here, the prefactor multiplying W is the total number of possible ways of taking N steps, so that n_1 steps are to the right and n_2 steps are to the left. Combining Equations (1.11) and (1.13)

$$P_N(m) = \frac{N!}{\left(\frac{N+m}{2}\right)!\left(\frac{N-m}{2}\right)!} p^{\frac{N+m}{2}} \cdot q^{\frac{N-m}{2}} \,. \tag{1.14}$$

Evidently, the probability is normalized, i.e.,

$$\sum_{n_1=0}^{N} P_N(n_1) = 1, \tag{1.15}$$

if we recall the binominal theorem that

$$(p+q)^N = \sum_{n_1=0}^{N} \frac{N!}{n_1!(N-n_1)!} p^{n_1} \cdot q^{N-n_1}, \tag{1.16}$$

and that

$$(p+q) = 1. \tag{1.17}$$

We can also easily calculate the mean and the fluctuation of n_1 from the following steps. First, the mean of n_1 is given by

$$< n_1 > = \sum_{n_1=0}^{N} n_1 \frac{N!}{n_1!n_2!} p^{n_1} \cdot q^{n_2} = p \frac{\partial}{\partial p} \sum_{n_1=0}^{N} \frac{N!}{n_1!n_2!} p^{n_1} \cdot q^{n_2}, \tag{1.18}$$

upon treating p and q as two independent variables. The right side of Equation (1.18), using the binomial theorem (1.16), can be formally written as

$$< n_1 > = p \frac{\partial}{\partial p} (p+q)^N = Np(p+q)^{N-1}. \tag{1.19}$$

Finally, we employ Equation (1.17) to arrive at

$$< n_1 > = Np. \tag{1.20}$$

Using similar tricks, the mean of n_1-squared is given by*

$$< n_1^2 > = < n_1^2 > + Npq. \tag{1.21}$$

Thus, the fluctuation in n_1^2 is

$$n_1^2 - < n_1^2 > = Npq. \tag{1.22}$$

Finally, from Equations (1.21) and (1.22) we can also evaluate the mean and the dispersion of m: the net displacement to the right. Equations (1.11) and (1.20) yield

$$< m > = 2 < n_1 > - N = 2N \ (p - 1/2). \tag{1.23}$$

* See Exercise 1.

Similarly*,

$$<(m-<m>)^2 = 4<(n_1-<n_1>)^2> = 4Npq. \tag{1.24}$$

Of special interest is the case when the right step and the left step are equally probable, i.e., there is no bias. We now have

$$p = q = \frac{1}{2}, \tag{1.25}$$

and hence,

$$<m> = 0, \ <m^2> = N. \tag{1.26}$$

Further, from Equation (1.14),

$$P_N(m) = \frac{N!}{\left(\frac{N+m}{2}\right)!\left(\frac{N-m}{2}\right)!}\left(\frac{1}{2}\right)^N. \tag{1.27}$$

Taking the logarithm of the left side and applying the Stirling formula:

$$\ln(n!) = \left(n+\frac{1}{2}\right)\ln(n) - n + \frac{1}{2}\ln(2\pi) + O\left(n^{-1}\right) \tag{1.28}$$

we obtain† (in the limit of large N, n_1 and n_2)

$$\ln(P_N(m)) \cong \ln\left(\sqrt{\frac{2}{\pi}}\right) + \left(N+\frac{1}{2}\right)\ln N - \frac{m}{2}[\ln(N+m) - \ln(N-m)]$$
$$-\frac{(N+1)}{2}[\ln(N+m) + \ln(N-m)]. \tag{1.29}$$

Because right and left steps occur with equal probability, it is reasonable to assume that the difference in the number of right and left steps has a magnitude $|m|$ that will be much smaller than the total number of N steps. Thus, we make use of the expansion:

$$\ln\left(1 \pm \frac{m}{N}\right) \approx \pm\frac{m}{N} - \frac{m^2}{2N^2} + \dots \quad .$$

---

* See Exercise 2.

† See Exercise 3.

From Equation (1.29) then*,

$$P_N(m) \approx \sqrt{\frac{2}{\pi N}} \exp\left(-\frac{m^2}{2N}\right). \tag{1.30}$$

Curiously, Equation (1.30) has the same structure as Equation (1.7) and we must ask whether there is a diffusion equation beneath the solution given in Equation (1.30). The answer is provided below by re-examining the random walk problem in terms of real position and time.

1.4 Random Walk on a Lattice and Its Continuum Limit

We consider a one-dimensional lattice with spacing l on which a random walker starts her journey from the middle (Figure 1.1). At any given instant, at an arbitrary site, after a mean waiting time ($= \lambda^{-1}$), the walker can jump to the nearest neighbor right or left site. The stay-put probability for the walker at the site m is given by the following rate equation:

$$\frac{d}{dt} P_m = -2\lambda P_m + \lambda(P_{m+1} + P_{m-1}) \ . \tag{1.31}$$

The first term on the right is the so-called loss-term for the leakage of probability to the two neigboring sites (and explains the factor of 2). The second term within round parentheses is the gain-term. Equation (1.31) is in the form of a difference equation. The transition from the discrete to the continuum can be effected by converting the difference equation (1.31) to a differential equation when the lattice spacing l becomes vanishingly small, by using the identity

$$\frac{\partial^2 P}{\partial x^2}(x,t) = \lim_{l \to 0} \frac{1}{l^2}[P_{m+1}(t) + P_{m-1}(t) - 2P_m(t)]. \tag{1.32}$$

Equation (1.31) now yields the diffusion equation

$$\frac{\partial}{\partial t} P(x,t) = D_s \frac{\partial^2}{\partial x^2} P(x,t), \tag{1.33}$$

that introduces the coarse-grained diffusion coefficient D_s:

$$D_s = \lim_{\bar{\lambda} \to \infty} \lim_{l \to 0} \left(\frac{\bar{\lambda} l^2}{2}\right), \quad \bar{\lambda} = 2\lambda, \tag{1.34}$$

* See Exercise 4.

where $(\bar{\lambda})^{-1}$ is the lifetime of the occupation of a site. We underscored the fact that we are dealing with stochastic diffusion here by inserting a subscript s under D.

It is interesting to note that while the random walk was analyzed earlier in Section 1.2 by a combinatorial analysis, the treatment above provides a temporal connotation. Indeed the two approaches can be brought to par by viewing N and n of Section 1.2 as continuous variables and deriving from Equation (1.30) the partial differential equation

$$\frac{\partial}{\partial N} P_N(m) = \frac{1}{2}\frac{\partial^2}{\partial m^2} P_N(m). \tag{1.35}$$

Equation (1.35) is exactly of the same form as the stochastic diffusion equation (1.33), yielding the following natural identification

$$N = 2D_S \frac{t}{l^2}. \tag{1.36}$$

In addition, the mathematical properties of Equation (1.33) [or Equation (1.35)] are the same as those of Fourier's physical diffusion equation (1.3) and all our earlier comments about boundary condition-specific solutions equally apply here.

It is pertinent to remark that the one-dimensional random walk, identical to the coin-tossing problem for which the probability that "heads" will occur m more times than "tails" among a total number of N trials, obeys the same equation as Equation (1.35). In fact, similar ideas influenced Rayleigh's treatise on the theory of sound (1894). Rayleigh showed that the average intensity resulting from a random mixing of acoustic waves of constant amplitude (or step size) and an arbitrary two-valued phase (or step direction to right or left) is indeed proportional to N in Section 1.2.

Further applications of the concepts of stochastic diffusion followed by formulation of the laws of errors in statistics (Edgeworth 1883) and in stock market prediction. Bachelier (1900) demonstrated that the time evolution of stock prices obeys a stochastic diffusion equation. The list would be incomplete if we failed to mention a fundamental principle of the probability theory of large numbers, namely Polya's central limit theorem (1920) that hinges on the Gaussian form [Equation (1.30)] of the underlying probability, such as $P_N(m)$.

Notwithstanding all these successes of the Laplacian concept of stochastic diffusion, Fourier's work on heat diffusion achieved a much wider following in the practical worlds of biology, chemistry, and geology. It is important also to understand the distinctions between physical and stochastic diffusion. While both $T(x, t)$ and $P(x, t)$ are intensive quantities, P is additive while T

is not! (P is additive because it can be envisaged as a fractional density.) (See Equation (1.46).) Moreover, stochastic diffusion is characterized by a single parameter D_s, whereas heat diffusion depends on two physical attributes such as specific heat c and thermal conductivity $\mathcal{K}$, measurable independently. The next section discusses a unification of physical and stochastic diffusion (Einstein 1905 and 1906).

1.5 Einstein on Brownian Motion

Here we review Einstein's famous paper (1905) on Brownian motion (Brown 1828, 1829), and his subsequent doctoral thesis (Einstein 1906). The thesis was an extraordinary and deep exposition of Einstein's penchant for unraveling real life phenomena—a new approach to the determination of Avogadro's number ($\mathcal{N}$) and the size (a) of a molecule. This approach perhaps contrasts the popular image of Einstein as a powerful mathematical physicist whose contribution was beyond the comprehension of mortals.

It is also curious to reflect back to the era prior to 1905 when chemists and physicists could not reach a consensus about the integrity of a molecule, i.e., molecular reality. Like all practitioners of physics, Einstein focused on estimating a number (such as $\mathcal{N}$) and *inter alia*, elucidating the natural phenomena of thermal fluctuations evident in daily life and as seen in experiments. His analysis was based on two separate, brilliantly reasoned ideas that led to two distinct relations between $\mathcal{N}$ and a, as detailed below.

1.5.1 Dressed Viscosity

Consider an extremely dilute concentration of solute molecules, such as sugar in a solvent like water. The solute particles are envisaged to be much bigger than the solvent particles, in the range of about 10^{-3} mm (~1000 nm), making Brownian motion a topically relevant aspect of nanofluids. Einstein surmised that since the addition of solute is expected to reduce the "elbow room" of the solvent particles and thereby decrease mobility, the new viscosity η^* must be larger than the original (bare) viscosity η of the solvent. Further, the difference (between η^* and η) should be small due to the diluteness of the solution and proportional to the fractional volume occupied by the solute per unit volume of the solvent. Thus

$$\frac{\eta^*}{\eta} = 1 + \frac{4}{3}\pi a^3 \frac{\mathcal{N}}{M}\rho. \tag{1.37}$$

where $(4/3)\pi a^3$ is the volume of each solute particle (assumed spherical) of radius a and ρ is the mass density of the solute. M is the molecular weight of the solute so that $M/\mathcal{N}$ measures the mass of each solute particle. Therefore, the product $M/\mathcal{N}.\rho$ specifies the number of solute particles per unit volume of the solvent.

As indicated earlier, the hypothesis in Equation (1.37) hinges on the assumption that the solute concentration is low. Indeed, detailed hydrodynamic arguments reveal that the second term on the right of Equation (1.37) is merely a leading order term in an expansion wherein the small parameter is proportional to ρ! In addition, Einstein recognized that the assumption of sphericity of a solute particle and the concomitant boundary condition for the flow of solvent past a solute was an oversimplification and that a more realistic analysis brings forth an additional prefactor of 5/2 in multiplying the second term. Thus, Equation (1.37) can be rewritten as

$$\frac{\eta^*}{\eta} = 1 + \frac{5}{2} \cdot \frac{4}{3} \pi a^3 \frac{\mathcal{N}}{M} \rho. \tag{1.38}$$

Before 1905 all other quantities barring $\mathcal{N}$ and a were experimentally known. Einstein made use of the available data and arrived at an estimate of the quantity $(\mathcal{N}a^3)$. In order to separately assess $\mathcal{N}$ and a, he needed one other equation to connect these two parameters. He derived it from an ingenious argument described below.

1.5.2 Synergy of Thermodynamics and Kinetics

Based on Brown's observations (1826 and 1829) of the random zig-zag motions of pollen particles under a microscope, Einstein was well aware that motion never stops even though the jar containing the water and the pollens in Brown's experiments were in equilibrium at constant temperature and pressure. He thus set up a gedanken measurement of the instantaneous force F on a tagged solute particle, say in the x-direction, for the sake of simplicity. Clearly F would be proportional to the local gradient $\partial\pi/\partial x$ of the pressure and the proportionality factor would be a volume required to make both sides dimensionally balanced. Thus

$$F = \frac{M}{\mathcal{N}} \frac{1}{\rho} \frac{\partial \pi}{\partial x}, \tag{1.39}$$

where the volume factor has already been introduced in connection with Equation (1.37). Einstein next relied on the Dutch Chemist Van't Hoff's remarkable hypothesis (1887) that the kinetics of a dilute solute in a solvent is governed by the ideal gas laws of Boyle and Charles. Hence

$$\frac{\partial \pi}{\partial x} = \frac{RT_B}{M} \frac{\partial p}{\partial x}, \tag{1.40}$$

where R is the gas constant. Combining Equations (1.39) and (1.40) yields

$$F = \frac{RT_B}{\mathfrak{N}} \frac{1}{\rho} \frac{\partial p}{\partial x}. \tag{1.41}$$

The next few steps serve as testimony to Einstein's open mindedness and broad appreciation of interdisciplinary science. He reverted to two key ideas of chemical engineering. The first was to recognize that force F is neither external nor mechanical. It has an intrinsic origin related to the viscous drag experienced by a solute particle due to incessant collisions with solvent particles. It is given by the Stokes law (Chaikin and Lubensky 2004)

$$F = -6\pi\eta a \upsilon, \tag{1.42}$$

where the negative sign indicates that F and the instantaneous velocity υ of the solute particle are in opposition. Finally, the velocity υ is given by another chemical engineering law developed by Fick (1855) (Chaikin and Lubensky 2004) that is the analogue of Ohm's law of electrodynamics:

$$\rho\upsilon = -D_p \frac{\partial \rho}{\partial x}. \tag{1.43}$$

The quantity D_p is a phenomenologically introduced diffusion coefficient. The p suffix highlights the fact that it represents physical diffusion (of Fourier type) of a macroscopic attribute such as density. Combining Equations (1.42) and (1.43),

$$F = D_p \frac{6\pi\eta a}{\rho} \frac{\partial \rho}{\partial x}. \tag{1.44}$$

It is amazing to realize that the fictitious force F of Einstein self-determines—it simply drops from a combination of Equations (1.43) and (1.44), leading to

$$D_p = \frac{RT}{6\pi\eta a \mathfrak{N}}. \tag{1.45}$$

Equation (1.45) is then the promised result that provides a measurement of another combination ($a\mathcal{N}$) from the experimented data on D_p. Einstein was then able to eliminate size a from the earlier determined factor $\mathcal{N}a^3$ and extract Avogadro's number. Einstein's estimated value of $\mathcal{N}$ was 6.6×10^{23}—very close to the presently accepted number (Einstein 1956).

1.5.3 Brownian Motion and Stochastic Diffusion

Eleven days after Einstein submitted his thesis, he published the famous paper (1905) in which he provided a microscopic insight into the analysis presented above. This work illustrates how profoundly Einstein was influenced by Boltzmann's idea of thermal fluctuations.

The basic premise was that concomitant with macroscopic diffusive motion of a region of a solution, the fundamental constituent (individual solute particle) performs a random walk in a continuum. Consider again one dimension in which $n(x,t)$ designates the number of solute particles per unit volume of solvent around x at time t. It is evident that $n(x,t)$ is also proportional to the probability of occupation (of the solute) around x:

$$n(x,t) = \bar{n}P(x,t), \tag{1.46}$$

where $\bar{n}$ is the mean density governed by

$$\bar{n} = \int dx \; n(x,t), \tag{1.47}$$

in consonance with the conservation of probability.

$$\int P(x,t)\,dx = 1. \tag{1.48}$$

Following Equation (1.33), $n(x,t)$ obeys the stochastic diffusion equation:

$$\frac{\partial}{\partial t} n(x,t) = D_s \frac{\partial^2 n(x,t)}{\partial x^2} . \tag{1.49}$$

Correspondingly, the solution of Equation (1.49) follows (1.7), yielding for open boundary condition:

$$n(x,t) = \frac{\bar{n}}{\sqrt{4\pi D_s t}} \exp -\left(\frac{x^2}{4D_s t}\right). \tag{1.50}$$

Equation (1.50) is consistent with the initial condition:

$$n(x,0) = \bar{n}\delta(x). \tag{1.51}$$

We can now consider the following experiment. Take a stop watch and focus on a tagged solute particle at time $t = 0$ at a point we define as the origin. The particle wanders around on a zig-zag path and moves to point x at time t. Continue to repeat the experiment and collate a large set of entries for x. While x will evidently average out to zero for an unbiased random walk the mean squared $< x^2 >$ will be given by

$$< x^2 > = \int dx \frac{n(x,t)}{\bar{n}} x^2. \tag{1.52}$$

From Equation (1.50),

$$< x^2 > = 2D_s t. \tag{1.53}$$

It is tempting to compare Equation (1.53) with (1.36) and realize that in the lattice picture we indeed have

$$< x^2 > = Nl^2 . \tag{1.54}$$

The extraordinary ingenuity of Einstein is the recognition that physical and stochastic diffusion are synonymous where Brownian motion is concerned. He was therefore led to take the remarkable step of equating D_s to the physical diffusion coefficient D_p derived earlier in Equation (1.45) to arrive at

$$< x^2 > = \frac{RT}{3\pi\eta a\mathcal{N}} t \tag{1.55}$$

Whether sugar molecules in water or polystyrene balls in a fluid medium, the mean-squared displacement is linearly proportional to time t. Further, Equation (1.55) provides a universal relation for the computation of Avogadro's number. Because the left side of Equation (1.55) measures a fluctuation property while the right side depends on the viscosity η that characterizes dissipation, Equation (1.55) became the first indication back in 1905 of the fluctuation–dissipation theorem (FDT)—a cornerstone of non-equilibrium statistical mechanics. The FDT was further expounded later by Nyquist (Wax 1954), Callen and Welton (1952), and Kubo (1957). Thus, it is not an exaggeration to infer that Einstein's work on Brownian motion and diffusion actually laid the foundation of relaxation phenomena of non-equilibrium statistical mechanics (Zwanzig 2001, Balakrishnan 2009).

1.6 Concluding Remarks

Einstein's work on diffusion led to a tremendous variety of applications in the context of

Construction industry (linked with the interdisciplinary topic of granular matter)

Dairy technology (based on colloidal properties of micelle suspension in milk)

Ecology (Brownian movement of aerosol suspensions in the atmosphere)

Pais wrote in his fascinating biography of Einstein (2008) titled *Subtle Is the Lord*: "Einstein might have enjoyed hearing this, since he was quite fond of applying physics to practical situations." In Einstein's own words: "... the theory of Brownian motion is of great importance since it permits an exact computation of $\mathfrak{N}$: the Avogadro number. The great significance is that one sees under the microscope part of the heat energy in the form of mechanical energy...." In 1919, four years after he proposed the general theory of relativity, Einstein further mused: "... because of the understanding of the essence of Brownian motion suddenly all doubts vanished about the correctness of Boltzmann's interpretation of thermodynamics."

Exercises

Derive:

1. Equation (1.21)
2. Equation (1.24)
3. Equation (1.29)
4. Equation (1.30)

References

Bachelier, L. 1900. Theorie de la speculation. *Annals de l' Ecole Normale Super ieure,* 3. Paris: Gaithier-Villars.

Balakrishnan, V. 2009. *Elements of Nonequilibrium Statistical Mechanics*. New Delhi: Ane Books.

Brown, R. 1828. A brief account of microscopical observations made in the months of June, July, and August, 1827, on the particles contained in the pollen of plants; and on the general existence of active molecules in organic and inorganic bodies. *Philos. Mag.* 4, 161.
Brown, R. 1829. Additional remarks on active molecules. *Philos. Mag.* 6, 161.
Callen, H.B. and R.F. Welton. 1951. Irreversibility and generalized noise. *Phys. Rev.* 83, 34.
Chaikin, P.M. and T.C. Lubensky. 2004. *Principles of Condensed Matter Physics.* Cambridge: Cambridge University Press.
Edgeworth, F. Y. 1883. The law of error. *Philos. Mag.* 16, 300.
Einstein, A. 1926. *Investigations on the Theory of the Brownian Motion.* New York: Dover.
Einstein, A. 1905. On the motion required by the molecular kinetic theory of heat, of particles suspended in fluids at rest. *Ann. der Physik* 17, 549.
Einstein, A. 1906. Zur theorie der Brownschen bewegung. *Ann. Physik,* 19, 371.
Fick, A. 1855. Ueber diffusion. *Ann. Der. Physik,* 94, 59; *Phil. Mag.* 10, 30 (in English).
Fourier, J. 1972. Theorie de al propagation de la Chaleur dans les solides, 21 December 1807. English translation: Cambridge: MIT Press, p. 516.
Kubo, R. 1957. Statistical–mechanical theory of irreversible processes. 1. General theory and simple applications to magnetic and conduction problems. *J. Phys. Soc. (Japan).* 12, 570.
Laplace, P. S. 1809. Théorie analytique des probabilities. *J. d'École Polytech.* VIII 235, 5.
Narasimhan, T.N. 2010. *Phys. Today.*
Narasimhan, T. N. 2009 (July). The dichotomous history of diffusion. *Phys. Today.* 48.
Narasimhan, T. N. 2010. On physical diffusion and stochastic diffusion. *Current Sci.* 98, 23.
Pais, A. 2008. *Subtle Is the Lord: The Science and Life of Albert Einstein.* Oxford: Oxford University Press.
Pearson, K. 1905. The problem of the random walk. *Nature* 72, 294.
Polya, G. 1920. Über den zentralen grenzwertsatz der wahrscheinlichkeitsrechnung und das momentenproblem. *Math. Z.* 8, 171.
Rayleigh, Lord. 1894. *The Theory of Sound.* London. Macmillan (reprinted 1929).
Wax, N., Ed. 1954. Selected papers. In *Noise and Stochastic Processes.* New York: Dover.
Zwanzig, R. 2001. *Nonequlibrium Statistical Mechanics.* New York: Oxford University Press.

2

Markov Processes

2.1 Genesis of Markov Concept

In a paper, Einstein (1906) provided a mathematical basis of his thesis work and went on to propound the idea of a Markovian stochastic process (Bharucha-Reid 1960, Gillespie 1991). Einstein's analysis of Brownian motion hinged on the simple premise that suspended pollen particles move independently of one another in a completely random manner.

He justified this assumption on the fact that the solute concentration was very low, and hence intersolute interaction could be neglected. Thus, if the density of solute particles within region x and $x + dx$ grows from $n(x, t)$ at time t to $n(x, t + \tau)$ at time $t + \tau$ (where $\tau << t$), the growth must have been at the expense of the density at the preceding step $(x - \Delta)$ (where $|\Delta| << |x|$) measured by $n(x - \Delta, t)$. Inherent in this concept is the premise of the random walk (Section 1.2). Furthermore, the dependence of $n(x, t)$ only on the density at the preceding step, i.e., $n(x - \Delta, t)$, is indeed the central assumption of a Markov process (see below). Einstein thus surmised that

$$n(x,t+\tau) = \int_{-\infty}^{\infty} n(x-\Delta,t)\phi(\Delta)d\Delta, \quad t >> \tau, \tag{2.1}$$

where $\phi(\Delta)\ d\Delta$ is the probability that a region is displaced through a point that lies between Δ and $\Delta + d\Delta$. Because Δ may lie anywhere within the domain of interest, it is integrated over. Again, it is pertinent to note that the joint probabilities of n and ϕ are written in a multiplicative form, implying an independence of the underlying processes; $\phi(\Delta)$ must satisfy the normalization condition:

$$\int_{-\infty}^{\infty} \phi(\Delta)\ d\Delta = 1. \tag{2.2}$$

In addition, since unbiased random walk means that right and left jumps are equally probable, we have the symmetry condition (Weiss 1994):

$$\phi(\Delta) = \phi(-\Delta). \tag{2.3}$$

We may now perform Taylor expansion of both sides of Equation (2.1) in accordance with

$$n(x,t+\tau) \approx n(x,t) + \tau \frac{\partial}{\partial t} n(x,t) + \ldots, \tag{2.4}$$

and

$$n(x-\Delta,t) \approx n(x,t) - \Delta \frac{\partial}{\partial x} n(x,t) + \frac{\Delta^2}{2} \frac{\partial^2 n(x,t)}{\partial x^2} + \ldots. \tag{2.5}$$

Substituting Equations (2.4) and (2.5) into (2.1) and using Equations (2.2) and (2.3), we obtain

$$\tau \frac{\partial n}{\partial t}(x,t) = \frac{\partial^2 n(x,t)}{\partial x^2} \cdot \int_{-\infty}^{\infty} \frac{\Delta^2}{2} \phi(\Delta)\, d\Delta. \tag{2.6}$$

Equation (2.6) is, of course, identical to the diffusion equation (1.49) if we identify the stochastic diffusion coefficient as

$$D_s = \frac{1}{2\tau} \int_{-\infty}^{\infty} \Delta^2 \phi\, (\Delta) d\Delta. \tag{2.7}$$

Evidently, Equation (2.7) is the microscopic analogue of the macroscopic formula (1.53) wherein the integral defines the second order jump moment, discussed in Chapter 3. We now provide a mathematical foundation of Equation (2.1) against the background of a stationary Markov process (SMP), but first we set up a few preliminaries.

2.2 Definition

A stochastic process may mean the momentum or displacement (or a combination of both) or a fluctuating voltage or current in a resistor. In a random process $\xi(t)$, the value of ξ has a dependence on the time t, an independent variable, unlike a deterministic situation. Thus in a different observation one gets other $\xi(t)$ functions. Hence, only certain probability distributions are directly observable (Cox and Miller 1994, Doob 1953, Feller 1972, Gardiner 1997, Grimmett and Strizaker 2001, Ito and McKean, 1985, Kac 1959, Lawler 2006, Lemons 2002, Mazo 2002, Montroll and Lebowitz 1987, Nelson 1967, Oppenheim et al. 1977, Soong 1973, Stratonovich 1981, van Kampen 1985, Wong 1983). Note that parameter t, while representing time in this chapter, can actually be any other attribute in a general stochastic process.

2.2.1 Joint and Conditional Probabilities

The following set of joint probability densities completely specify a stochastic process $\xi(t)$ (Dattagupta 1987, Gardiner 1985, van Kampen 1981, Wax 1954):

$P_1(\xi_1,t_1)d\xi_1$ = probability of finding ξ in the range $\xi_1, \xi_1 + d\xi_1$ at time t_1.

$P_2(\xi_1,t_1;\xi_2 t_2)d\xi_1 d\xi_2$ = joint probability of finding ξ in the range ξ_1, $\xi_1 + d\xi_1$ at time t_1 and in the range $\xi_2, \xi_2 + d\xi_2$ at time t_2.

$P_3(\xi_1,t_1;\xi_2 t_2;\xi_3 t_3)d\xi_1, d\xi_2 d\xi_3$ = joint probability of finding a set of three values of ξ in the range $d\xi_1$, $d\xi_2$, and $d\xi_3$ at times t_1, t_2 and t_3 etc.

Evidently, the Ps satisfy

$$\text{Positivity}: \; P_n \geq 0,$$

Symmetry: $P_n(\xi_1,t_1;\xi_2 t_2 \ldots \xi_n,t_n)$ is a symmetric function of the set of variables $(\xi_1,t_1),(\xi_2,t_2),\ldots(\xi_n,t_n)$, i.e. $P_2(\xi_1,t_1;\xi_2 t_2) = P_2(\xi_2,t_2;\xi_1 t_1)$,

$$\text{Projection: } P_m(\xi_1,t_1;\ldots\xi_m,t_m) = \int\!\!\int d\xi_{m+1}\ldots d\xi_n P_n(\xi_1,t_1;\ldots\xi_n,t_n), \text{ for } m<n. \quad (2.8)$$

Along with joint probabilities, it is useful to define what are called conditional probabilities. For instance, the third order joint probability can be written as

$$P_3(\xi_1,t_1;\xi_2,t_2;\xi_3,t_3) = P_2(\xi_1,t_1;\xi_2,t_2)P(\xi_1,t_1;\xi_2,t_2 \mid \xi_3,t_3). \quad (2.9)$$

The second term on the right side is obviously interpreted as a conditional probability that the stochastic process $\{\xi,t\}$ takes a value between ξ_3 and $\xi_3 + d\xi_3$ at time t_3, given that it had the value ξ_1 at t_1 and ξ_2 at t_2 $(t_3 > t_2 > t_1)$. Note that a vertical bar is used to define a conditional probability. More generally, $P_3(\xi_1,t_1;\xi_2,t_2;\ldots;\xi_{n-1},t_{n-1} \mid \xi_n t_n$ specifies the conditional probability that ξ lies in the interval $\xi_n, \xi_n + d\xi_n$ at time t_n, given that ξ is equal to $\xi_1, \xi_2 \ldots \xi_{n-1}$ at the times $t_1, t_2 \ldots t_{n-1}$ (where $t_n > t_{n-1} > \ldots > t_2 > t_1$).

2.3 Stationary Markov Process

For practical applications to problems in biology, chemistry, geology, and physics, a class of stochastic methods called Markov processes assumes significance. For a Markov process, the entire nature of the stochastic variables is embedded in P_1 and P_2. A process is Markovian if the conditional probability $P(\xi_1,t_1;\xi_2,t_2;\ldots,\xi_{n-1}\,|\,\xi_n,t_n)$ depends only on ξ_{n-1} at time t_{n-1}, thus

$$P(\xi_1,t_1;\xi_2,t_2;\ldots;\xi_{n-1},t_{n-1}\,|\,\xi_n,t_n) = P(\xi_{n-1},t_{n-1}\,|\,\xi_n,t_n). \tag{2.10}$$

Evidently, once P_1 and P_2 are known, the remaining P_n's can be evaluated in terms of P_1 and P_2. For example, from (2.9) and (2.10),

$$P_3(\xi_1,t_1;\xi_2,t_2\xi_3,t_3) = P_2(\xi_1,t_1;\xi_2,t_2)P(\xi_2,t_2\,|\,\xi_3,t_3) = P_2(\xi_1,t_1;\xi_2,t_2)\frac{P_2(\xi_2,t_2;\xi_3,t_3)}{P_1(\xi_2,t_2)}, \tag{2.11}$$

and so on. We may also express the right side of Equation (2.11) in terms of conditional probabilities:

$$P_3(\xi_1,t_1;\xi_2,t_2;\xi_3,t_3) = P_1(\xi_2,t_2)P(\xi_1,t_1\,|\,\xi_2,t_2)P(\xi_2,t_2\,|\,\xi_3,t_3). \tag{2.12}$$

Generalizing,

$$P_n(\xi_1,t_1;\xi_2,t_2;\xi_n,t_n) = P_1(\xi_2,t_2)P(\xi_1,t_1\,|\,\xi_2,t_2)\ldots P(\xi_{n-1},t_{n-1}\,|\,\xi_n,t_n). \tag{2.13}$$

Thus, the a priori probability function, along with a set of conditional probabilities, completely specifies the n-point joint probabilities, via the above chain rule reminiscent of the transfer matrix of equilibrium statistical mechanics (Huang 1987).

For facilitating practical calculations it is useful to devise a differential equation for the two point conditional probability P_2, as shown below. Recall that by employing the projection property [Equation (2.8)], P_2 can be written as

$$P_2(\xi_1,t_1;\xi_3,t_3) = \int d\xi_2 P_3(\xi_1,t_1;\xi_2,t_2;\xi_3,t_3) = P_1(\xi_1,t_1)\int d\xi_2 P(\xi_1,t_1\,|\,\xi_2,t_2)P(\xi_2,t_2\,|\,\xi_3,t_3),$$

from Equation (2.13). Conversely, the left side also equals $P_1(\xi_1,t_1)P(\xi_1,t_1\,|\,\xi_3,t_3)$, yielding

$$P(\xi_1,t_1\,|\,\xi_3,t_3) = \int d\xi_2 P(\xi_1,t_1\,|\,\xi_2,t_2)P(\xi_2,t_2\,|\,\xi_3,t_3),\quad t_2 \in [t_1,t_3]. \tag{2.14}$$

The above equation embodies an important property of Markov processes in that the transition probability for traversing from ξ_1 to ξ_2 via $\xi_3$is the probability for going from ξ_1 to ξ_2 times the probability for going from ξ_2 to ξ_3, much like transition amplitudes in the Green function formulation of quantum mechanics. The above equation suggesting that successive transitions are statistically independent, in the sense that the probabilities are multiplicative, is known as the Smoluchowski-Chapman-Kolmogorov (SCK) equation.

We now convert the SCK equation into an integro-differential equation by employing a trick inherent in Einstein's formulation of Equation (2.1). Assume that the epochs t_3 and t_2 are infinitesimally close to each other, i.e., $t_3 = t_2 + \Delta t$, where Δt is vanishingly small. We may then Taylor-expand $P(\xi_2, t_2 \mid \xi_3, t_3)$ in Δt where the zeroth-order term, for which t_3 exactly matches t_2, must yield a distribution that is a delta function in ξ_2 and ξ_3. Thus

$$P(\xi_2, t_2 \mid \xi_3, t_3) \approx \delta(\xi_2 - \xi_3) + (\Delta t)\frac{\partial}{\partial t_2} P(\xi_2, t_2 \mid \xi_3, t_3)\Big|_{t_3 = t_2} + \vartheta(\Delta t^2), \quad (2.15)$$

where the first term on the right is the probability density of no jump. Since conservation of probability implies that

$$\int d\xi_3 P(\xi_2, t_2 \mid \xi_3, t_3) = 1, \quad (2.16)$$

Equation (2.15) yields

$$\int d\xi_3 \frac{\partial}{\partial t_2} P(\xi_2, t_2 \mid \xi_3, t_3)\Big|_{t_3 = t_2} = 0 \ . \quad (2.17)$$

We now make use of the identity:

$$\lambda(\xi_2, t_2) - \int \lambda(\xi_3, t_2)\ \delta(\xi_2 - \xi_3)\ d\xi_3 = 0 \ . \quad (2.18)$$

Substituting this for the right side of Equation (2.15), we obtain

$$\int d\xi_3 \left[\frac{\partial}{\partial t_2} P(\xi_2, t_2 \mid \xi_3, t_3\Big|_{t_3 = t_2} + \lambda(\xi_3, t_2)\delta(\xi_3 - \xi_2) \right] = \lambda(\xi_2, t_2), \quad (2.19)$$

where λ must have the dimension of a rate (frequency). It is evident that the integrand is a function of ξ_2, ξ_3, and t_2, and can be denoted as $W(\xi_2 \mid \xi_3; t_2)$. It is also clear that, for $\xi_2 \neq \xi_3$,

$$W(\xi_2 \mid \xi_3; t_2) = \frac{\partial}{\partial t_2} P(\xi_2, t_2 \mid \xi_3, t_3)\Big|_{t_3 = t_2} . \quad (2.20)$$

Hence, the left side of Equation (2.20) has the obvious interpretation of being the probability per unit time that ξ "jumps" (instantaneously) from ξ_2 to ξ_3 at time t_2. Also, because,

$$\int d\xi_3 W(\xi_2 \mid \xi_3; t_2) = \lambda(\xi_2, t_2), \tag{2.21}$$

λ is the total probability per unit time that a transition took place. We may now use the above results to rewrite Equation (2.15) as

$$P(\xi_2, t_2 \mid \xi_3, t_3) \approx \delta(\xi_2 - \xi_3)[1 - \lambda(\xi_3, t_2)\Delta t] + \Delta t \ W(\xi_2 \mid \xi_3; t_2). \tag{2.22}$$

We are then ready to substitute the left side of Equation (2.22) in (2.14) to obtain

$$\begin{aligned} P(\xi_1, t_1 \mid \xi_3, t_2 + \Delta t) = P(\xi_1, t_1 \mid \xi_3, t_2)[1 - \lambda(\xi_3, t_2)\Delta t] \\ + \Delta t \int d\xi_2 P(\xi_1, t_1 \mid \xi_2, t_2) \ W(\xi_2 \mid \xi_3; t_2). \end{aligned} \tag{2.23}$$

Rearranging terms, dividing throughout by the small parameter Δt and taking the latter to zero, we have the integro-differential equation:

$$\frac{\partial}{\partial t} P(\xi_1, t_1 \mid \xi, t) = \int d\xi' [P(\xi_1, t_1 \mid \xi', t) \ W(\xi' \mid \xi, t) - P(\xi_1, t_1 \mid \xi, t) \ W(\xi \mid \xi' t)], \tag{2.24}$$

where, in the last term, we exploited Equation (2.21). Furthermore, if we choose a given initial condition, we may suppress the dependence of P on (ξ_1, t_1) and rewrite Equation (2.24):

$$\frac{\partial}{\partial t} P(\xi, t) = \int d\xi' [P(\xi', t) \ W(\xi' \mid \xi, t) - P(\xi, t) \ W(\xi \mid \xi', t)], \tag{2.25}$$

where

$$P(\xi, t = 0) = \delta(\xi - \xi_1). \tag{2.26}$$

The above has the familiar form of a master equation in which the first term under the integral is the gain for jumping into ξ, whereas the second term is the loss for jumping out of ξ. For most physical applications—and diffusion is definitely physical—the stochastic process is both Markovian and stationary. The stationary status implies that results of measurements are independent of time translation and hence the origin of time has no special significance. This means

$$P_n(\xi_1,t_1+\tau;\xi_2,t_2+\tau;\ldots;\xi_n,t_n+\tau)=P_n(\xi_1,t_1;|\xi_2,t_2;\ldots;\xi_n,t_n), \qquad (2.27)$$

which further implies

$$P_1(\xi_1,t_1)=P_1(\xi_1),\ \ P_2(\xi_1,t_1;\xi_2,t_2)=P_2(\xi_1,\xi_2,t), \qquad (2.28)$$

independent of t_1 and t_2 and dependent only on the time difference $t(=t_2-t_1)$.
From the basic definition of probability it is clear that for a stationary process

$$P_2(\xi_1,\xi_2,t)=p(\xi_1)P(\xi_1\,|\,\xi_2,t), \qquad (2.29)$$

where we have denoted the single point probability hitherto defined as $P_1(\xi_1)$ to be simply $p_1(\xi_1)$ which measures the a priori probability of finding ξ in the range ξ_1 and $\xi_1+d\xi_1$ initially. The conditional probability must obviously satisfy

$$\text{Positivity:}\quad P(\xi_1\,|\,\xi_2,t)\geq 0, \qquad (2.30)$$

$$\text{Conservation:}\quad \int d\xi_2 P(\xi_1\,|\,\xi_2,t)=1, \qquad (2.31)$$

$$\text{Projection:}\quad \int d\xi_1 p(\xi_1)P(\xi_1\,|\,\xi_2,t)=p(\xi_2). \qquad (2.32)$$

The master equation (2.25) for a stationary Markov process (SMP) assumes the structure

$$\frac{\partial}{\partial t}P(\xi_1\,|\,\xi,t)=\int d\xi'[P(\xi_1\,|\,\xi',t)\ W(\xi'\,|\,\xi)-P(\xi_1\,|\,\xi,t)\ W(\xi\,|\,\xi')], \qquad (2.33)$$

where the jump probability W and concomitantly the rate constant λ [Equation (2.19)] are time-independent.

References

Bharucha-Reid, A.T. 1960. *Elements of the Theory of Markov Processes and Their Applications*. New York: McGraw Hill.

Cox, D.R. and H.D. Miller 1994. *The Theory of Stochastic Processes*. London: Chapman & Hall.

Dattagupta, S. 2011. *Relaxation Phenomena in Condensed Matter Physics*, 2nd ed. Kolkata: Levant.
Dattagupta, S. 1987. *Relaxation Phenomena in Condensed Matter Physics*. Orlando, FL: Academic Press.
Doob, J.L. 1953. *Stochastic Processes*. New York: John Wiley & Sons.
Einstein, A. 1956. *Investigations on the Theory of the Brownian Motion*. New York: Dover.
Einstein, A. 1906a. Eine neue bestimmung der moleküldimensionen. *Ann. Physik* 17, 549.
Einstein, A. 1906b. Zur theorie der Brownschen bewegung. *Ann. Physik*, 19, 371.
Feller, W. 1972. *An Introduction to Probability Theory and Its Applications*, Vols. 1 and 2. New Delhi: Wiley Eastern.
Gardiner, C.W. 1997. *Handbook of Stochastic Processes for Physics, Chemistry and the Natural Sciences*. Berlin: Springer.
Gillespie, D.T. 1991. Markov Processes: *An Introduction for Physical Scientists*. New York: Academic Press.
Grimmett, G. and D. Strizaker. 2001. *Probability and Random Processes*, 3rd ed. Oxford: Oxford University Press.
Huang, K. 1987. *Statistical Mechanics*, 2nd ed. New York: John Wiley & Sons.
Ito, K. and H.P. McKean. 1965. *Diffusion Processes and Their Sample Paths*. Berlin: Springer.
Kac, M. 1959. *Probability and Related Tropics in Physical Sciences*. New York: Wiley-Interscience.
Lawler, G.F. 2006. *Introduction to Stochastic Processes*, 2nd ed. London: Chapman & Hall.
Lemons, D.S. 2002. *An Introduction to Stochastic Processes*. Baltimore: Johns Hopkins University Press.
Mazo, R.M. 2002. *Brownian Motion: Fluctuations, Dynamics and Applications*. Oxford: Clarendon Press.
Montroll, E.W. and J.L. Lebowitz, Eds. 1987. *Fluctuation Phenomena*. Amsterdam: North–Holland.
Nelson, E. 1967. *Dynamical Theories of Brownian Motion*. Princeton, NJ: Princeton University Press.
Oppenheim, I., K. Shuler and G. Weiss. 1977. *Stochastic Processes in Chemical Physics: The Master Equation*. Cambridge: MIT Press.
Soong, T.T. 1973. *Random Differential Equations in Science and Engineering*. New York: Academic Press.
Stratonovich, R.L. 1981. *Topics in the Theory of Random Noise*, Vol. I. New York: Gordon & Breach.
van Kampen, N.G. 1981. *Stochastic Process in Physics and Chemistry*. Amsterdam: North-Holland.
Wax, N., Ed. 1954. Selected papers. In *Noise and Stochastic Processes*. New York: Dover.
Weiss, G.H. 1994. *Aspects and Applications of the Random Walk*. Amsterdam: North-Holland.
Wong, E. 1983. *An Introduction to Random Processes*. New York: Springer.

3

Gaussian Processes

3.1 Introduction

The probability distribution expressed in Equation (1.30) or Equation (1.50) is ubiquitous in nature and considered Gaussian. It is not surprising then that stochastic diffusion and Gaussian stochastic processes go hand in hand. Since Equation (1.30) represents the probability density of a single variable m, the Gaussian nature is attributed merely to a single point probability $P_1(\xi,t)$ in the sense of Equation (2.8).

Because a stochastic process must be specified in terms of all joint probability densities, it is not sufficient simply on the basis of Equation (1.30) or (1.50) to call stochastic diffusion a Gaussian process. We must delve deeper into a proper definition of Gaussian processes. This is what we attempt to do below and in Chapter 4, and *inter alia* address whether stochastic diffusion is a Gaussian process and a Markov process as well. We also remark on the issue of stationarity. However, before we deal with Gaussian stochastic processes, it is expedient to discuss Gaussian random variables (Chaturvedi 1983).

A random variable is said to be Gaussian if the range of values it can take extends from $-\infty$ to $+\infty$ and if the probability distribution over this range is Gaussian, with a mean $\langle\xi\rangle$ (Feller 1966, van Kampen 1981):

$$P(\xi) = \frac{1}{\sigma\sqrt{2\pi}} \exp\left[\frac{-(\xi - \langle\xi\rangle)^2}{2\sigma^2}\right], \tag{3.1}$$

where σ is called the variance. We now generalize to n variables $\xi_1, \xi_2, \xi_3, \ldots \xi_n$ that may be denoted compactly as a column vector ξ:

$$\begin{pmatrix} \xi_1 \\ \xi_2 \\ \xi_3 \\ \cdot \\ \cdot \\ \cdot \\ \xi_N \end{pmatrix} \tag{3.2}$$

The joint probability distribution is then

$$P_n(\xi_1,\ \xi_2,\ldots\xi_n)=\frac{(\det\underline{\underline{A}})^{1/2}}{(2\pi)^{n/2}}\exp\left[-\frac{1}{2}(\underline{\xi}-\langle\underline{\xi}\rangle)^T.\underline{\underline{A}}.(\underline{\xi}-\langle\underline{\xi}\rangle)\right], \tag{3.3}$$

where the superscript T denotes "transpose" and $\underline{\underline{A}}$ is a positive definite symmetric matrix. What is the significance of the covariance matrix $\underline{\underline{A}}$? We show that its inverse is given by what is called the second cumulant, defined below. Introducing

$$\eta_i=\xi_i-\langle\xi_i\rangle, \tag{3.4}$$

we note that by definition,

$$\langle\eta_i\eta_j\rangle=\frac{(\det\underline{\underline{A}})^{1/2}}{(2\pi)^{n/2}}\int_{-\infty}^{\infty}d\underline{\eta}\ (\eta_i\eta_j)\exp\left[-\frac{1}{2}\underline{\eta}^T.\underline{\underline{A}}.\eta\right]. \tag{3.5}$$

Because $\underline{\underline{A}}$ is a real, symmetrical matrix, it can be diagonalized by an orthogonal transformation:

$$\underline{\underline{\Lambda}}=\underline{\underline{S}}^T\underline{\underline{A}}\underline{\underline{S}},\quad \underline{\underline{S}}^T\underline{\underline{S}}=1, \tag{3.6}$$

where $\underline{\underline{\Lambda}}$ is diagonal. We set

$$\underline{\eta}=\underline{\underline{S}}.\underline{\zeta}, \tag{3.7}$$

and use the fact that the Jacobian of the transformation (3.7) is unity to obtain

$$\langle\eta_i\eta_j\rangle=\frac{(\det\underline{\underline{A}})^{1/2}}{(2\pi)^{n/2}}\int_{-\infty}^{\infty}d\underline{\zeta}(\underline{\underline{S}}.\underline{\zeta})_i\ (\underline{\underline{S}}.\underline{\zeta})_j\exp\left[-\frac{1}{2}\underline{\zeta}^T.\underline{\underline{\Lambda}}.\underline{\zeta}\right]$$

$$=\frac{(\det\underline{\underline{A}})^{1/2}}{(2\pi)^{n/2}}\sum_{kl}S_{ik}S_{jl}\int_{-\infty}^{\infty}d\underline{\zeta}\zeta_k\zeta_l\exp\left[-\frac{1}{2}\sum_{m-1}^{n}\underline{\underline{\Lambda}}_m\zeta_m^2\right] \tag{3.8}$$

The right side of Equation (3.8) clearly equals* $(\underline{\underline{A}}^{-1})_{ij}$. Restoring the form of n_i [cf. Equation (3.4)], we conclude

$$\langle(\xi_i-\langle\xi_i\rangle)(\xi_j-\langle\xi_j\rangle)\rangle=(\underline{\underline{A}}^{-1})_{ij}\ . \tag{3.9}$$

* See Exercise 1.

Note that Equation (3.9) can be equivalently expressed as

$$\langle \xi_i \xi_j \rangle - \langle \xi_i \rangle \langle \xi_j \rangle = (\underline{\underline{A}}^{-1})_{ij}. \tag{3.10}$$

3.1.1 Moments and Cumulants

The entity $\langle \xi_i \xi_j \rangle$ is known as the second moment of the set of random variables $\underline{\xi}$. We therefore find that the inverse of the $\underline{\underline{A}}$ matrix can be identified with the variance matrix using the analogy with the single point distribution (3.1). Finally, the quantity on the left side of Equation (3.9) is called the second order cumulant. We will denote it by double brackets:

$$\langle\langle \xi_i \xi_j \rangle\rangle = \langle \xi_i \xi_j \rangle - \langle \xi_i \rangle \langle \xi_j \rangle. \tag{3.11}$$

A special (and frequently encountered) case is the zero mean (or first moment) for which Equation (3.11) can be rewritten as

$$\langle\langle \xi_i \xi_j \rangle\rangle = \langle \xi_i \xi_j \rangle. \tag{3.12}$$

A distinctive signature of Gaussian random variables is that all cumulants higher than the second vanish. Equivalently, all moments higher than the second can be rewritten in terms of the second moment. To prove this statement we must first establish the following important property of the multivariate Gaussian distribution of Equation (3.3):

$$\langle \xi_i f(\underline{\xi}) \rangle = \sum_j \langle \xi_i \xi_j \rangle \left\langle \frac{df(\underline{\xi})}{d\xi_j} \right\rangle, \tag{3.13}$$

where $f(\underline{\xi})$ is a polynomial function of ξ_is. Using Equations (3.10) and (3.12), we find that the right side of Equation (3.13) equals $\sum_j (\underline{\underline{A}}^{-1})_{ij} \left\langle \frac{df(\underline{\xi})}{d\xi_j} \right\rangle$. Therefore, proving Equation (3.13) is tantamount to showing that

$$\left\langle \frac{df(\underline{\xi})}{d\xi_i} \right\rangle = \sum_j \underline{\underline{A}}_{ij} \langle \xi_j f(\underline{\xi}) \rangle. \tag{3.14}$$

It is now easy to establish Equation (3.14) by rewriting the right side in terms of the integral over the probability distribution [Equation (3.3)]. If the first moment is zero, all higher odd moments are also zero, as can be proven with

Equation (3.12). What about the even moments higher than two? Consider for instance the fourth moment $\langle \xi_i \xi_j \xi_k \xi_l \rangle$. By identifying the polynomial $f(\underline{\xi})$ with $\langle \xi_j \xi_k \xi_l \rangle$, we may write from Equation (3.13),

$$\begin{aligned}
\langle \xi_i \xi_j \xi_k \xi_l \rangle &= \sum_m \langle \xi_i \xi_m \rangle \left\langle \frac{d}{d\xi_m} (\xi_j \xi_k \xi_l) \right\rangle \\
&= \sum_m \langle \xi_i \xi_m \rangle \; (\delta_{mj} \langle \xi_k \xi_l \rangle + \delta_{mk} \langle \xi_j \xi_l \rangle + \delta_{ml} \langle \xi_j \xi_k \rangle) \\
&= \langle \xi_i \xi_j \rangle \langle \xi_k \xi_l \rangle + \langle \xi_i \xi_k \rangle \langle \xi_j \xi_l \rangle + \langle \xi_i \xi_l \rangle \langle \xi_j \xi_k \rangle.
\end{aligned} \tag{3.15}$$

Thus the fourth order moment can be completely expressed in terms of second order moments. Equation (3.15) further implies that the fourth order cumulant, defined as

$$C_4 = \langle \xi_i \xi_j \xi_k \xi_l \rangle - \langle \xi_i \xi_j \rangle \langle \xi_k \xi_l \rangle - \langle \xi_i \xi_k \rangle \langle \xi_j \xi_l \rangle - \langle \xi_i \xi_l \rangle \langle \xi_j \xi_k \rangle, \tag{3.16}$$

is identically zero. Finally, rewriting Equation (3.15) in the compact form:

$$\langle \xi_i \xi_j \xi_k \xi_l \rangle = \sum_{pairs} \langle \xi_p \xi_q \rangle \langle \xi_r \xi_s \rangle,$$

we can generalize as

$$\langle \xi_i \xi_j \xi_k \xi_l \ldots \rangle = \sum_{pairs} \langle \xi_p \xi_q \rangle \langle \xi_r \xi_s \rangle \ldots \tag{3.17}$$

Therefore, the even moments of Gaussian random variables (with zero means) factorize in terms of second moments. It is interesting to note that the converse statement is also true, i.e., if the moments of a probability distribution factorize as in Equation (3.18), the distribution is Gaussian, rendering Equation (3.18) as a necessary and sufficient condition for Gaussian status.

3.1.2 Characteristic Function

A useful way of computing the moments of a probability density is to evaluate the characteristic function, defined as

$$C(\underline{k}) = \langle \exp(i\underline{k}.\underline{\xi}) \rangle = \sum_{m_1=0}^{\infty} \sum_{m_2=0}^{\infty} \ldots \sum_{m_n=0}^{\infty} \frac{(ik_1)^{m_1}}{m_1!} \frac{(ik_2)^{m_2}}{m_2!} \ldots \langle \xi_1^{m_1} \xi_2^{m_2} \ldots \xi_n^{m_n} \rangle. \tag{3.18}$$

For a Gaussian with zero mean we may write

$$C(\underline{k}) = \frac{(\det \underline{\underline{\mathcal{A}}})^{1/2}}{(2\pi)^{n/2}} \int d\underline{\xi} \exp\left[-\frac{1}{2}\underline{\xi}^T \cdot \underline{\underline{\mathcal{A}}} \cdot \underline{\xi} + \frac{i}{2}\underline{k}^T \cdot \underline{\xi} + \frac{i}{2}\underline{\xi}^T \cdot \underline{k}\right]$$

$$= \frac{(\det \underline{\underline{\mathcal{A}}})^{1/2}}{(2\pi)^{n/2}} \int d\underline{\xi} \exp\left[-\frac{1}{2}(\underline{\xi} - i\underline{\underline{\mathcal{A}}}^{-1}\underline{k})^T \underline{\underline{\mathcal{A}}}(\underline{\xi} - i\underline{\underline{\mathcal{A}}}^{-1}\cdot \underline{k}) - \frac{1}{2}\underline{k}\cdot^T \underline{\underline{\mathcal{A}}}^{-1}\cdot \underline{k}\right] \quad (3.19)$$

$$= \exp\left(-\frac{1}{2}\underline{k}^T \cdot \underline{\underline{\mathcal{A}}}^{-1} . \underline{k})\right).$$

Employing Equation (3.9), we may further write Equation (3.20) as

$$C(\underline{k}) = \exp\left[-\frac{1}{2}\sum_{ij} k_i \langle \xi_i \xi_j \rangle k_j\right]. \quad (3.20)$$

It is interesting to note that for a single random variable ξ, the characteristic function is simply the Fourier transform of the probability distribution (1.50). The value of the characteristic function, as is apparent from the right side of Equation (3.18), is that by differentiating $C(\underline{k})$ with respect to k_i an appropriate number of times, and setting $k_i = 0$, we can evaluate an arbitrary moment. The latter, as further implied by Equation (3.20), is given entirely in terms of the second moment. It is also easy to see that for a Gaussian with non-zero mean

$$C(\underline{k}) = \exp\left[-\frac{1}{2}\underline{k}^T . \underline{\underline{\mathcal{A}}}^{-1} . \underline{k} + i\underline{k}^T . \langle \underline{\xi} \rangle\right]$$

$$= \exp\left[i\sum_j k_j \langle \xi_j \rangle - \frac{1}{2}\sum_{ij} k_i << \xi_i \xi_j >> k_j\right]. \quad (3.21)$$

3.1.3 Cumulant Generating Function

The cumulant generating function $K(\underline{k})$, another useful entity for Gaussian random variables and Gaussian random processes discussed below, is obtainable from the logarithm of the characteristic function. Thus,

$$C(\underline{k}) = \exp(K(\underline{k})), \quad (3.22)$$

where

$$K(\underline{k}) = i\sum_j k_j \langle \xi_j \rangle - \frac{1}{2}\sum_{ij} k_i << \xi_i \xi_j >> k_j. \tag{3.23}$$

An alternate way of expressing the cumulant generating function is

$$K(\underline{k}) = \sum_{m_i=0}^{\infty} \frac{(ik_1)^{m_1}}{m_1!} \frac{(ik_2)^{m_2}}{m_2!} \ldots << \xi_1^{m_1} \xi_2^{m_2} \ldots >>. \tag{3.24}$$

A comparison of the right sides of Equations (3.23) and (3.24) is yet another substantiation of the fact that cumulants higher than the second vanish.

Another important theorem from Marcinkiewiez (1939) states that the cumulant generating function $K(k)$ is at best quadratic in k or contains all powers of $\underline{k}$. Correspondingly, all but the first two cumulants of a probability distribution vanish or the number of non-vanishing cumulants is infinite. This theorem has a relation to the Kramers–Moyal expansion in the derivation of the Fokker–Plank equation discussed in Chapter 5.

3.2 Gaussian Stochastic Processes

With this background to Gaussian random variables, we are set to introduce Gaussian processes. In general, a stochastic process is a function of random variables ω and t, wherein t is usually time in dynamical systems (Chaturvedi 1983):

$$\xi_\omega(t) = f(\omega, t). \tag{3.25}$$

In the two-parameter space of ω and t, *we have* two distinct ways of viewing a stochastic process f: (1) for a fixed ω, $\xi_\omega(t)$ is an ordinary function of time t, as in a deterministic system, and is called a realization of the process; (2) for a given value of t, $\xi_\omega(t)$ is a stochastic variable in the sense of Section 3.1. The latter is now a function of the random variable ω. The stochastic process $\xi_\omega(t)$ is then a continuum of random variables, one for each t.

As discussed in Section 2.2.1, we need all joint probabilities P_n to completely define a stochastic process. In analogy with Equation (3.3), a stochastic process is said to be Gaussian if all the joint probabilities are Gaussian and expressible in terms of the second order cumulant:

$$P_n(\xi_1,t_1;\xi_2,t_2;\ldots\xi_n,t_n)=\frac{(\det \underline{\underline{A}})^{1/2}}{(2\pi)^{n/2}}\exp\left[-\frac{1}{2}\sum_{ij}(\xi(t_i)-\langle\xi(t_i)\rangle)\underline{\underline{A}}_{ij}(\xi(t_j)-\langle\xi(t_j)\rangle)\right], \tag{3.26}$$

where the matrix $\underline{\underline{A}}$ is defined through its inverse:

$$(A^{-1})_{ij} = << \xi(t_i)\xi(t_j) >> = \langle(\xi(t_i)-\langle\xi(t_i)\rangle)\ (\xi(t_j)-\langle\xi(t_j)\rangle)\rangle \tag{3.27}$$

In the context of non-equilibrium statistical mechanics, the right side of Equation (3.27) is customarily referred to as a correlation function (see Chapter 4). It is natural to expect that all the formulae we derived earlier for Gaussian random variables will equally apply to Gaussian stochastic processes, with the proviso that partial derivatives will be replaced by functional derivatives and summation over t by integration. We enumerate all these formulae below.

3.2.1 Novikov Theorem

For a polynomial functional $f[\xi]$ of the stochastic process $\xi(t)$, we have (Novikov, 1964, 1965)

$$\langle\xi(t)f[\xi]\rangle=\int dt'\ \langle\xi(t)\xi(t')\rangle\left\langle\frac{\delta f(\xi)}{\delta\xi(t')}\right\rangle, \tag{3.28}$$

where $\left\langle\frac{\delta f(\xi)}{\delta\xi(t')}\right\rangle$, is the functional derivative of $f(\xi)$ with respect to $\xi(t')$.

3.2.2 Moment Theorem

For a Gaussian stochastic process with zero mean, i.e., $\langle\xi(t)\rangle=0$, the odd moments vanish and the even moments factorize pairwise.

$$\langle\xi(t_1)\xi(t_2)\ldots\rangle=\sum_{pairs}\langle\xi(t_p)\xi(t_q)\rangle\langle\xi(t_r)\xi(t_s)\rangle. \tag{3.29}$$

3.2.3 Characteristic Functional

The characteristic functional is defined by, in analogy with Equation (3.18), as

$$C(\underline{k}) = \exp\left(i\int_{-\infty}^{\infty} dt\ k(t)\ \xi(t)\right)$$

$$= \sum_{m=0}^{\infty} \frac{i^m}{m!} \int_{-\infty}^{\infty} dt_1 ... \int_{-\infty}^{\infty} dt_n k(t_1)...k(t_m)\ \langle \xi(t_1)...\xi(t_m)\rangle. \quad (3.30)$$

Further, for a Gaussian stochastic process with zero mean,

$$C(\underline{k}) = \exp\left[-\frac{1}{2}\int_{-\infty}^{\infty} dt_1 \int_{-\infty}^{\infty} dt_2 k(t_1)\ k(t_2)\langle \xi(t_1)\xi(t_2)\rangle\right], \quad (3.31)$$

and for $\langle \xi(t) \neq 0, \rangle$

$$C(\underline{k}) = \exp\left[i\int dt_1 k(t_1)\langle \xi(t_1)\rangle - \frac{1}{2}\ dt_1 \int dt_2 k(t_1) k(t_2) << \xi(t_1)\xi(t_2) >>\right], \quad (3.32)$$

where the double angular brackets have been defined in Equation (3.27). Interestingly, the characteristic function introduced in Equation (3.1) as a mathematical entity is directly related to the time-dependent structure factor, as can be measured in gas and liquid-phase spectroscopy. Finally, the Marcinkiewiez theorem mentioned at the end of Section 3.1 holds for stochastic processes as well.

3.3 Stationary Gauss–Markov Process

We saw earlier that for a Markov process, the joint n-point probability density is expressed entirely in terms of the two-point conditional probability [cf. Equation (2.13)]. Conversely, a Gaussian process is determined fully in terms of the second order cumulant [cf. Equation (3.26)]. Therefore, a Gaussian process is not in general a Markov process; nor is a Markov process generally a Gaussian stochastic process. However, in nature, including Brownian motion, we find a class of Gaussian processes that are Markovian and stationary. They are governed by what is known as Doob's theorem (Wax 1954).

3.3.1 Doob's Theorem

A stationary Gaussian process is Markovian if and only if the autocorrelation function, i.e., the correlation of the same stochastic process at two different times, is an exponential of the difference between the two epochs (Doob 1953):

$$<< \xi(t_1)\xi(t_2) >> \; = \; C\exp[-\gamma\,|\,t_1 - t_2\,|\,], \tag{3.33}$$

where γ *is* a rate parameter and C a constant. For a multi-component stochastic process for which $\xi(t\ )$ is a column vector, $\xi(t\ ), \gamma$ is replaced by a rate matrix Γ. The proof is sketched below for the case in which $\langle\xi(t\ )\rangle = 0$, but can be easily generalized to the situation* in which $\langle\xi(t\ )\rangle \neq 0$. Our starting point is the defining equation [cf. Equation (2.9)] for the two-point conditional probability:

$$P(\xi_1, t_1 \,|\, \xi_3, t_3) = \frac{P_2(\xi_1, t_1; \xi_3, t_3)}{P_1(\xi_1, t_1)}. \tag{3.34}$$

The quantities on the right side for a Gaussian process are given by Equations (3.1) and (3.26):

$$P(\xi_1, t_1) = \frac{1}{\sqrt{2\pi\langle\xi_1^2\rangle}} \exp\left(-\frac{\xi_1^2}{2\langle\xi_1^2\rangle}\right), \tag{3.35}$$

(read ξ_1 as $\xi(t_1)$)

$$P_2(\xi_1, t_1; \xi_3, t_3) = \frac{(\det \underline{\underline{\mathcal{A}}})^{1/2}}{2\pi} \exp\left[-\frac{1}{2}\left(\mathcal{A}_{11}\xi_1^2 + \mathcal{A}_{33}\xi_3^2 + 2\mathcal{A}_{13}\xi_1\xi_3\right)\right]. \tag{3.36}$$

Recall that the inverse of $\underline{\underline{\mathcal{A}}}$ is given by

$$\underline{\underline{\mathcal{A}}}^{-1} = \begin{pmatrix} \langle\xi_1^2\rangle & \langle\xi_1\xi_3\rangle \\ \langle\xi_1\xi_3\rangle & \langle\xi_3^2\rangle \end{pmatrix}. \tag{3.37}$$

It is clear that the determinant is

$$\mathcal{D}_{13} = \det \underline{\underline{\mathcal{A}}} = \langle\xi_1^2\rangle\langle\xi_3^2\rangle - \langle\xi_1\xi_3\rangle^2. \tag{3.38}$$

* See Exercise 2.

Further, the $\mathcal{A}$ matrix is evidently

$$\underline{\underline{\mathcal{A}}} = \frac{1}{\mathcal{D}_{13}} = \begin{pmatrix} \langle \xi_3^2 \rangle & -\langle \xi_1 \xi_3 \rangle \\ -\langle \xi_1 \xi_3 \rangle & \langle \xi_1^2 \rangle \end{pmatrix} . \tag{3.39}$$

Substituting Equations (3.35) and (3.36) into (3.34), we obtain

$$\begin{aligned} P(\xi_1, t_1 \mid \xi_3, t_3) &= \frac{\langle \xi_1^2 \rangle}{2\pi \mathcal{D}_{13}} e^{\frac{\xi_1^2}{2\langle \xi_1^2 \rangle}} .\exp\left\{ -\frac{1}{2\mathcal{D}_{13}} \left(\langle \xi_3^2 \rangle \xi_1^2 + \langle \xi_1^2 \rangle \xi_3^2 - 2 \langle \xi_1 \xi_3 \rangle \xi_1 \xi_3 \right) \right\} \\ &= \frac{\langle \xi_1^2 \rangle}{2\pi \mathcal{D}_{13}} \exp\left\{ -\frac{\langle \xi_1^2 \rangle}{2\mathcal{D}_{13}} \left(\xi_3 - \xi_1 \frac{\langle \xi_1 \xi_3 \rangle}{\xi_1^2} \right)^2 \right\}. \end{aligned} \tag{3.40}$$

With this result for a Gaussian process, i.e., Equation (3.40), we first evaluate the conditional average of $\xi = \xi_3$ at time t_3, given that the value of ξ at an earlier time t_1 was ξ_1. The required expression is

$$\begin{aligned} \langle \xi_3 \rangle_{\xi_1(t_1)=\xi_1} &= \int d\xi_3 \xi_3 P(\xi_1, t_1 | \xi_3, t_3), \; t_3 > t_1 \\ &= \xi_1 \frac{\langle \xi_1 \xi_3 \rangle}{\langle \xi_1^2 \rangle} . \end{aligned} \tag{3.41}$$

We now calculate the left side of Equation (3.41) anew based only on the assumption that the stochastic process $\xi(t)$ is Markovian. For the latter we have [cf. Equation (2.14)]

$$< \xi_3 >_{\xi_1(t_1)=\xi_1} = \int \int d\xi_2 d\xi_3 \;\; \xi_3 \;\; P(\xi_1, t_1 \mid \xi_2, t_2) \;\; P(\xi_2, t_2 \mid \xi_3, t_3). \tag{3.42}$$

However, if the process is also a Gaussian,

$$\int d\xi_3 \;\; \xi_3 \;\; P(\xi_2, t_2 \mid \xi_3, t_3) = \xi_2 \frac{< \xi_2 \xi_3 >}{< \xi_2^2 >}, \tag{3.43}$$

using the result of Equation (3.41). Substituting then in Equation (3.42),

$$< \xi_3 >_{\xi_1(t_1)=\xi_1} = \int d\xi_2 \cdot \;\; \xi_2 \frac{< \xi_2 \xi_3 >}{< \xi_2^2 >} \cdot \;\; P(\xi_1, t_1 \mid \xi_2, t_2),$$

which, with a further use of Equation (3.41), yields

$$\langle \xi_3 \rangle_{\xi_1(t_1)=\xi_1} = \frac{\langle \xi_2 \xi_3 \rangle}{\langle \xi_2^2 \rangle} . \xi_1 \frac{\langle \xi_1 \xi_2 \rangle}{\langle \xi_1^2 \rangle}. \tag{3.44}$$

We can now combine the results of (3.41) and (3.44) to conclude that for a process that is both Gaussian and Markovian, we have the important equality:

$$\frac{<\xi_1\xi_3>}{\sqrt{<\xi_1^2><\xi_3^2>}} = \frac{<\xi_1\xi_2>}{\sqrt{<\xi_1^2><\xi_2^2>}} \cdot \frac{<\xi_2\xi_3>}{\sqrt{<\xi_2^2><\xi_3^2>}}, \quad t_3 > t_2 > t_1. \tag{3.45}$$

Our final step in the quest for establishing Doob's theorem is to assume further that the process is stationary as well as Gauss–Markovian. For a stationary process

$$\langle \xi_1^2 \rangle = \langle [\xi(t_1)]^2 \rangle = \langle \xi^2 \rangle, \tag{3.46}$$

independent of time t_1. In addition, the auto-correlation function depends only on the time difference of the two epochs. For instance,

$$<\xi_1\xi_2> = <\xi(t_1)\xi(t_2)> = <\xi(0)\xi(t_2 - t_1)>, \quad t_2 > t_1. \tag{3.47}$$

Combining with Equation (3.42),

$$\langle \xi(0)\xi(t_3 - t_1) \rangle = \langle \xi(0)\xi(t_3 - t_2) \rangle. \ \langle \xi(0)\xi(t_2 - t_1) \rangle. \tag{3.48}$$

The factorization implied in Equation (3.48) is both a necessary and sufficient condition for a one-dimensional Gaussian process to be Markovian. For the factorization to be valid, we must have

$$\langle \xi(0) \ \xi(t_3 - t_1) \rangle = \langle \xi^2 \rangle \exp[-\gamma \,|\, t_3 - t_1 \,|\,], \tag{3.49}$$

hence the proof of Doob's theorem.

Exercises

1. Derive Equation (3.9).
2. Prove the statement following Equation (3.34).

References

Chaturvedi, S. 1983. In Agarwal, G.S. and S. Dattagupta, Eds., *Stochastic Processes: Formalism and Applications,* Vol. 184. Berlin: Springer.

Doob, J.L.1953. *Stochastic Process.* New York: John Wiley & Sons.

Feller, W. 1966. *Introduction to Probability Theory and Its Applications,* Vol. 2. New York: John Wiley & Sons.

Marcinkiewiez, J. 1939. Sur une propriete de la loi de Gauss. *Math. Z.* 44, 612.

Novikov, E. A. 1965. Functionals and the random-force method in turbulence theory. *Sov. Phys.* JETP 20, 1290.

Novikov, E.A. 1964. zh. Eksp. Teor. Fiz. 47, 1919 [*Sov. Phys. JETP*, 1990].

van Kampen, N.G. 1981. *Stochastic Process in Physics and Chemistry.* Amsterdam: North-Holland.

Wax, N., Ed. 1954. Selected papers. In *Noise and Stochastic Processes.* New York: Dover.

4

Langevin Equations

4.1 Introduction

At the beginning of Chapter 3, we questioned whether diffusion in the sense of Brownian motion discussed in Chapter 1 is a Gaussian–Markov process. The answer remains unsettled as we have been merely able to connect the one-point Gaussian distribution in Equation (1.50) to the one-dimensional position coordinate x of the Brownian particle.

The purpose of this chapter is to pursue this issue further by examining the mathematical structure of Brownian motion in detail (Chandrasekhar 1943) with the aid of Langevin equations (Langevin 1908). The simplest of these equations refers to the stochastic dynamics of a free particle under the fluctuating effects of its environment, much like Brown's pollen grains undergoing incessant collisions with the surrounding water molecules. Before we discuss Langevin equations, it is useful to recall some of the well-known results of the classical kinetic theory of ideal gases.

An ideal gas is a collection of low density molecules that are too far apart to interact until they are in direct contact when a collision occurs. The kinetic theory makes a tacit assumption of timescale separation in that the time of impact during a collision is taken to be effectively zero or much smaller than the mean time T_m (mean free time) between collisions. This assumption is known as impact approximation (Huang 1987). The resultant kinetic theory based on the Boltzmann transport equation admits a stationary (or equilibrium) solution for a probability distribution of the momentum p—the well-known Maxwell distribution:

$$P_{eq}(p) = \left(\frac{\beta}{2\pi m}\right)^{1/2} \exp\left(-\frac{\beta p^2}{2m}\right), \tag{4.1}$$

where $\beta = 1/K_B T$, K_B is the Boltzmann constant, T *is* the temperature, and m is the mass of the molecule. Note that Equation (4.1) is a Gaussian distribution of a single random variable p, in the sense of Chapter 3. The task in front of a stochastic model is to set up a time-dependent formulation that has Equation (4.1) as a natural stationary solution.

Our dynamical formulation throughout this book will be based on Hamiltonian mechanics operating in the phase space of position and momentum variables. Initially, we keep the discussion simple by considering only one dimensional motion for which the Hamiltonian equations of motion can be written as

$$\begin{aligned} \frac{dx}{dt} &= \frac{\partial \mathcal{H}}{\partial p} = \frac{p}{m} \\ \frac{dp}{dt} &= -\frac{\partial \mathcal{H}}{\partial x} = -\frac{\partial V}{\partial x}, \end{aligned} \tag{4.2}$$

where $\mathcal{H}$ is the Hamiltonian given by

$$\mathcal{H} = \frac{p^2}{2m} + V(x), \tag{4.3}$$

$V(x)$ represents the potential energy. The reader may recall the curious asymmetry of nature that only for harmonic motion ($V(x) \propto x^2$) is the set of Equations (4.2) in x and p linear, while the equation of motion for x is linear in p, that for the latter generally has non-linear terms in x! Non-linearity implies that usual matrix algebra methods or integral transforms of the Fourier or Laplace type do not work. However, in all cases, the first equation in (4.2) is integrable and yields

$$x(t) = x(0) + \frac{1}{m}\int_0^t dt' p(t'). \tag{4.4}$$

Thus, for a given initial value $x(0)$, $x(t)$ is entirely determined in terms of $p(t)$. In that sense, $p(t)$is the driver whereas $x(t)$ is the driven process.

4.2 Free Particle in Momentum Space

We now assume that the system described by Equation (4.2) is embedded in an environment of fluctuating forces. The driver process takes a very simple form when the system at hand is only a free particle, as in ideal gas kinetic theory. The second part of Equation (4.2) then assumes the form for $V(x) = 0$ (Balakrishnan 2009, Dattagupta and Puri 2004)

$$\frac{dp}{dt} = F_e(t), \tag{4.5}$$

where the right side is the force due to the environment. Inasmuch as we can invoke the time scale separation of ideal gas kinetics for which the impact approximation holds, we are allowed to perform coarse graining in time for $t >> \tau_{im}$ where the latter term is the time of impact

$$F_e(t) = F_s + F_r(t), \tag{4.6}$$

where F_s is the systematic component of the force that results from the average cumulative effect of environmental collisions while $F_r(t)$ is the residual random force. This concept is entirely in tune with Einstein's formulation of Brownian motion (Chapter 1). The genius of Langevin (who was almost a contemporary of Einstein) was modeling systematic force as a linear function of instantaneous momentum

$$F_s = -\gamma p(t), \tag{4.7}$$

where the constant of proportionality γ (having the dimension of a frequency) is called the coefficient of friction and the negative sign shows that the systematic force is in opposition to the motion. Again the perceived scenario is in conformity with real experience of dropping a metal ball in an oil jar and seeing the ball slow down due to upward viscous force. Indeed F_s can be identified with the Stokes force of Equation (1.42).

We can now model the random force $F_r(t)$. As a first step. it is evident that the average force taken over many equivalent realizations of the environment (denoted by angular brackets, below), must vanish, i.e.,

$$\langle F_r(t) \rangle = 0. \tag{4.8}$$

Further, if we take the average of the product of the random forces at two different times t_1 and t_2, that must also disappear unless t_1 and t_2 are the same, i.e.,

$$\langle F_r(t_1) F_r(t_2) \rangle = 0, \quad \text{for} \quad t_1 \neq t_2. \tag{4.9}$$

In the sense of the theory of distribution, we can write

$$\langle F_r(t_1) F_r(t_2) \rangle = \Gamma \delta(t_1 - t_2), \tag{4.10}$$

where δ is the Dirac delta function and Γ is a parameter of dimension: $(force)^2 \times time$. Combining all these we have the free-particle Langevin equation as

$$x(t) = x(0) + \frac{1}{m}\int_0^t dt' p(t'), \tag{4.11a}$$

$$\frac{dp(t)}{dt} = -\gamma p(t) + F_r(t), \tag{4.11b}$$

$$\langle F_r(t) \rangle = 0, \tag{4.11c}$$

$$\langle F_r(t) F_r(t') \rangle = \Gamma \delta(t - t'). \tag{4.11d}$$

We have thus set up a hierarchy of stochastic processes, the most primitive of these is $F_r(t)$ that governs the driver process $p(t)$ via Equation (4.11b). The latter further drives the driven process $x(t)$ through Equation (4.11a).

What kind of a stochastic process is $F_r(t)$? First note that its correlation in Equation (4.10) implies that it is a stationary process. Also, because $\delta(t) = \delta(-t)$, Equation (4.10) can be rewritten as

$$\langle F_r(t) F_r(t_2) \rangle = \Gamma \delta(|t_1 - t_2|). \tag{4.12}$$

Because the power spectrum of the process defined by

$$S_\omega = \frac{1}{2\pi}\int_{-\infty}^{\infty} d\tau \; e^{i\omega\tau} \langle F_r(0) F_r(\tau) \rangle, \qquad \tau \equiv |t_2 - t_1|, \tag{4.13}$$

is a constant (= Γ), in view of the delta function nature of the correlation, the stationary process defined by Equation (4.12) is called a white noise. However, the presence of the delta function also implies that the process is not stochastic in the strictest mathematical sense. On the other hand, because the delta function has the following exponential representation:

$$\delta(t) = \underset{\nu \to \infty}{Lim} \frac{\nu}{2} e^{-\nu|t|}, \tag{4.14}$$

the process $F_r(t)$ is used to model a large number of phenomena with very rapid (i.e., $\nu \to \infty$) fluctuations. Indeed this limiting form of the exponential correlation makes $F_r(t)$ a Gaussian–Markov process as well, in the sense of Doob's theorem [cf. Equation (3.33)].

The Gaussian nature of $F_r(t)$ can also be established by showing that all the odd order correlation functions of $F_r(t)$ vanish. For instance, the three-point correlation can be written as $< F_r(t_1)\ F_r(t_2)\ F_r(t_3) >$, which is of course zero if $t_1 \neq t_2 \neq t_3$, purely from the randomness of $F_r(t)$ but, even when $t_1 = t_2 = t_3$, the correlation still is zero, from inversion (i.e., parity) symmetry. Next, the four-point correlation of $F_r(t)$ can be first factored into a symmetric, equally weighted linear combination of products of all the possible two-point correlations, i.e.,

$$< F_r(t_1)F_r(t_2)F_r(t_3)F_r(t_4) > = < F_r(t_1)F_r(t_2) >< F_r(t_3)F_r(t_4) > + < F_r(t_1)F_r(t_3) >< F_r(t_2)F_r(t_4) > \\ + < F_r(t_1)F_r(t_4) >< F_r(t_2)F_r(t_3) > . \tag{4.15}$$

From Equation (4.10),

$$< F_r(t_1)F_r(t_2)F_r(t_3)F_r(t_4) > \; = \; \Gamma[\delta(t_1 - t_2)\delta(t_3 - t_4) + \delta(t_1 - t_3)\delta(t_2 - t_4) + \delta(t_1 - t_4)\delta(t_2 - t_3)]. \tag{4.16}$$

Similar factorization holds for all even-order correlation functions—the hallmark of a Gaussian process.* Having established the nature of $F_r(t)$, we now focus on the momentum $p(t)$ governed by Equation (4.11b). The solution of Equation (4.11b) is evidently

$$p(t) = p(0)\ e^{-\gamma t} + \int_0^t dt' e^{-\gamma(t-t')} F_r(t'). \tag{4.17}$$

For a fixed initial momentum $p(0)$ and upon averaging over $F_r(t)$ per Equation (4.11c), we have

$$\langle p(t) \rangle = p(0) e^{-\gamma t}, \tag{4.18}$$

which is a solution of the constitutive equation:

$$\frac{d}{dt}\langle p(t) \rangle = -\gamma \langle p(t) \rangle. \tag{4.19}$$

The inverse of $\gamma (= \gamma^{-1})$ is interpreted as the relaxation time for momentum fluctuations—sometimes also called Smoluchowski time. Note that time-reversal symmetry is manifestly broken in Equation (4.19), indicative of the presence of dissipation, which is why γ is also known as the friction coefficient.

* See Exercise 1.

To examine momentum fluctuations, we first consider the mean squared momentum which, from Equations (4.17), (4.11c), and (4.11d), can be written as

$$< p^2(t) > = p^2(0)\ e^{-2\gamma t} + \int_0^t dt' \int_0^t dt''\ e^{-\gamma(t-t')-\gamma(t-t'')} < F_r(t')\ F_r(t'') > , \quad (4.20)$$

because the cross terms vanish in view of Equation (4.11c) and causality. (The momentum is the effect while the noise is the cause; hence the initial momentum must be decoupled from the noise.) Use of the delta function in the correlation [cf. Equation (4.11d)] then yields

$$\langle p^2(t)\rangle = p^2(0)e^{-2\gamma t} + \frac{\Gamma}{2\gamma}(1-e^{-2\gamma t}). \quad (4.21)$$

We now need to fix the parameter Γ. Recall Brown's experiment in which the liquid jar was always in thermal equilibrium at a temperature T. Equations (4.18) and (4.21) measure how the average momentum and its fluctuation evolve over time when a given tagged particle is identified with an initial momentum $p(0)$. As expected, the average momentum relaxes to zero as the time t approaches infinity whereas the fluctuation becomes

$$\underset{t\to\infty}{Lim}\langle p^2(t)\rangle = \frac{\Gamma}{2\gamma}. \quad (4.22)$$

However, we expect that at infinite times the tagged particle (in Brown's case, the pollen grain) would have reached thermal equilibrium of the environment. Hence

$$\underset{t\to\infty}{Lim}\langle p^2(t)\rangle = \langle p^2\rangle_{eq}, \quad (4.23)$$

where the latter must be governed by the momentum distribution in Equation (4.1). Clearly,

$$< p^2 >_{eq} = \int_{-\infty}^{\infty} dp\ \ p^2 P_{eq}(p) = mK_B T, \quad (4.24)$$

which is also the statement of the equipartition theorem. From Equation (4.22),

$$\Gamma = 2m\gamma K_B T. \quad (4.25)$$

Two comments regarding Equation (4.25) are now in order. First, it vindicates the separation of the environmentally induced force $F_e(t)$ of Equation (4.6) into the systematic force F_s governed by γ [(cf. Equation (4.7)] and the random force F_r measured by Γ [cf. Equation (4.10)], thus implying that the origins of γ and Γ are the same. They both belong to the environment.

Second, γ is a parameter that characterizes dissipation, whereas Γ, being proportional to the power spectrum of the random force [cf. Equation (4.13)], describes fluctuations. We shall therefore call Equation (4.25) the second fluctuation–dissipation theorem and reserve the first fluctuation–dissipation theorem for fluctuations of the driver process, i.e., $x(t)$ [cf. Equation (4.11a)] implicit in Einstein's work on Brownian motion.

Based on the above results, we may draw the inevitable conclusion that dissipation is an essential ingredient of a system's ability to reach equilibrium. Indeed the connecting formula between Γ and γ via Equation (4.25) is an expression of inner consistency of what Einstein called the "correctness of Boltzmann's interpretation of thermodynamics" (see Section 1.6 of Chapter 1).

We now turn our attention to the autocorrelation of momentum from Equation (4.17):

$$< p(t)p(\tau) > = p_0^2 \; e^{-\gamma(t+\tau)} + \int_0^t dt' \int_0^\tau dt'' e^{-\gamma(t-t')-\gamma(\tau-t'')} < F_r(t')F_r(t'') > . \tag{4.26}$$

Again, the cross terms vanish because of Equation (4.11c). Further use of the delta function in the integrand for $t > \tau$ leads to

$$< p(t)p(\tau) >= p_0^2 e^{-\gamma(t+\tau)} + mK_B T(e^{-\gamma(t-\tau)} - e^{-\gamma(t+\tau)}), \tag{4.27}$$

where we also used Equation (4.25). We now have two routes to obtaining autocorrelation in equilibrium. The first is to let both t and τ run to infinity while keeping their difference $(= t - \tau)$ fixed, thus yielding

$$< p(t)p(\tau) >_{eq} = mK_B T \; e^{-\gamma(t-\tau)}, \quad t > \tau. \tag{4.28}$$

Alternatively, we can average p_0 in Equation (4.27) over the equilibrium momentum distribution of Equation (4.1) and the resultant equipartition theorem yields the same result (4.28). Finally, the assumption of $t > \tau$ does not imply a loss of generality. The reverse case of $\tau > t$ would have produced the same answer as Equation (4.28) by simply interchanging t and τ. Hence we have the compact expression:

$$< p(t)p(\tau) >_{eq} = \; < p^2)_{eq} e^{-|t-\tau|}. \tag{4.29}$$

Thus, although the random force is delta-correlated, the momentum is exponentially correlated. What about the higher order correlations of the momentum? Since all odd correlations of $F_r(t)$ are zero, it is easy to show that

$$< p(t_1)p(t_2)p(t_3) >_{eq} = 0. \tag{4.30}$$

Next, using Equations (4.16) and (4.17), we can prove that*

$$< p(t_1)p(t_2)p(t_3)p(t_4) >_{eq} = (mK_BT)^2 \{e^{-\gamma|t_1-t_2|-\gamma|t_3-t_4|} + e^{-\gamma|t_1-t_3|-\gamma|t_2-t_4|} + e^{-\gamma|t_1-t_4|-\gamma|t_2-t_3|}\}. \tag{4.31}$$

This result implies the factorization

$$\langle p(t_1)p(t_2)p(t_3)p(t_4)\rangle_{eq} = \langle p(t_1)p(t_2)\rangle_{eq}\langle p(t_3)p(t_4)\rangle_{eq} + \langle p(t_1)p(t_3)\rangle_{eq}\langle p(t_2)p(t_4)\rangle_{eq}$$
$$+\langle p(t_1)p(t_4)\rangle_{eq}\langle p(t_2)p(t_3)\rangle_{eq}, \tag{4.32}$$

that further establishes the Gaussian nature of the stochastic process $p(t)$.

The form of the autocorrelation function in Equation (4.29) additionally implies that $p(t)$ is a stationary Gaussian–Markov process. The latter property ensures that all joint probability densities are expressible in terms of the single-point probability and the two-point conditional probability [cf. Chapter 3; see Equations (3.36) and (3.41)]. These are given by

$$P_1(p_1) = \frac{1}{\sqrt{2\pi(< p_1^2 >_{eq}}} \exp\left(- \frac{p_1^2}{2 < (p_1^2)_{eq} >}\right) \tag{4.33}$$

and

$$P(p_1, t_1 \mid p_2, t_2) = \left[\frac{< p_1^2 >_{eq}}{2\pi(< p_1^2 >_{eq} < p_2^2 >_{eq} - < p(t_1)p(t_2) >_{eq}^2}\right]^{1/2}$$

$$.\exp\{ - \frac{< p_1^2 >_{eq}}{2 << p_1^2 >_{eq} < p_2^2 >_{eq} - < p(t_1)p(t_2) >_{eq}^2}(p_2 - p_1 \frac{< p(t_1)p(t_2) >}{< p_1^2 >_{eq}})^2\}, t_2 > t_1. \tag{4.34}$$

---

* See Exercise 2.

Using the equipartition theorem of Equation (4.24), Equation (4.33) reduces to Equation (4.1). Further, Equation (4.28) implies that

$$P(p_1,t_1 \mid p_2,t_2) = \left[\frac{1}{2\pi m K_B T(1-e^{-2\gamma(t_2-t_1)})}\right]^{1/2}$$
$$.\exp\left\{-\frac{1}{2\pi m K_B T(1-e^{-2\gamma(t_2-t_1)})}\left(p(t_2)-p(t_1)e^{-\gamma(t_2-t_1)}\right)^2\right\}, \quad t_2 > t_1 \tag{4.35}$$

The above distributions constitute an Ornstein–Uhlenbeck process (Uhlenbeck and Ornstein 1930). We should remark, however, that while the Langevin equation for momentum is in the form of an equation of motion, like the Heisenberg formulation of quantum mechanics, the associated probability distributions of Equations (4.33) and (4.34) are in the Schrödinger picture (Dirac 1958). This distinction (and also the correspondence) will be amplified later.

4.3 Free Particle in Position Space

Having discussed in detail the properties of the momentum process for a free particle, we now concentrate on the stochastic process associated with the coordinate $x(t)$. This will enable us to make contact between the Langevin equation formulation and Einstein's analysis of Brownian motion discussed in Chapter 1. By combining Equations (4.11a) and (4.17), it is easy to see that

$$x(t) = x(0) + \frac{p(0)}{m\gamma}(1-e^{-\gamma t}) + \int_0^t dt' \int_0^{t'} dt'' e^{-\gamma(t'-t'')} F_r(t''). \tag{4.36}$$

Clearly,

$$\langle x(t)\rangle = x(0) + \frac{p(0)}{m\gamma}(1-e^{-\gamma t}), \tag{4.37}$$

which for short times ($\gamma t << 1$) behaves as

$$\langle x(t)\rangle \approx x(0) + \frac{p(0)}{m}t + \ldots, \tag{4.38}$$

in accordance with Newtonian mechanics. The next higher order term is interesting to inspect:

$$\langle x(t)\rangle \approx x(0)+\frac{p(0)}{m}t-\frac{1}{2}\frac{\gamma\, p(0)}{m}t^2+\ldots \tag{4.39}$$

As expected, the motion is decelerated by friction where the deceleration equals $\frac{\gamma p(0)}{m}$. Physically, deceleration results from the retarding Stokes force appropriate to the initial momentum $p(0)$. To estimate the fluctuation in $x(t)$, it is more convenient rewrite Equation (4.4) as

$$X(t)=\frac{1}{m}\int_0^t dt' p(t'), \tag{4.40}$$

where

$$X(t)=x(t)-x(0). \tag{4.41}$$

Evidently,

$$\langle X^2(t)\rangle_{eq}=\frac{1}{m^2}\int_0^t dt'\int_0^t dt''\langle p(t')p(t'')\rangle_{av}. \tag{4.42}$$

Because the autocorrelation of the momentum depends only on the time difference $t'-t''$and is also a symmetric function of this time difference, i.e., of $|t'-t''|$, the region of integration [originally a square in the (t',t'') plane (Figure 4.1) can be halved to include only the triangle in which $t''<t'$, while at the same time the integrand is doubled. Hence,

$$\langle X^2(t)\rangle_{eq}=\frac{2}{m^2}\int_0^t dt'\int_0^{t'} dt''<p(t)p(t'')>_{eq}. \tag{4.43}$$

Substituting Equation (4.29), we obtain

$$\begin{aligned}\langle X^2(t)\rangle_{eq}&=\frac{2K_BT}{m\gamma^2}\int_0^t dt'\int_0^{t'} dt'' e^{-\gamma(t'-t'')}\\&=\frac{2K_BT}{m\gamma^2}(\gamma t-1+e^{-\gamma t}).\end{aligned} \tag{4.44}$$

Figure 4.2 shows a plot of Equation (4.44).

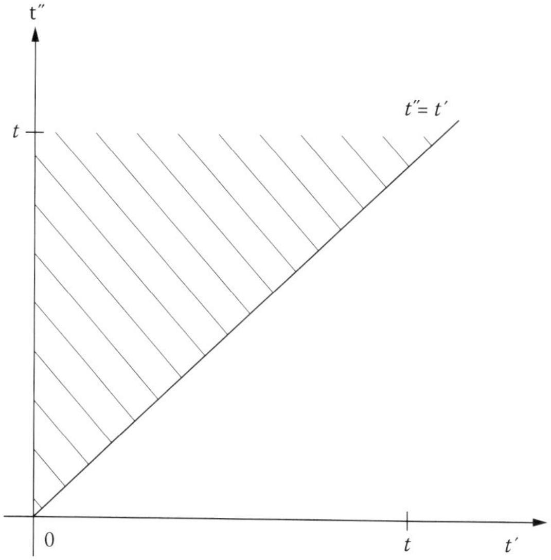

FIGURE 4.1
Relevant integration region.

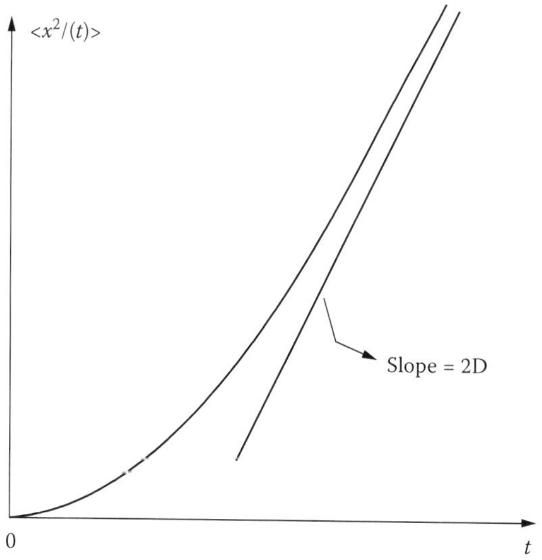

FIGURE 4.2
Plot of Equation (4.45) depicting mean squared displacement of a free Brownian particle as a function of time t. As expected at short times $\gamma t \ll 1$, we see ballistic behavior $\sim t^2$ while at long times $\gamma t \gg 1$ diffusive motion sets in, characterized by a straight line with slope 2D.

If we want the mean-squared displacement to follow the result in Equation (1.53) for Brownian motion, we must consider the time regimes in which $\gamma t >> 1$, in which case

$$\left\langle X^2(t)\right\rangle_{eq} = \frac{2K_BT}{m\gamma}t. \tag{4.45}$$

Comparing with Equation (1.53),

$$Ds = \frac{K_BT}{m\gamma}. \tag{4.46}$$

Further, comparison with Equation (1.55) yields

$$\gamma = \frac{6\pi a\eta}{m}, \tag{4.47}$$

which indeed identifies the systematic force F_s in Equation (4.7) with Einstein's use of the Stokes relation [Equation (1.42)].

One other important remark: because the mean squared displacement of a Brownian particle is explicitly dependent on time $t, x(t)$ is not a stationary stochastic process although its driver process $p(t)$ is. The autocorrelation of the displacement in equilibrium may be ascertained from Equation (4.37):

$$\left\langle X(t)X(t')\right\rangle_{eq} = \frac{1}{m^2}\int_0^t dt_1 \int_0^{t'} dt_2 < p(t_1)\; p(t_2) >_{eq}. \tag{4.48}$$

From Equation (4.29),

$$\left\langle X(t)X(t')\right\rangle_{eq} = \frac{K_BT}{m}\int_0^t dt_1 \int_0^{t'} dt_2 \; e^{-\gamma|t_1-t_2|}. \tag{4.49}$$

Assuming first that $t > t'$, the integral to the right can be split into two parts as follows:

$$\left\langle X(t)X(t')\right\rangle_{eq} = \frac{K_BT}{m}\int_0^{t'} dt_1 \int_0^{t'} dt_2 \; e^{-\gamma(t_1-t_2)} + \frac{K_BT}{m}\int_{t'}^{t} dt_1 \int_0^{t'} dt_2 \; e^{-\gamma(t_1-t_2)}, \tag{4.50}$$

where we removed from the second term the vertical bars in the exponent because t_1 is always greater than t_2, in view of the extant domains of integration. The first term can be further manipulated by employing the technique mentioned below Equation (4.42). Thus

$$\langle X(t)X(t')\rangle_{eq} = \frac{2K_BT}{m}\int_{t'}^{t'} dt_1 \int_{0}^{t_1} dt_2\, e^{-\gamma(t_1-t_2)} + \frac{K_BT}{m}\int_{t'}^{t} dt_1 \int_{0}^{t'} dt_2\, e^{-\gamma(t_1-t_2)}, \quad (4.51)$$

where the vertical bars in the exponent have been removed altogether. We finally arrive at

$$< X(t)X(t') >_{eq} = \frac{K_BT}{m\gamma^2}[2\gamma\, t' - 1 + e^{-\gamma t} + e^{-\gamma t'} - e^{-\gamma(t-t')}] \quad (t > t'). \quad (4.52)$$

The expression for $t < t'$ is obtained simply by interchanging t and t' in Equation (4.53). We also note that for $t = t'$ we recover Equation (4.44) as expected. With this form of the correlation function, it is evident that $X(t)$, in general, is not a Markov process. However, if we examine the behavior of Equation (4.53) in the high-friction (diffusive) regime $\gamma t >> 1$, $\gamma t' >> 1$, we find

$$\langle X(t)X(t')\rangle_{eq} = \frac{2K_BT}{m\gamma}t'. \quad (t > t'). \quad (4.53)$$

Using Equation (4.44), we have in general,

$$\langle X(t)X(t')\rangle_{eq} = 2\mathcal{D}_s \min(t,t') = 2\mathcal{D}_s[t\theta(t-t') + t'\theta(t'-t)], \quad (4.54)$$

where $\theta(t)$ is the Heaviside theta function. Again, the correlation function explicitly depends on both the time arguments t and t', underscoring the non-stationary nature of the process $X(t)$.

4.3.1 High-Friction Brownian Regime

Our picture of Brownian motion based on the Langevin equation conforms to the timescale separation indicated systematically in Figure 4.3. The Langevin equations are valid in the regime $t >> \tau_m$. The latter time indicates mean time between collisions. In addition, for $t > \tau_s$, where τ_s is the Smoluchowski time ($= \gamma^{-1}$), Brownian motion appears, in conformity with Equation (4.44). The concomitant single-point probability distribution is given by [cf. Equation (1.50)]

$$P(x,t) = \frac{1}{\sqrt{4\pi D_s t}} e^{-\frac{x^2}{4D_s t}}. \quad (4.55)$$

FIGURE 4.3
Various timescales of diffusive processes.

In relation to Equations (4.45) and (4.53) we also have

$$\frac{< X(t_1)X(t_3) >}{\sqrt{< X^2(t_1) > < X^2(t_3) >}} = \sqrt{\frac{t_1}{t_3}}, \quad t_3 > t_1. \tag{4.56}$$

Equation (3.45) is exactly satisfied, rendering the Brownian motion in $x(t)$ a Gaussian–Markov process, albeit a non-stationary one. Hence, the two-point conditional probability can be easily constructed from Equation (3.41):

$$P(x_1, t_1 | x_2, t_2) = \sqrt{\frac{1}{4\pi D_s(t_2 - t_1)}} e^{-\frac{(x_2 - x_1)^2}{4D_s(t_2 - t_1)}}. \tag{4.57}$$

The process described by Equations (4.55) and (4.57) is known as the Wiener process. We have an alternate physical way of looking at the Wiener process. Guided by the timescale separation indicated in Figure 4.3, the momentum is completely averaged over, i.e., thermalized. In this domain therefore, dp/dt in Equation (4.11b) is nearly equal to zero and implies the approximate identity:

$$p(t) \approx \frac{1}{\gamma} F_r(t). \tag{4.58}$$

Substituting in Equation (4.11a),

$$X(t) \approx \frac{1}{m\gamma} \int_0^t dt' \;\; F_r(t'). \tag{4.59}$$

Thus the Wiener process may be regarded as an integral of the Gaussian white noise. Clearly,

$$\langle X(t) \rangle_{eq} = 0, \tag{4.60}$$

while

$$\langle X^2(t)\rangle_{eq} = \frac{1}{m^2\gamma^2}\int_0^t\int_0^t dt'\,dt''\langle F_r(t')\;F_r(t'')\rangle$$

$$= \frac{\Gamma}{m^2\gamma^2}\int_0^t\int_0^t dt'dt''\delta(t'-t'') \tag{4.61}$$

$$= \frac{2KT}{m\gamma}t,$$

using the second fluctuation dissipation theorem [Equation (4.25)]. Next, we consider the unequal time correlation function given by

$$\langle X(t)X(t')\rangle_{eq} = \frac{1}{m^2\gamma^2}\int_0^t dt_1\int_0^{t'} dt_2 < F_r(t_1)\;F_r(t_2) >,$$

which can be further decomposed as for $t > t'$ [as in the steps below Equation (4.48)],

$$\langle X(t)X(t')\rangle = \frac{1}{m^2\gamma^2}\left[\left(\int_{t'}^t dt_1\int_0^{t'} dt_2 + \int_0^{t'} dt_1\int_0^{t'} dt_2\right)\langle F_r(t_1)F_r(t_2)\rangle\right], \quad (t > t'). \tag{4.62}$$

The contribution of the first term within the square parentheses is identically zero due to delta-correlated random force. The second term, after doing the integrals over the delta function in t_1 and t_2, yields

$$\langle X(t)X(t')\rangle = \frac{\Gamma}{m^2\gamma^2}t', \quad (t > t'). \tag{4.63}$$

In general,

$$\langle X(t)X(t')\rangle_{eq} = 2D_s \min\,(t,t'), \tag{4.64}$$

as in Equation (4.53). We shall return to a further discussion of the high-friction limit in the next chapter from the view of the Fokker–Planck equation.

4.3.2 Summary of Results

We enumerate below the salient features of the Langevin model.

1. The effect of the heat bath on the system of interest is modeled as a systematic drag force plus a completely random (stochastic) force.
2. The systematic force is assumed to be proportional to the instantaneous momentum of the system (specifically, the Brownian particle) and is endowed with a negative sign that underscores the point that it is always in opposition to the motion. The resultant equation of motion explicitly breaks time reversal symmetry.
3. The random force is taken to be a delta-correlated white noise, which is a stationary Gaussian process.
4. Since the random force governs the time variation of the momentum, the latter is also a stochastic process, albeit exponentially correlated in time, for a free particle. In fact, the free-particle momentum is a stationary Gaussian–Markov process governed by an Ornstein–Uhlenbeck distribution.
5. Finally, the position coordinate x, driven by momentum through its defining form based on Hamilton's equation of motion, is, of course, also a stochastic process. It is the time integral of the Ornstein–Uhlenbeck process. However, x turns out to be a non-stationary, non-Markovian, but Gaussian process.
6. Only in the limit of large friction (i.e., over-damped limit) does the position coordinate (Wiener process) become Markovian while retaining its non-stationarity. Historically, this was the limiting situation analyzed by Einstein in the context of Brownian motion and called diffusion. Clearly the Langevin model allows us to generalize the notion of diffusion in various angles, some of which will be discussed later.
7. The Langevin equations are formulated on physically motivated separation of various timescales (Figure 4.3). First, the equations are valid only for times much larger than the impact time τ_{im} and the time τ_m between collisions that play no further role in the discussion. The next timescale in the hierarchy is the so-called Smoluchowski time τ_s, which is the inverse of the friction coefficient. Only for times much larger than τ_s does one enter the diffusive regime of Einstein.
8. It is then instructive to estimate the timescales τ_{im}, τ_m, and τ_s. For all practical purposes and in conformity with the impact approximation of kinetic theory of dilute gases, τ_{im} may be set to be zero. (The impact time may be assumed to be an order of magnitude greater than the timescale over which electromagnetic interactions occur; thus τ_{im} may be taken to be of the order of 10^{-13} sec).

Next, for estimating τ_m we need the mean distance $\bar{a}$ between the Brownian particles and the water molecules and the RMS speed υ_{RMS} of a molecule, because $\tau_m = \bar{a}/\upsilon_{RMS}$. The quantity $\bar{a}$ is essentially also the mean intermolecular distance and hence can be estimated from the mass density ρ of water ($\sim 10^3$ kg/m^3), the molecular weight M of water (18), and Avogadro's number $\mathfrak{N}(\approx 6\times 10^{23})$. Thus $\bar{a} = (M10^{-3}/\rho\mathfrak{N})^{1/3} \approx 3\times 10^{-10} m$. Next, $\upsilon_{RMS} = (K_B T/m_{H_2O})^{1/2}$, where the mass of a molecule is again given by $m_{H_2O} = M\times 10^{-3}/\mathfrak{N} \approx 3x10^{-26} kg$. Taking T as the room temperature, $\upsilon_{RMS} \approx 4\times 10^2 m/s$, $\tau_m \approx 10^{-12} s$. While being exceedingly small, τ_m is about 10 times larger than τ_{im} at room temperature. It is then a moot point to fix the noise correlation time ν^{-1} of Equation (4.14) as τ_{im} or τ_m and it could be either. In summary, for times shorter than τ_m, the entire Langevin description breaks down.

Finally, we come to estimate the Smoluchowski time $\tau_s (= \gamma^{-1} = m/6\pi a\eta)$ given by Equation (4.47), in which the mass m is that of the particle. Assuming the density of the particle to be the same as that of water (recall that Brown's pollen grains floated in water) and assuming a radius of 1000 nm for the particle, $m \approx 410^{-15} kg$. We now need the viscosity η of water which, at room temperature, is $\approx 10^{-3}$ *Newton* $-$ sec$/m^2$. Thus $\tau_s \approx 10^{-7} s$, much larger than τ_{im} or τ_m. Interestingly, diffusion sets in for times longer than τ_s.

4.4 Harmonic Oscillator

While spreading (or diffusion) and bounded motion (in a harmonic potential) may sound like oxymorons, the harmonic oscillator occupies such a special position in physics that it is useful to consider its Langevin dynamics in some detail. The free-particle Langevin equations (4.11a) and (4.11b) can now be modified as

$$\frac{dx}{dt} = p \tag{4.65}$$

$$\frac{dp}{dt} = -\gamma p - m\omega_0^2 x + F_r(t), \tag{4.66}$$

where ω_0 is the frequency of the oscillator. These two equations can be written in a compact form as

$$\frac{d}{dt}\underline{x} = \underline{\underline{\Lambda}}.\underline{x} + \frac{1}{m}\underline{F}(t), \tag{4.67}$$

$$\underline{x} = \begin{pmatrix} x(t) \\ p(t)/m \end{pmatrix}, \quad F(t) = \begin{pmatrix} 0 \\ F_r(t) \end{pmatrix}, \quad \underline{\underline{\Lambda}} = \begin{pmatrix} 0 & 1 \\ -\omega_0^2 & -\gamma \end{pmatrix}. \tag{4.68}$$

The solution of Equation (4.69) is

$$\underline{x}(t) = e^{\underline{\Lambda}t}\underline{x}(0) + \frac{1}{m}\int_0^t dt' e^{\underline{\Lambda}(t-t')}\underline{F}(t'). \tag{4.69}$$

Using properties of 2×2 matrices we can show

$$e^{\underline{\Lambda}t} = \frac{e^{-\gamma t/2}}{\omega}\begin{pmatrix} \omega\cos(\omega t) + \frac{\gamma}{2}\sin(\omega t) & \sin(\omega t) \\ -\omega_0^2 & \omega\cos(\omega t) - \frac{\gamma}{2}\sin(\omega t) \end{pmatrix}, \tag{4.70}$$

where

$$\omega = \sqrt{\omega_0^2 - \frac{1}{4}\gamma^2}. \tag{4.71}$$

From Equation (4.61) then

$$\begin{aligned} x(t) = {} & \frac{e^{-\gamma t/2}}{\omega}\left[x(0)\left(\omega\cos\ \omega t + \frac{\gamma}{2}\sin\ \omega t\right) + \frac{p(0)}{m}\sin\ \omega t\right] \\ & + \frac{1}{m\omega}\int_0^t dt' e^{-\frac{\gamma(t-t')}{2}} F_v(t')\ \sin\omega(t-t'), \end{aligned} \tag{4.72}$$

and

$$\begin{aligned} p(t) = {} & \frac{e^{-\gamma t/2}}{\omega}\left[-m\omega_0^2(0)\sin\ \omega t + p(0)\left(\omega\cos\ \omega t + \frac{\gamma}{2}\sin\ \omega t\right)\right] \\ & + \frac{1}{\omega}\int_0^t dt' e^{-\frac{\gamma(t-t')}{2}} F_v(t')\left[\omega\cos\omega(t-t') - \frac{\gamma}{2}\sin\omega(t-t')\right]. \end{aligned} \tag{4.73}$$

Clearly,

$$\langle x(t)\rangle = \frac{e^{-\gamma t/2}}{\omega}\left[x(0)\left(\omega\cos\ \omega t + \frac{\gamma}{2}\sin\ \omega t\right) + \frac{p(0)}{m}\sin\ \omega t\right]. \tag{4.74}$$

and

$$\langle p(t)\rangle = \frac{e^{-\gamma t/2}}{\omega}\left[-m\omega_0^2 x(0)\ \sin\omega t + p(0)\left(\omega\cos\omega t - \frac{\gamma}{2}\sin\omega t\right)\right]. \tag{4.75}$$

Further, the fluctuations in $x(t)$ and $p(t)$ are respectively given by

$$\langle (x(t) - \langle x(t)^2\rangle\rangle = \frac{K_B T}{m\omega_0^2}\left[1 - \frac{e^{-\gamma t}}{\omega^2}\left(\omega_0^2 + \frac{1}{2}\gamma\omega\ \sin 2\omega t - \frac{\gamma^2}{4}\cos 2\omega t\right)\right], \tag{4.76}$$

$$\langle (p(t) - \langle p(t)^2\rangle\rangle = mK_B T\left[1 - \frac{e^{-\gamma t}}{\omega^2}\left(\omega_0^2 - \frac{1}{2}\gamma\omega\ \sin 2\omega t - \frac{\gamma^2}{4}\cos 2\omega t\right)\right]. \tag{4.77}$$

The cross correlation is

$$\langle (x(t) - \langle x(t)\rangle)\rangle = (p(t) - \langle p(t)\rangle) = \frac{\gamma K_B T}{\omega^2}\ e^{-\gamma t}\sin^2\omega t. \tag{4.78}$$

We shall revisit this problem in the next chapter. However, it is important to note that in the long time limit ($\gamma t >> 1$), the mean squared displacement now equilibrates [cf. (4.68)] in contrast to the free particle case because the harmonic potential always keeps the motion bounded.

4.5 Diffusive Cyclotron Motion

Consider a particle with an electric charge q, moving under the influence of the Lorentz force due to a constant magnetic field $\vec{B}$ and a dissipative heat bath that causes damping. The corresponding Langevin equation (Balakrishnan 2009, Singh and Dattagupta 1996) can be written

$$m\dot{\vec{\upsilon}} = q\frac{(\vec{\upsilon}\times\vec{B})}{c} - m\gamma\vec{\upsilon} + \vec{\eta}(t). \tag{4.79}$$

The vector $\vec{\eta}(t)$ is a stationary Gaussian noise with zero mean and correlation given by

$$\langle \eta_i(t)\ \eta_j(t')\rangle = \Gamma\delta_{ij}\ \ \delta(t - t'), \tag{4.80}$$

where *i and j* indicate Cartesian directions. The coefficient Γ is given by the usual fluctuation–dissipation relation.

$$\Gamma = 2m\gamma\ K_B T. \tag{4.81}$$

In writing Equations (4.79) and (4.80), we tacitly used the fact that a Lorentz force normal to $\vec{\upsilon}$ means that the magnetic field $\vec{B}$ does not work on the particle. As a result, the heat bath is characterized by the systematic damping term and the noise is not affected. This assumption is also in conformity with the fact that the $\vec{B}$ field, appearing via the vector potential in the partition function, can be integrated out and the equilibrium Gibbs measure remains unaffected by $\vec{B}$. Choosing the $\vec{B}$ field in an arbitrary direction defined by the unit vector $\hat{n}$, i.e.,

$$\vec{B} = B\ \hat{n}, \tag{4.82}$$

the *jth* Cartesian component of the velocity $\vec{\upsilon}$ can be written from Equation (4.79) as

$$m\dot{\upsilon}_j(t) \ = \ qB \in_{jkl} \upsilon_k(t)\ n_l - m\gamma\ \ \upsilon_j(t) + \eta_j(t), \tag{4.83}$$

where $\in_{jkl}$ is the totally antisymmetric third rank tensor called the Levi-Civita symbol. Introducing now a time-independent matrix $\underline{\underline{M}}$, with elements given by

$$M_{kj} \ = \ \in_{kjl} n_l, \tag{4.84}$$

and employing the antisymmetry property of the Levi-Civita symbol, we may rewrite Equation (4.84) as

$$\dot{\upsilon}_j(t) = -\gamma\upsilon_j(t) - \omega_c\upsilon_k(t)M_{kj} + \frac{1}{m}\eta_j(t), \tag{4.85}$$

where the cyclotron frequency ω_c is given by

$$\omega_c = qB/mc. \tag{4.86}$$

Because the equilibrium distribution remains Maxwellian, the velocity of the particle remains a *stationary* process and the fluctuation–dissipation relation $\Gamma = 2m\gamma K_B T$ also remains intact. Hence, the correlation function $<\upsilon_i(t')\ \upsilon_j(t)>$ is a function of the time difference $|t\text{–}t'|$ alone for every pair of indices *i* and *j*. For convenience, we define the velocity correlation matrix

$\underline{\underline{C}}\ (t-t')$ as a 3×3 matrix with elements obtained from the correlation functions by dividing them by $^{K_BT}/_{m}$, the mean squared value of the velocity:

$$C_{ij}(t-t') \;=\; \frac{<\upsilon_i(t')\ \upsilon_j(t)>eq}{^{K_BT}/_{m}} \;=\; \frac{<\upsilon_i(o)\ \upsilon_j(t-t')>eq}{^{K_BT}/_{m}} \quad . \tag{4.87}$$

The differential equation for $C_{ij}(t)$ is found simply by multiplying Equation (4.85) by $\upsilon_i(o)$ from the left, and performing a complete average, namely an average over the realizations of the noise and an average over the initial velocity $\upsilon_i(o)$ with respect to the equilibrium distribution, thus yielding

$$C_{ij}(t') = -\,\omega_c C_{ik}(t) M_{kj} - \gamma C_{ij}(t), \quad (t>0), \tag{4.88}$$

wherein we employed causality, i.e.,

$$\overline{\upsilon_i(o)\ \eta_j(t)} \;=\; \overline{\upsilon_i(o)}\ \overline{\eta_j(t)} = 0, \tag{4.89}$$

based on the premise that noise is the cause and velocity is the effect. Equation (4.88) can be written in a matrix form:

$$\dot{\underline{\underline{C}}}(t) = -\underline{\underline{C}}(t)\ (\underline{\underline{I}}\ \gamma + \underline{\underline{M}}\ \omega_c)\ , \tag{4.90}$$

where $\underline{\underline{I}}$ is the 3×3 unit matrix. The solution of Equation (4.90) can be expressed as

$$\underline{\underline{C}}(t) = \ \underline{\underline{C}}(o)\ \exp\,[-(\underline{\underline{I}}\gamma + \underline{\underline{M}}\omega_c)t] = \ e^{-\gamma t}\,\underline{\underline{C}}(o)\ e^{-\underline{\underline{M}}\omega_c t}, \tag{4.91}$$

for all $t > o$. It is evident that $\underline{\underline{C}}$ (o) is just the unit matrix, because

$$C_{ij}(o) \;=\; \frac{<\upsilon_i(o)\ \upsilon_j(0)>eq}{^{K_BT}/_{m}} \;=\; \delta_{ij}. \tag{4.92}$$

It remains to calculate the exponential in Equation (4.91) by employing the properties of 3×3 matrices (Messiah 1961)*:

$$\exp\,[-\ \underline{\underline{M}}\omega_c t] = \ \underline{\underline{I}}\ -\ \underline{\underline{M}}\sin(\omega_c t) + \ \underline{\underline{M}}^2(1-\cos\omega_c t) \quad . \tag{4.93}$$

Further, for the given $\underline{\underline{M}}$ at hand,

$$(\underline{\underline{M}}^2)_{kj} = \in_{klp} n_p \in_{ljq} n_q = n_k n_j - \delta_{kj} \tag{4.94}$$

---

* See Exercise 3.

Substituting Equations (4.92) through (4.94) into Equation (4.91) we find

$$C_{ij}(t) = e^{-\gamma t}[n_i n_j + (\delta_{ij} - n_i n_j)\ \cos(\omega_c t) - \epsilon_{ijk}\ n_k \sin(\omega_c t)]. \tag{4.95}$$

Further insight into Equation (4.95) can be had by choosing the B field along the space fixed z axis, which implies $(n_1, n_2, n_3) = (0, 0, 1)$. We find

$$< v_z(o)\ v_z(t) >_{eq} = \frac{K_B T}{m}\ e^{-\gamma t}, \tag{4.96a}$$

$$< v_z(o)\ v_x(t) >_{eq} = < v_y(o)\ v_y(t) >_{eq} = \frac{K_B T}{m}\ e^{-\gamma t} \cos(\omega_c t), \tag{4.96b}$$

$$< v_x(o)\ v_y(t) >_{eq} = < v_y(o)\ v_x(t) >_{eq} = -\frac{K_B T}{m}\ e^{-\gamma t} \sin(\omega_c t) \tag{4.96c}$$

$$< v_x(o)\, v_z(t) >_{eq} = < v_y(o)\ v_z(t) >_{eq} = < v_z(o)\, v_x(t) >_{eq} = < v_z(o)\, v_y(t) >_{eq} = 0. \tag{4.96d}$$

Exercises

Derive:

1. Equation (4.16).
2. Equation (4.31).
3. Equation (4.93).

References

Balakrishnan, V. 2009. *Elements of Nonequilibrium Statistical Mechanics*. New Delhi: Ane Books Pvt. Ltd.

Chandrasekhar, S. 1943. Stochastic problems in physics and astronomy. *Rev. Mod. Phys.* 15, 1.

Dattagupta, S. and S. Puri 2004. *Dissipative Phenomena in Condensed Matter: Some Applications*. New York: Springer.

Dirac, P.A.M. 1958. *The Principles of Quantum Mechanics*, 4th Ed. London: Oxford University Press.

Huang, K. 1987. *Statistical Mechanics*, 2nd ed. New York: John Wiley & Sons.

Langevin, P. 1908. Comptes Rendus. *Acad. Sci. Paris* 146, 530.

Messiah, A. 1961. *Quantum Mechanics*. Amsterdam: North-Holland.

Singh, J. and S. Dattagupta. 1996. *Pramana J. Phys.* 47, 199.

5

Fokker–Planck Equation

5.1 Introduction

The aim of this chapter is to continue the discussion of stochastic and physical diffusion related to Brownian motion discussed in Chapter 1 and mathematical foundations based on Langevin equations introduced in Chapter 4. We also remind the reader that the classical diffusion processes in the position–space expounded by Einstein can be derived from the high-friction limit of the full phase space Langevin equations. The latter describe a composite stochastic process in a combined position–momentum space which is stationary, Gaussian, and Markovian. However, both stationarity and Markovian nature are lost when the motion is projected into the position space alone. Although this was demonstrated for a free particle in Section 4.3, the concept that a Markov process projected into a lower dimensional space can lead to non-Markovian behavior is general.

We again note that Langevin equations written for observables (– operators in quantum mechanics) conform to the Heisenberg view. On the other hand, an entity that embodies the wave function of the Schrödinger picture is the von Neumann density operator, usually denoted by $\rho(t)$, whose classical counterpart is the probability distribution function (*pdf*) discussed earlier. It is helpful to introduce a formalism for the *pdf* completely at par with the Langevin description. The resultant Fokker–Planck (FP) equations serve as the subjects of this chapter. The simplest FP equation can be derived from the master equation (2.25) for a stationary Markov process, as shown below (Agarwal 1983, Risken 1996).

We assume that only small jumps are allowed such that the final state ξ in the probability rate $W(\xi'|\xi)$ can differ from the initial state ξ' by an infinitesimal amount $\zeta(\zeta = \xi - \xi')$. In terms of ξ and ζ, the master equation (2.25) can be rewritten as

$$\frac{\partial}{\partial t}P(\xi,t) = \int d\zeta P(\xi-\zeta,t)W(\xi-\zeta\,|\,\zeta) - P(\xi,t)\int d\zeta W(\xi\,|\,\xi-\zeta)\ . \tag{5.1}$$

Furthermore, using the fact that jumps are symmetric, i.e., ζ and $-\zeta$ jumps are equally likely; see Equation (2.3),

$$W(\xi - \zeta) = W(\xi\,|+\zeta). \tag{5.2}$$

Because $P(\xi - \zeta, t)$ is tacitly assumed to be a slowly varying function of ξ, we may use the Taylor expansion:

$$P(\xi-\zeta,t)W(\xi-\xi\zeta) \approx P(\xi,t)\ W(\xi\zeta) - \zeta\frac{\partial}{\partial\xi}[P(\xi,t)W(\xi|\zeta)] + \frac{\zeta^2}{2}\frac{\partial^2}{\partial\xi^2}[P(\xi,t)W(\xi\zeta)]. \tag{5.3}$$

Substituting Equation (5.3) in Equation (5.1) and defining the jump moments (per unit time) as

$$C_1(\xi) = \int d\zeta\zeta W(\xi|\zeta), \quad \text{and} \quad C_2(\xi) = \int d\zeta\zeta^2 W(\xi|\zeta), \tag{5.4}$$

we obtain the desired FP equation

$$\frac{\partial}{\partial t}P(\xi,t) = \frac{\partial}{\partial\xi}[C_1(\xi)P(\xi,t)] + \frac{1}{2}\frac{\partial^2}{\partial\xi^2}[C_2(\xi)P(\xi,t)]. \tag{5.5}$$

The terms on the right of Equation (5.5) may be regarded as the first two terms of the so-called Kramers–Moyal expansion:

$$\frac{\partial}{\partial t}P(\xi,t) = \sum_{n=1}^{\infty}\frac{(-1)^n}{n!}\frac{\partial^n}{\partial\xi^n}[C_n(\xi)P(\xi,t)]. \tag{5.6}$$

If the Kramers-Moyal expansion terminates, only C_1 and C_2 are nonzero while all other higher moments $C_n(n>2)$ vanish (the condition under which the FP equation is valid). Otherwise all C_n s are non-vanishing. Recalling that for a Gaussian process all cumulants higher than the second are zero (Chapter 3), it is clear that the FP equation and Gaussian stochastic processes go hand in hand. This point will be amplified later.

5.2 FP Equation in Velocity

In line with our rudimentary discussion of the Langevin equations for a free particle, it is helpful to consider the corresponding FP equation. The entity ξ, generally a multivariable stochastic process, is taken here to be a one-dimensional momentum variable p. The first jump moment $C_1(p)$ [cf. Equation (5.4)] is the coarse-grained rate with which the momentum p varies with time and from Equation (4.7) is given by $-\gamma p$. Thus Equation (5.5) yields what is called the Rayleigh equation, written as

$$\frac{\partial}{\partial t}P(p,t) = \gamma\frac{\partial}{\partial p}[p\ P(p,t] + \frac{1}{2}\frac{\partial^2}{\partial p^2}[C_2(p)P(p,t)]. \tag{5.7}$$

The second moment $C_2(p)$ that measures small fluctuations in p must be positive definite even in the limit $p \to 0$ and must satisfy

$$C_2(p) \approx C_2(0) + \vartheta(p^2) \approx C_2(0). \tag{5.8}$$

How do we fix the second moment $C_2(0)$? The procedure is very similar to our discussion in Chapter 4 in relation to the second fluctuation–dissipation theorem cited in Equation (4.25). We demand that Equation (5.7) yield the thermal equilibrium distribution of Equation (4.1). If the latter must be a solution of Equation (5.7) we can check by direct differentiation that

$$C_2(0) = 2m\gamma K_B T. \tag{5.9}$$

The second moment is just the strength of the Langevin white noise [cf. Equation (4.11d)] captured by the power spectrum [cf. Equation (4.13)]. Combining Equations (5.7) through (5.9) we arrive at

$$\frac{\partial}{\partial t} P(p,t) = \gamma \frac{\partial}{\partial p}[p\, P(p,t)] + m\gamma K_B T \frac{\partial^2}{\partial p^2} P(p,t). \tag{5.10}$$

It is easy to verify that the solution of Equation (5.10) is the Ornstein–Uhlenbeck distribution (Equation (4.35), rewritten here for the sake of clarity:

$$P(p,t) = \left[\frac{1}{2\pi m K_B T(1-e^{-2\gamma t})}\right]^{1/2} .\exp\left\{-\frac{1}{2\pi m K_B T(1-e^{-2\gamma t})}\right\}(p - p_0 e^{-\gamma t})^2,$$

with

$$P(p,t=0) = \delta(p - p_0). \tag{5.11}$$

This result underscores the fact that the FP equation is yet another description for a stationary Gaussian–Markov process.

5.3 FP Equation in Position: Stochastic Diffusion

The power of the FP equation (5.5) can be demonstrated by showing that it is open to a flexible interpretation of the process ξ, even when it is non stationary. In Section 5.2, we took ξ as the momentum p. Now we consider the case when ξ is viewed as the position x of a Brownian particle discussed in Chapter 1. From the random walk picture (Section 1.2) in which the left and right jumps occur with equal probability, it is clear that the first moment

$$C_1(x) = 0. \tag{5.12}$$

The second moment $C_2(x)$ [cf. Equation (5.4)] is the mean squared jump length times the jump rate and is therefore twice the coefficient for stochastic diffusion D_s [cf. Equation (1.34)]. The differential equation then reads

$$\frac{\partial}{\partial t}P(x,t) = D_s\frac{\partial^2 P}{\partial x^2}(x,t), \tag{5.13}$$

our old friend—the diffusion equation of Einstein [cf. Equation (1.33)].

5.4 FP Equation in Force Field: Kramers Equation

In this subsection we deviate from the free-particle case and consider the more important example of a particle moving in a potential field $V(x)$ [cf. Equation (5.3)] (Kramers 1940). The corresponding Langevin equations constitute a generalized version of Equation (4.11b):

$$\frac{dx(t)}{dt} = \frac{p(t)}{m}, \tag{5.14a}$$

$$\frac{dp(t)}{dt} = -\gamma\; p(t) - \frac{\partial V}{\partial x} + F_r(t). \tag{5.14b}$$

The full phase space FP equation can then be written from Equation (5.5) as

$$\begin{aligned}\frac{\partial}{\partial t}P(x,p,t) = &-\frac{\partial}{\partial x}[C_{11}(x,p)P(x,p,t)] - \frac{\partial}{\partial p}[C_{12}(x,p)P(x,p,t)]\\ &+\frac{1}{2}\frac{\partial^2}{\partial x^2}[(C_2)_{11}(x,p)P(x,p,t)] + \frac{1}{2}\frac{\partial^2}{\partial p^2}[(C_2)_{22}(x,p)P(x,p,t)]\\ &+\frac{\partial^2}{\partial x\,\partial p}[(C_2)_{12}(x,p)P(x,p,t)].\end{aligned} \tag{5.15}$$

Recalling that W in Equation (5.4) describes mean jump rates, the first two moments are obtained from the mean of the equations of motion in (5.14a) and (5.14b) and thus

$$\begin{aligned}C_{11}(x,p) &= \frac{p}{m},\\ C_{12}(x,p) &= -\gamma p - \frac{\partial V}{\partial x}.\end{aligned} \tag{5.16}$$

To calculate the second jump moments, we choose a timescale Δt over which x changes by Δx and p by Δp. Thus

$$(C_2)_{11}(x) = \left\langle \frac{(\Delta X)^2}{\Delta t} \right\rangle = \left\langle \frac{(\Delta X)^2}{\Delta t} \right\rangle \Delta t = p^2 \Delta t,$$

$$(C_2)_{12}(x,p) = \left\langle \frac{(\Delta X)(\Delta p)}{\Delta t} \right\rangle = \left\langle \left(\frac{\Delta X}{\Delta t} \right) \left(\frac{\Delta p}{\Delta t} \right) \right\rangle \Delta t = p \left(-\gamma p - \frac{\partial V}{\partial x} \right) \Delta t, \quad (5.17)$$

which all go to zero in the limit $\Delta t \to 0$. We are finally left with $(C_2)_{22}(p)$ which is assumed for small jumps to be independent of x and p and replaced by a constant C. The derived FP equation is then

$$\frac{\partial}{\partial t} P(x,p,t) = \left\{ -\frac{p}{m} \frac{\partial}{\partial x} + \frac{\partial}{\partial p} \left(\gamma p + \frac{\partial V}{\partial x} \right) + \frac{C}{2} \frac{\partial^2}{\partial p^2} \right\} P(x,p,t). \quad (5.18)$$

For determining the constant C, we employ the same argument used to evaluate the corresponding quantity for a Rayleigh particle [see sequel to Equation (5.8)]. We insist that the equilibrium Boltzmann distribution,

$$P_{eq}(x,p) = \left(\frac{\beta}{2\pi m} \right)^{1/2} \frac{e^{-\beta\left(\frac{p^2}{2m} + V(x)\right)}}{\int dx\, e^{-\beta V(x)}}, \quad (5.19)$$

is also a solution of Equation (5.18). Imposition of this condition is sometimes known as demanding a detailed balance. We find*

$$C = 2m\gamma K_B T, \quad (5.20)$$

as in Equation (5.9). The final form of Equation (5.18) is then

$$\frac{\partial}{\partial t} P(x,p,t) = \left\{ -\frac{p}{m} \frac{\partial}{\partial x} + \frac{\partial}{\partial p} \left(\gamma p + \frac{\partial V}{\partial x} \right) + m\gamma K_B T \frac{\partial^2}{\partial p^2} \right\} P(x,p,t). \quad (5.21)$$

Kramers studied this equation extensively in connection with escape over a barrier (discussed in Section 5.7), and hence Equation (5.21) is known as the Kramers equation. It is interesting to note that when $V(x) = 0$, Equation (5.21) does not reduce to the Rayleigh equation (5.10) but has a full phase space description, though we are now dealing with a free particle. The corresponding equation reads

$$\frac{\partial}{\partial t} P(x,p,t) = \left\{ -\frac{p}{m} \frac{\partial}{\partial x} + \frac{\partial}{\partial p} (\gamma p) + m\gamma K_B T \frac{\partial^2}{\partial p^2} \right\} P(x,p,t). \quad (5.22)$$

* See Exercise 1.

5.5 FP Equation in Force Field in High-Friction Limit: Smoluchowski Equation

It may be recalled that the Wiener process can be recovered from the full set of Langevin equations by going to the high-friction ($\gamma t >> 1$) limit (Section 4.3.1). A natural question is to ask what form the Kramers equation takes for ($\gamma t >> 1$). As in Section 4.3.1, we assume that the momentum p equilibrates to zero so that the Equation (5.14b) simplifies as

$$p \approx -\frac{1}{\gamma}\frac{\partial V}{\partial x} + \frac{1}{\gamma}F_r(t), \tag{5.23}$$

which, when substituted in Equation (5.14a) yields

$$\frac{dx}{dt} = -\frac{1}{m\gamma}\frac{\partial V}{\partial x} + \frac{1}{m\gamma}F_r(t). \tag{5.24}$$

Equation (5.24) is a now Langevin equation (for the displacement x) with a renormalized damping $(m\gamma)^{-1}$ and renormalized noise $(m\gamma)^{-1}F_r(t)$. Correspondingly, the FP equation (5.5) is now an equation for the probability density equation of a single variable x:

$$\frac{\partial}{\partial t}P(x,t) = -\frac{\partial}{\partial x}[C_1(x)P(x,t)] + \frac{1}{2}\frac{\partial^2}{\partial x^2}[C_2(x)P(x,t)]. \tag{5.25}$$

The first moment is clearly

$$C_1(x) = \left\langle \frac{\Delta x}{\Delta t} \right\rangle. \tag{5.26}$$

Thus from Equation (5.23),

$$C_1(x) = -\frac{1}{m\gamma}\frac{\partial V}{\partial x}. \tag{5.27}$$

Conversely, the second moment is obtained from the Einstein relation [cf. Equations (2.7) and (4.46)]

$$C_2(x) = \left\langle \frac{(\Delta X)^2}{\Delta t} \right\rangle = 2D_s = \frac{2K_BT}{m\gamma}. \tag{5.28}$$

We finally have

$$\frac{\partial}{\partial t}P(x,t) = \frac{1}{m\gamma}\left\{\frac{\partial}{\partial x}\left(\frac{\partial V}{\partial x}P(x,t)\right) + K_BT\frac{\partial^2}{\partial x^2}P(x,t)\right\}. \tag{5.29}$$

Clearly, when the potential is zero (i.e., we are dealing with a free particle), we recover from above the familiar diffusion equation (5.13).

Equation (5.29) was derived by Smoluchowski (1906) who, like Langevin, was another contemporary of Einstein. The heuristic argument presented above in arriving at the Smoluchowski equation can be made more rigorous by a systematic expansion of the Kramers equation in powers of γ^{-1} (van Kampen 1981).

Recall that the diffusion equation (5.13) was derived from the simple random walk model of Section 1.3 when the walker had no bias. Appropriately, the first term on the right of Equation (5.29) is known as drift and represents biased flow under a force field, in order to distinguish it from diffusion—a term naturally reserved for the second order derivative.

5.6 FP Equation for Damped Harmonic Oscillator

The Kramers equation (5.21) is not amenable to an explicit closed form solution unless the potential is (i) zero (free particle), (ii) linear in x, e.g., for a particle falling under gravity, as in the case of sedimentation, and (iii) quadratic in x, e.g., for harmonic motion. We studied the Langevin equations and their solutions for a harmonic oscillator in Section 4.4. We now complete that discussion through a comparative study relative to the Kramers equation. For harmonic motion, Equation (5.21) reads

$$\frac{\partial}{\partial t}P(x,p,t)=\left\{-\frac{p}{m}\frac{\partial}{\partial x}+\frac{\partial}{\partial p}\left(\gamma p+m\omega_0^2 x\right)+m\gamma K_B T\frac{\partial^2}{\partial p^2}\right\}P(x,p,t). \quad (5.30)$$

We can easily check that the bivariate (in x and p) Gaussian distribution [cf. Equation (3.3)]:

$$P(x,p,t)=\frac{(\det \underline{\underline{A}})^{1/2}}{2\pi}\exp\left[-\frac{1}{2}(\underline{x}-<\underline{x}>)^T.\underline{\underline{A}}.(\underline{x}-<\underline{x}>)\right], \quad (5.31)$$

solves Equation (5.30) with the initial condition:

$$P(x,p,0)=\delta(x-x_0)\delta(p-p_0), \quad (5.32)$$

where x_0 and p_0 are the initial values of the position and momentum respectively, and $\langle x(t)\rangle$ and $\langle p(t)\rangle$ are given by Equations (4.65) and (4.66). Recall from Equation (3.9) that the inverse of the covariance matrix is given by

$$\underline{\underline{A}}^{-1}=\begin{pmatrix} <(x-<x(t)>)^2> & <(x-<x(t)>)\ (p-<p(t)>)> \\ <(x-<x(t)>)\ (p-<p(t)>)> & <p-<p(t)>)^2> \end{pmatrix},$$

where the entries are given by Equations (4.74) through (4.78). Therefore,

$$\underline{\underline{A}} = \frac{1}{\det \underline{\underline{A}}} \begin{pmatrix} <(p-<p(t)>)^2> & <(x-<x(t)>)\ (p-<p(t)>)> \\ -<(x-<x(t)>)\ (p-<p(t)>)> & <x-<x(t)>)^2> \end{pmatrix}, \tag{5.33}$$

where

$$\det \underline{\underline{A}} = <(x-<x(t)>)^2><(p-<p(t)>)^2> - <(x-<x(t)>)(p-<p(t)>)>^2. \tag{5.34}$$

The treatment above is complementary to the one presented in Section 4.4 and firmly establishes that the stochastic process associated with a damped harmonic oscillator is Gaussian and Markovian. We can also derive the solution for the Kramers equation (5.22), appropriate to a free particle by taking the $\omega_0 \to 0$ limit in the above equation. The solution is again given by Equation (5.31), except now [cf. Equations (4.74) through (4.78)]:

$$<x(t)> = x_0 + \frac{p_0}{m\gamma}(1-e^{-\gamma t}), \tag{5.35}$$

$$\langle p(t) \rangle = p_0 e^{-\gamma t}, \tag{5.36}$$

$$<(x(t)-<x(t)>)^2> = \frac{K_B T}{m\gamma^2}[2\gamma t - 3 + 4e^{-\gamma t} - e^{-2\gamma t}], \tag{5.37}$$

$$<(p(t)-<p(t)>)^2> = mK_B T[1-e^{-2\gamma t}], \tag{5.38}$$

$$<(x(t)-<x(t)>)(p(t)-<p(t)>)> = \frac{K_B T}{\gamma} e^{-\gamma t} \sinh^2\left(\frac{1}{2}\gamma t\right). \tag{5.39}$$

5.7 FP Equation for Cyclotron Motion

We discussed in Section 4.5 the Langevin description of charged particle dynamics in a magnetic field. We would like now to conduct a complementary analysis with the help of the FP equation (Balakrishnan 2009). We can utilize Equation (5.10) by adding an extra term (third term on the right of the equation below) that corresponds to the Lorentz force,

$$\dot{P}(\vec{\upsilon},t) = \nabla_\upsilon \bullet [\gamma \vec{\upsilon} P(\vec{\upsilon},t)] + \frac{\gamma}{m} KT \nabla_\upsilon^2 P(\vec{\upsilon},t) - \nabla_\upsilon \bullet \left[\frac{q}{mc}(\vec{\upsilon} \times \vec{B})\ P(\vec{\upsilon},t)\right]. \tag{5.40}$$

Unlike the original Stokes law-type drift term that is dissipative, the additional Lorentz term is deterministic. The solution of Equation (5.40) can also be read out from the Ornstein–Uhlenbeck distribution [Equation (5.11)] by noting certain facts. The Gaussian in $\vec{\upsilon}-\vec{\upsilon}_0 e^{-\gamma t}$ arises from the fact that in the absence of the magnetic field, the average velocity vector is given by $\vec{\upsilon}_0 e^{-\gamma t}$ and the Gaussian describes the diffusive spread of the deviation from the mean velocity. On the other hand, if friction were absent and the particle moved only under the influence of the magnetic field, the velocity at time t, starting from an initial velocity $\vec{\upsilon}_0$, would have evolved in accordance with Equation (4.77) sans the friction and noise terms.

$$\dot{\upsilon}_j(t) = -\omega_c \upsilon_k(t) \underline{\underline{M}}_{kj}. \tag{5.41}$$

The above equation allows the physical interpretation that the longitudinal component of the velocity remains unaltered whereas the transverse component rotates (akin to Larmor precession) around the $\vec{B}$ field by an angle $(\omega_c t)$. Because the stochastic process remains additive even in the presence of a magnetic field, the mean velocity may be obtained simply by riding on the rotated frame and exponentially damping the apparently static velocity by the term $e^{-\gamma t}$, thus yielding $\vec{u}(t,\vec{\upsilon}_0)e^{-\gamma t}$. Hence the modified Ornstein-Uhlenbeck solution can now be expressed as

$$\mathrm{P}(\vec{\upsilon},t) = \left[\frac{m}{2\pi KT(1-e^{-2\gamma t})}\right]^{3/2} \exp\left\{-\frac{m\left(\vec{\upsilon}-\vec{u}(t,\vec{\upsilon}_0)e^{-\gamma t}\right)^2}{2KT(1-e^{-2\gamma t})}\right\}, \tag{5.42}$$

with the initial condition

$$P(\vec{\upsilon},0) = \delta(\vec{\upsilon}-\vec{\upsilon}_0). \tag{5.43}$$

In equilibrium ($t \to \infty$), the above reduces to the usual Maxwellian, with no contribution from the magnetic field, as expected. The velocity correlation matrix and the associated correlation functions are already covered in Equations (4.95) through (4.96).

5.8 Diffusion across Barrier

We noted earlier that the Smoluchowski equation in an arbitrary potential is not analytically tractable. However, under certain simplifying limiting conditions, this equation lends itself to an approximate treatment for what may be called an escape rate that finds innumerable applications to various branches of science, some of which will be discussed later. For now, we mention the conditions under which the method is applicable, with relation

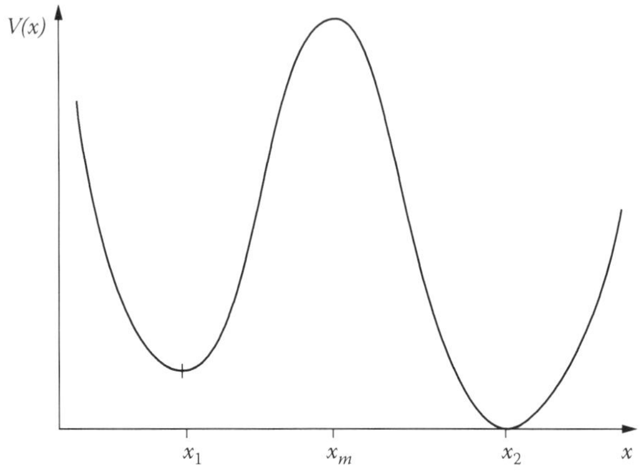

FIGURE 5.1
Asymmetric double well.

to Figures 5.1 and 5.2 (Chandrasekhar 1943). The figures depict a one-dimensional mechanical potential $V(x)$ that has two minima (at x_1 and x_2) in Figure 5.1 and a single minimum (*at* x_1) in Figure 5.2. In both cases the maximum is taken to occur at (x_m)

Imagine now a bunch of non-interacting Brownian particles moving in the potential $V(x)$ in the background of a fluctuating thermodynamic system (comprised of the surrounding molecules, in Brown's experiment) at a fixed temperature T. At any point in time, the particles are found all over, but if $V(x_m) >> V(x_1, x_2)$ most of the particles are expected to be lumped around x_1 and x_2. In particular, in thermal equilibrium, the momentum distribution

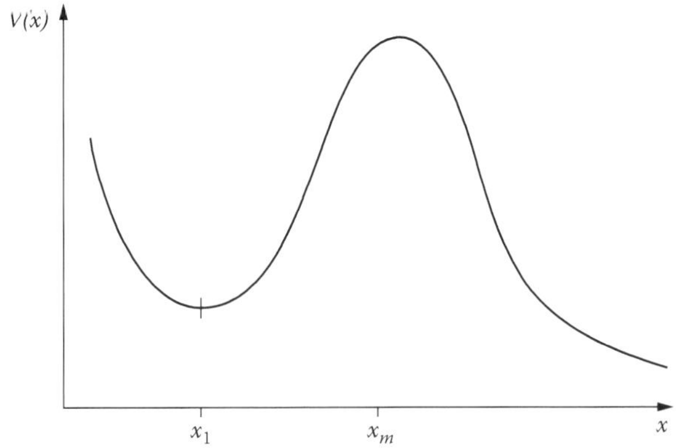

FIGURE 5.2
Barrier crossing.

is given by Equation (4.1) whereas position distribution is governed by the Boltzmann factor:

$$P^{eq}(x) = \frac{e^{-\beta V(x)}}{\int dx' e^{-\beta V(x')}}. \tag{5.44}$$

It is clear that $P^{eq}(x)$ is expected to sharply peak around x_1 and x_2. The difference between the two situations in Figures 5.1 and 5.2 is that in the latter, the particles that escape to $x \to +\infty$ never return. The particles in Figure 5.1 diffuse back and forth, maintaining a higher population at x_2 on the average. Because Figure 5.1 depicts a more general solution, we will focus only on that. For a classical system at zero temperature, particles that lock around x_1 and x_2 can never escape. At a finite temperature, statistical mechanics takes over, and the Brownian "kicks" from the surrounding molecules enable the particles to surmount the steep barrier at x_m.

Although this process is expected to be very slow, especially if $\beta(V(x_m) - V(x_{1,2})) >> 1$, the rate of escape is finite and we will estimate it below with the aid of the Smoluchowski equation. (Recall that evaporation means that a dish of water keeps losing its molecules to the vapor state although the ambient temperature is much lower than the boiling point of water.) It is clear then that the background of this discussion is one in which the friction is large, i.e., we are in the diffusive regime of Chapter 1. First note that the Smoluchowski equation (5.29) can be written as a continuity equation:

$$\frac{\partial}{\partial t} P(x,t) + \frac{\partial}{\partial x} j(x,t) = 0. \tag{5.45}$$

where the current density $j(x,t)$ is

$$j(x,t) = -(m\gamma)^{-1}\left(\frac{\partial V}{\partial x} + K_B T \frac{\partial}{\partial x}\right) P(x,t). \tag{5.46}$$

We consider a near-equilibrium or a mildly non-equilibrium situation in which the probability $P(x,t)$ is nearly time independent and the current density j is finite but independent of x [cf. Equation (5.45)]. The corresponding quantity in this quasi-static regime is denoted by $j^{qs}(t)$ and is given by

$$j^{qs}(t) = -(m\gamma)^{-1}\left(\frac{\partial V}{\partial x} + K_B T \frac{\partial}{\partial x}\right) P^{qs}(x,t), \tag{5.47}$$

where the right side must conspire to be independent of x! The significance of $j^{qs}(t)$ is that it represents the rate of accumulation of particles at x_2 at the expense of depletion at x_1:

$$j^{qs}(t) = -\dot{n}_1 = \dot{n}_2, \tag{5.48}$$

where n_1 and n_2 denote the number of particles around x_1 and x_2, given by

$$n_{1,2} = \int_{x_{1,2}-\Delta x_{1,2}}^{x_{1,2}+\Delta x_{1,2}} dx\ P^{qs}(x,t), \tag{5.49}$$

where $\Delta x_{1,2}$ is a small region around x_1 and x_2 respectively. Thus, in a high barrier–weak noise limit, we reduced the Smoluchowski problem to an effective two-state problem, examples of which abound in nature. Our strategy is to derive a rate equation for this two-state system whose natural outcome is expected to be the escape rate itself. Note that the current density in the quasi-static region can be re-expressed from Equation (5.47) as

$$j^{qs}(t) = -\frac{(m\gamma)^{-1}}{\beta} e^{-\beta V(x)} \frac{\partial}{\partial x}(e^{+\beta V(x)} P^{qs}(x,t)). \tag{5.50}$$

Transferring the Boltzmann factor to the left side and exploiting the x-independence of $j^{qs}(t)$, we can determine the latter by integrating from x_1 to x_2, thus

$$j^{qs}(t)\int_{x_1}^{x_2} dx e^{\beta V(x)} = -\frac{(m\gamma)^{-1}}{\beta}\left[e^{\beta V(x_2)} P^{qs}(x_2,t) - e^{\beta V(x_1)} P^{qs}(x_1,t)\right]. \tag{5.51}$$

To express the right side in terms of n_1 and n_2 so that Equation (5.48) takes a proper rate equation-like structure, we use the sharp-peaking argument discussed earlier to rewrite Equation (5.51):

$$n_{1,2} \approx P^{qs}(X_{1,2},t) \int_{x_{1,2}-\Delta x_{1,2}}^{x_{1,2}+\Delta x_{1,2}} dx e^{-\beta(V(x)-V(x_{1,2}))}. \tag{5.52}$$

Combining Equations (5.48), (5.51), and (5.52), we arrive at the rate equation:

$$\dot{n}_1 \approx -\dot{n}_2 \approx \frac{(m\gamma)^{-1}}{\beta I_3}\left(\frac{n_2}{I_2} - \frac{n_1}{I_1}\right), \tag{5.53}$$

where the integrals are

$$I_{1,2} = \int_{x_{1,2}-\Delta x_{1,2}}^{x_{1,2}+\Delta x_{1,2}} dx e^{-\beta V(x)}, \tag{5.54}$$

and

$$I_3 = \int_{x_1}^{x_2} dx e^{\beta V(x)}. \tag{5.55}$$

The escape or depletion rate is then given by

$$\upsilon_{12} = \frac{(m\gamma)^{-1}}{\beta}\frac{1}{I_1 I_3}, \tag{5.56}$$

where

$$I_1 \approx \sqrt{\frac{2\pi}{\beta V^{11}(x_1)}}e^{-\beta V(x_1)}, \tag{5.57}$$

and

$$I_3 \approx \sqrt{\frac{2\pi}{\beta V^{11}(x_m)1}}e^{+\beta V(x_m)} \tag{5.58}$$

The integrals were evaluated by Gaussian approximation around the minimum and maximum. The corresponding accumulation rate υ_{21} can be obtained by interchanging 2 and 1 in the above expressions. Combining Equations (5.56) through (5.58) we have the Arrhenius formula:

$$\upsilon_{12} = \frac{1}{2\pi m\gamma}\sqrt{V^{11}(x_1)\,|\,V^{11}(x_m)}e^{-\beta(V(x_m)-V(x_1))}. \tag{5.59}$$

Exercise

1. Prove Equation (5.20).

References

Agarwal, G. S. 1983. In Agarwal, G.S. and S. Dattagupta, Eds., *Stochastic Processes: Formalism and Applications*, Vol. 184. Berlin: Springer.

Agarwal, G.S. and C.V. Kunasz. 1983. *Phys. Rev.* A27, 996.

Balakrishnan, V. 2009. *Elements of Nonequilibrium Statistical Mechanics.* New Delhi: Ane Books.

Chandrasekhar, S. 1943. Stochastic problems in physics and astronomy. *Rev. Mod. Phys.* 15, 1.

Kramers, H. A. 1940. Brownian motion in a field of force and the diffusion model of chemical reactions. *Physica. (Utrecht)* 7, 284.

Risken, H. 1996. In Haken, H., Ed., *The Fokker-Planck Equation: Methods of Solution and Applications.* Series in Synergetics 18. Berlin: Springer.

Smoluchowski, M. 1912. Vorträge über die kinetische theorie der Materie und Elektrizität. *Phys. Zeit.* 12, 1069.

Smoluchowski, M. 1906. Zur kinetischen theorie der Brownschen molekularbewegung und der suspensionen. *Ann Phys.* 326, 576.

van Kampen, N.G. 1981. *Stochastic Process in Physics and Chemistry.* Amsterdam: North-Holland.

6

Jump Diffusion

6.1 Introduction

Our discussion of diffusive processes in Chapters 1 to 5 considered cases in which position and momentum are viewed as continuous variables. The basic entity is the momentum that acquires randomness because of heat bath-induced noise, as in the Langevin picture. The driven variable, i.e., position coordinate, is also endowed with randomness and at the high-friction limit, merges with the Einstein definition of Brownian motion. In conformity with the continuous nature of a random process, the Smoluchowski equation can be written as a linear, second other differential operator.

$$\frac{\partial}{\partial t}P(x,t) = W_x P(x,t), \tag{6.1}$$

where

$$W_x = \frac{1}{m\gamma}\frac{\partial}{\partial x}\left(\frac{\partial V}{\partial x} + K_B T\frac{\partial}{\partial x}\right). \tag{6.2}$$

While the above-mentioned formalism is quite appropriate for gaseous and liquid states, we need an alternate approach for fluctuation phenomena in solids. A solid consists of a discrete array of atoms localized at lattice sites. At finite temperatures, the atoms perform vibratory motions around their equilibrium lattice positions which, at moderate temperatures, may be viewed as harmonic. However, at random time instants and aided by background fluctuations, an atom can jump into an available vacancy or jump into an empty spot at an interstitial site. Indeed, interstitial jumps in an empty lattice can be viewed as an uncorrelated random walk (Section 1.3).

On the other hand, the jumps of atoms from a normal lattice site (not an interstice) or a point defect at a substitutional site to a nearby vacancy constitute a correlated random walk governed by the so-called Bardeen Herring factor (Flynn 1972). In such cases, and simpler ones involving only a finite number of sites, we need a formulation that is distinct from the ones presented.

Vibrations of atoms around lattice sites occur on timescales comparable to the phonon (Debye) time scale which is of the order of 10^{-12} sec whereas the mean time between successive jumps is about 10^{-8} sec. Thus about 10^4 cycles of vibration are completed before the atom decides to jump. The concomitant timescale separation allows us to treat vibrations and jump motions as statistically uncorrelated. We will focus only on the latter and expand the diffusion concept to include jump diffusion. Like continuous diffusion, jump diffusion is also a stationary Markov process but in a discrete space. The central equation for studying jump diffusion is the discrete version of the master equation (2.25) written as

$$\frac{\partial}{\partial T} P(\xi,t) = \sum_{\xi'}{}^{'} [P(\xi',t)\ W(\xi'|\xi) - P(\xi,t)\ W(\xi|\xi')], \tag{6.3}$$

where the prime in the summation implies that the term $\xi' = \xi$ is excluded. Evidently the jump matrix W is now not a differential operator, unlike in Equation (2.25). Interestingly we have come full circle. From the discrete random walk in Section 1.3, we went to the continuous diffusive limit and we are now back to discussing a discrete stochastic process in accordance with Equation (6.3) from which the one-dimensional random walk of Section 1.3 follows as a limiting case!

6.2 Operator Notation

It is convenient to regard Equation (6.3) as an operator equation of motion (much like the Heisenberg equation in quantum mechanics) in which $\hat{P}$ and $\hat{W}$ (now adorned by overhead caps) are operators in the linear vector space spanned by the allowed values of the stochastic process ξ. Correspondingly, we introduce a notation $|\xi)$ (like the Dirac ket) to represent a stochastic state in which the random variable assumes the value ξ. The conditional probability $P(\xi_0 \,|\, \xi,t)$ can then be written as a matrix (Dattagupta 1987):

$$P(\xi_0 \,|\, \xi,t) = (\xi \,|\, \hat{P}(t) \,|\, \xi_0). \tag{6.4}$$

The probability matrix $\hat{P}(t)$ evidently satisfies the initial condition:

$$\hat{P}(0) = 1, \tag{6.5}$$

where 1 is the unit operator. Thus

$$(\xi \mid \hat{P}(t) \mid \xi_0) = \delta(\xi - \xi_0), \tag{6.6}$$

where the states $|\xi)$ are assumed to form an orthonormal complete set, satisfying the closure property:

$$\sum_{\xi} |\xi)(\xi| = 1.$$

The master equation (6.3) can now be expressed in the compact form:

$$\frac{\partial}{\partial t}\hat{P}(t) = \hat{W}\hat{P}(t), \tag{6.7}$$

reflecting the Heisenberg structure, in which the jump matrix W is defined as

$$(\xi \mid \hat{W} \mid \xi_0) = W(\xi_0 \mid \xi). \tag{6.8}$$

The above represents an instantaneous probability per unit time for a jump of the stochastic process from the state $|\xi_0)$ to the state $|\xi)$. It is easy to demonstrate how Equation (6.7) reduces to the master equation (6.3). We have

$$\frac{\partial}{\partial t}(\xi \mid \hat{P}(t) \mid \xi_0) = \sum_{\xi'} (\xi \mid \hat{W} \mid \xi')\ (\xi' \mid \hat{P}(t) \mid \xi_0), \tag{6.9}$$

where we have made use of Equation (6.7). We further note from Equations (6.7) and (6.5)

$$\hat{W} = \frac{\partial \hat{P}(t)}{\partial t}\Big|_{t=0}. \tag{6.10}$$

Because the total probability is conserved we must have

$$\sum_{\xi'} (\xi' \mid \hat{P}(t) \mid \xi) = 1, \tag{6.11}$$

from which follows that

$$\sum_{\xi'} \left(\xi \mid \hat{W} \mid \xi'\right) = 0. \tag{6.12}$$

$$\sum_{\xi'}{}' (\xi' | \hat{W} | \xi) = -(\xi | \hat{W} | \xi). \tag{6.13}$$

Equation (6.9) then yields

$$\frac{\partial}{\partial t}\left(\xi | \hat{P}(t) | \xi_0\right) = \sum_{\xi'}{}' [(\xi | \hat{W} | \xi')\ (\xi' | \hat{P}(t) | \xi_0) - (\xi' | \hat{W} | \xi)\ (\xi | \hat{P}(t)\ \xi_0)]. \tag{6.14}$$

Reverting to the conditional probability we find

$$\frac{\partial}{\partial t}\hat{P}(\xi_0 | \xi, t) = \sum_{\xi'}{}' [\hat{W}(\xi' | \xi)\ \hat{P}(\xi_0 | \xi', t) - \hat{W}(\xi\ | \xi')\ \hat{P}(\xi_0 | \xi, t)], \tag{6.15}$$

which, by suppressing the dependence on the initial state $|\xi_0)$, is identical to Equation (6.13). As we saw in the case of various avatars of the Fokker–Planck equation (Rayleigh, Kramers, or Smoluchowski equation), the probability matrix must be such that asymptotically, i.e. as $t \to \infty$, equilibrium must obtain, which means

$$\lim_{t \to \infty} (\xi | P(\hat{t}) | \xi_0) = p(\xi), \tag{6.16}$$

where $p(\xi)$ is the equilibrium one-point probability appropriate to the state $|\xi)$ sometimes also called the *a priori* probability. Equation (6.14) then implies, since the left side is zero in equilibrium:

$$0 = \sum_{\xi'}{}' [(\xi | \hat{W} | \xi)\ p(\xi') - (\xi | \hat{W} | \xi')\ p(\xi)]. \tag{6.17}$$

A sufficient (but not necessary) condition to satisfy Equation (6.17) is that

$$p(\xi')\ (\xi' | \hat{W} | \xi') = p(\xi)\ (\xi' | \hat{W} | \xi). \tag{6.18}$$

This establishes the so-called detailed balance relation. Summarizing, the stationary Markov process for jump diffusion is founded on a probability operator:

$$\hat{P}(t) = \exp(\hat{W} t). \tag{6.19}$$

The operator $\hat{W}$ is a finite dimensional matrix whose eigenvalues and eigenfunctions completely specify the stochastic process at hand. One eigenvalue of W must be zero, appropriate to the stationary (equilibrium) distribution attained at $t \to \infty$ while all other eigenvalues must be negative to ensure boundedness. The $\hat{W}$ matrix, also called the jump matrix or relaxation matrix, must satisfy the conservation of probability in Equation (6.12) and the detailed balance relation of Equation (6.18).

6.3 Two-State Jump and Telegraph Processes

The simplest example of jump diffusion is a two-state process, also known as a dichotomic Markov or telegraph process (Feller 1972, Dattagupta 1987). While it is difficult to imagine this process to represent jump diffusion in a lattice, which was the motivation for introducing jump processes at the outset, in certain restricted conditions of diffusion in a double well, a spin one half impurity, or a two-level atom, the two-state process provides a handy model. The W matrix is evidently a 2×2 arrangement defined by

$$|I) = \begin{pmatrix} 1 \\ 0 \end{pmatrix}, \quad |2) = \begin{pmatrix} 1 \\ 0 \end{pmatrix}. \tag{6.20}$$

Thus

$$W = \begin{pmatrix} -w & w \\ w & -w \end{pmatrix}, \tag{6.21}$$

where the diagonal elements are fixed by the conservation condition. w is the jump rate from state $|1)$ to $|2)$ and vice versa. The detailed balance condition of Equation (6.18) assumes a simple form:

$$p_1(2|\hat{W}|1) = p_2(1|\hat{W}|2), \tag{6.22}$$

where

$$p_1 = p_2 = ½, \tag{6.23}$$

i.e., the two states are equally probable a priori. Although the $\hat{P}(t)$ matrix in Equation (6.19) is easily calculable by exploiting properties of 2×2 exponential matrices, we will employ a trick that may be easily generalized later to multilevel situations. The W matrix can be written as

$$\hat{W} = \lambda(\hat{\mathfrak{J}} - 1), \tag{6.24}$$

where $\hat{\mathfrak{J}}$ is also a 2×2 matrix with special properties:

$$\mathfrak{J}_{ij} = 1/2, \quad i, j = 1, 2, \tag{6.25}$$

and λ, the relaxation rate, is given by

$$\lambda = 2w. \tag{6.26}$$

Clearly, $\hat{\mathfrak{J}}$ is idempotent:

$$(\hat{\mathfrak{J}})^2 = \hat{\mathfrak{J}}, \tag{6.27}$$

as can be easily verified from Equation (6.25). Using this property, the $\hat{P}(t)$ matrix reduces to a simple structure,

$$\hat{P}(t) = \frac{1}{2}\begin{pmatrix} (1+e^{-\lambda t}) & (1-e^{-\lambda t}) \\ (1-e^{-\lambda t}) & (1+e^{-\lambda t}) \end{pmatrix}. \tag{6.28}$$

Once we have $\hat{P}(t)$, all physical properties can be derived. For instance, the mean of ξ is given by

$$\langle \xi(t) \rangle = \sum_{ij} p_i (i \,|\, \hat{P}(t) \,|\, j) \xi_j, \tag{6.29}$$

which, from Equations (6.28) and (6.23), vanishes if ξ is assumed to take two values $+\xi$ and $-\xi$. On the other hand, the autocorrelation is ξ given by

$$C(t) = \langle \xi(o)\xi \,|\, (t) \rangle = \sum_{ij} p_i \quad \xi_i (i \,|\, \hat{P}(t) \,|\, j) \; \xi_j = \langle \xi^2 \rangle e^{-\lambda t}. \tag{6.30}$$

The fact that the correlation function is a single exponential although the underlying process is not Gaussian is, of course, not a violation of Doob's theorem mentioned earlier. The present process is a discontinuous jump process and is thus outside the ambit of Doob's theorem. The correlation time τ_c is defined by

$$\tau_C = \frac{1}{C(0)} \int_0^\infty dt \; C(t) \tag{6.31}$$

τ_C turns out to be precisely λ^{-1} in the present case.

6.4 Multi-State Jump Process

The above treatment for a two-state jump process can be easily generalized to a multistate situation though the W matrix can continue to be expressed as Equation (6.24) where now,

$$\lambda = Nw \quad \text{and} \quad \mathfrak{F}_j = \frac{1}{N}, \text{ independent of } i \text{ and } j. \tag{6.32}$$

Thus, $\hat{\mathfrak{F}}$ continues to be an idempotent matrix that allows the $\hat{P}(t)$ matrix to be expressed as

$$(i \mid P(t) \mid j) = \frac{1}{N} + \left(\delta_{ij} - \frac{1}{N}\right)e^{-\lambda t}, \quad i, j = 1, 2, \ldots N, \tag{6.33}$$

where δ_{ij} is the Kronecker delta. The mean and correlations are now given by:*

$$\langle \xi(t) \rangle = \frac{1}{N}\sum_{n=1}^{N} \xi_n, \tag{6.34}$$

$$C(t) = \langle \xi \rangle^2 + \left(\langle \xi^2 \rangle - \langle \xi \rangle\right)^2 e^{-\lambda t}, \tag{6.35}$$

where

$$\langle \xi^2 \rangle = \frac{1}{N}\sum_{n=1}^{N} \xi_n^2. \tag{6.36}$$

6.5 Kubo–Anderson Process

The Kubo–Anderson process that has myriad applications in spectroscopy (discussed later) can be considered a straightforward generalization of the two-state jump process (Anderson 1954, Blume 1968, Kubo 1954). Instead of assuming that both the states are equally probable a priori as in Equation (6.23) we now assume that the elementary jump probability depends only on the final state. Thus

$$(i \mid \hat{\mathfrak{F}} \mid j) = f_i, \tag{6.37}$$

* See Exercise 1.

independent of j, where f_i is an arbitrary function of the state i. Now the detailed balance relation of Equation (6.18) reads:

$$p_j(i|\hat{\mathfrak{I}}|j) = p_i(j|\hat{\mathfrak{I}}|i), \tag{6.38}$$

which implies that

$$f_i = p_i, \tag{6.39}$$

once we use Equation (6.37) and also the conservation probability:

$$\sum_i p_i = 1, \quad \sum_i (i|\hat{\mathfrak{I}}|j) = 1. \tag{6.40}$$

Hence,

$$(i|\hat{\mathfrak{I}}|j) = p_i. \tag{6.41}$$

Equation (6.41) as a model is applicable to a multistate case in which the indices i, j run from $1,2, \ldots N$. Once again the $\mathfrak{I}$ matrix is idempotent*:

$$(\mathfrak{I})^n = \ldots\ldots = (\mathfrak{I})^{n-1} = \hat{\mathfrak{I}}. \tag{6.42}$$

Using this, we obtain

$$(i|\hat{P}(t)|j) = e^{-\lambda t}(\delta_{ij} - p_i + p_i \; e^{\lambda t}). \tag{6.43}$$

The correlation function now is

$$C(t) = \langle \xi \rangle^2 + \left(\langle \xi^2 \rangle - \langle \xi \rangle^2 \right) \; e^{-\lambda t}, \tag{6.44}$$

where

$$\langle \xi \rangle = \sum_{n=1}^{N} p_n \xi_n, \tag{6.45}$$

$$\langle \xi^2 \rangle = \sum_{n=1}^{N} p_n \xi_n^2. \tag{6.46}$$

* See Exercise 2.

From the multistate case we can, of course, move to a situation in which $N \to \infty$. Thus we are led to a W matrix whose dimension is (countable) infinity. The process ξ may now be regarded as almost continuous for which the correlation function still has the structure of Equation (6.44) but now the summations in Equations (6.45) and (6.46) are replaced by integrals, i.e.,

$$\langle \xi \rangle = \int d\xi' p(\xi')\xi', \quad \text{and} \quad \langle \xi^2 \rangle = \int d\xi' \; p(\xi') \; (\xi')^2. \tag{6.47}$$

The jump matrix $\hat{W}$ has the form [cf. Equations (6.24) and (6.25)].

$$(\xi \,|\, \hat{W} \,|\, \xi_0) = \lambda\left[p(\xi) - \delta(\xi - \xi_0)\right]. \tag{6.48}$$

The corresponding conditional probability matrix can be directly read from Equation (6.43). It is

$$(\xi \,|\, P(t) \,|\, \xi_0) = e^{-\lambda t}\delta(\xi - \xi_0) + p(\xi)(1 - e^{-\lambda t}). \tag{6.49}$$

Viewing ξ as the momentum variable of a Brownian particle, the Kubo–Anderson process depicts a situation diametrically opposite to that described by the Rayleigh process (cf. Section 5.2). Here the probability of jump (due to a collision) from the momentum p_0 to p is independent of the latter. This implies that a collision is so strong that the memory of the initial momentum is completely wiped out afterward. Hence, the Kubo–Anderson process is also known as the strong collision approximation (Dattagupta and Blume 1974).

This is in total contrast to the Langevin picture in which the effect of a collision alters the momentum by only a tiny bit, justifying the linear damping term in the Langevin equation (4.7). Naturally, the Rayleigh process is called the weak collision approximation (Rautian and Sobelman 1967). We shall discuss the spectroscopic context in detail later.

6.6 Interpolation Model

In many situations concerning rotational diffusion of molecules in a liquid or gas phase, an improvement on the strong collision approximation occurs during modeling of the collision operator $\hat{\mathfrak{I}}$ when a bit more of the remembrance of the initial state is injected. Thus, Equation (6.41) is generalized (Dattagupta and Sood 1979, Balakrishnan 1979, Fixman and Rider 1969) to

$$(\xi \,|\, \hat{\mathfrak{I}} \,|\, \xi_0) = r \;\; p(\xi) + (1 - r) \;\; \delta(\xi - \xi_0), \tag{6.50}$$

when the primary transition operator is an interpolation between the initial and asymptotic distribution, r being an interpolation parameter $(0 \leq r \leq 1)$. Evidently for $r = 1$, Equation (6.50) reduces to the strong collision approximation (designated the J diffusion model in rotational spectroscopy) whereas for $r = 0$ we have the so-called M diffusion model (Gordon 1966). Writing the transition operator $\hat{\mathfrak{T}}$ as

$$\mathfrak{T} = r\ \hat{\mathfrak{T}}_0 + (1-r)\hat{I}, \tag{6.51}$$

and the W matrix can be expressed in the same generic form applicable to the strong collision approximation (Equation 6.24) in Equation (6.48):

$$\hat{W} = \nu(\hat{\mathfrak{T}}_0 - \hat{1}), \tag{6.52}$$

where, however,

$$\nu = r\lambda. \tag{6.53}$$

Thus, the interpolation model simply alters the relaxation rate λ to a reduced value ν, thereby slowing the relaxation process. The corresponding conditional probability now reads [cf. Equation (6.49)]:

$$(\xi \mid \hat{P}(t) \mid \xi_0) = e^{-\nu t}\ \ \delta(\xi - \xi_0) + p(\xi)(1 - e^{-\nu t}). \tag{6.54}$$

Through this simple mechanism a larger variety of experimental data on Raman and infrared studies of rotational relaxation can be satisfactorily investigated.

6.7 Kangaroo Process

As noted earlier, the probability of an elementary jump, say for a Rayleigh particle, is considered completely independent of the initial momentum for a Kubo–Anderson process [Equation (6.48)]. This reflects in the constancy of the jump rate λ. However, because the effect of collisions is to ultimately equilibrate the probability distribution, we could argue that the effective rate of collisions may be chosen to be lower for particles whose momenta are further away from the most probable momentum of the equilibrium distribution in Equation (6.1) (Brissaud and Frisch 1974, 1971). In that case, λ will have to be a function of the precollision momentum and this functional dependence must be tuned so that the goal of hastening equilibration cited above can be achieved. We may then extend Equation (6.48) to

$$(\xi\,|\,\hat{W}\,|\,\xi_0) = \lambda(\xi_0)[q(\xi) - \delta(\xi - \xi_0)], \tag{6.55}$$

where $q(\xi)$ has to be naturally distinct from the equilibrium distribution $p(\xi)$ but must satisfy

$$0 = \int d\xi(\xi\,|\,\hat{W}\,|\,\xi_0) = \lambda(\xi_0)\int d\xi \;\; q(\xi) - \lambda(\xi_0). \tag{6.56}$$

In other words, $q(\xi)$ must also be normalized

$$\int d\xi \;\; q(\xi) = 1. \tag{6.57}$$

Furthermore, the detailed balance relation (6.18) and Equation (6.50) imply

$$p(\xi_0)\cdot\lambda(\xi_0)\cdot q(\xi) = \;\; \hat{p}(\xi)\cdot\lambda(\xi)\cdot q(\xi_0) \tag{6.58}$$

Integrating Equation (6.58) over ξ_0 and using Equation (6.53) we obtain

$$q(\xi) = \lambda(\xi)p(\xi)\,/\,\bar{\lambda}(\xi) \tag{6.59}$$

where

$$\bar{\lambda}(\xi) = \int d\xi_0 \;\;\; p(\xi_0) \;\; \lambda(\xi_0). \tag{6.60}$$

Clearly, when $\lambda(\xi_0)$ is independent of ξ_0, $\bar{\lambda}(\xi)$ reduces to λ and we are back to the Kubo–Anderson process. Thus the latter may be seen as a special case of Equation (6.50) we designated the kangaroo process. Although the form of $\hat{W}$ as in Equation (6.24) and the analytic structure of $\hat{P}(t)$ as in Equation (6.49) are not applicable for a kangaroo process, we may still write:

$$\hat{W} = (\hat{\mathfrak{I}} - \hat{1})\,\hat{\Lambda} \tag{6.61}$$

where the matrix of $\hat{\mathfrak{I}}$ is now [cf. Equation (6.55)]:

$$(\xi\,|\,\hat{\mathfrak{I}}\,|\,\xi_0) = q(\xi), \tag{6.62}$$

and $\hat{\Lambda}$ is another matrix that is diagonal and is given by

$$(\xi\,|\,\hat{\Lambda}\,|\,\xi_0) = \lambda(\xi_0)\;\delta(\xi - \xi_0). \tag{6.63}$$

Evidently, Equations (6.61) to (6.63) are consistent with Equation (6.50). The question now is how to obtain $\hat{P}(t)$ from Equation (6.61). The process is facilitated by considering the Laplace transform defined by

$$\tilde{P}(s) = \int_0^\infty dt\;\; e^{-st}\;\; \hat{P}(t). \tag{6.64}$$

From Equations (6.19) and (6.64), formally.

$$\tilde{P}(s) = \frac{1}{(s + \hat{\wedge}) - \hat{\mathfrak{I}}\hat{\Lambda}}. \tag{6.65}$$

The above equation can be written as a geometric series.

$$\tilde{P}(s) = \frac{1}{s+\hat{\Lambda}}\left[1 + (\hat{\mathfrak{I}}\hat{\Lambda})\frac{1}{s+\hat{\Lambda}} + (\hat{\mathfrak{I}}\hat{\Lambda})\frac{1}{s+\hat{\Lambda}}(\hat{\mathfrak{I}}\hat{\Lambda})\frac{1}{s+\hat{\Lambda}} + \dots\right]. \tag{6.66}$$

Hence,

$$\begin{aligned}(\xi\,|\,\tilde{P}(s)\,|\,\xi_0) = \frac{1}{s+\lambda(\xi_0)}[\delta(\xi - \xi_0) + \frac{1}{s+\lambda(\xi)}(\xi\,|\,\hat{\mathfrak{I}}\hat{\Lambda}\,|\,\xi_0) \\ + \int d\xi' \frac{1}{s+\lambda(\xi)}(\xi\,|\,(\hat{\mathfrak{I}}\hat{\Lambda})\,|\,\xi')\frac{1}{s+\lambda(\xi')}(\xi'\,|\,(\hat{\mathfrak{I}}\hat{\Lambda})\,|\,\xi_0) + \dots]\end{aligned} \tag{6.67}$$

where we have used the closure property

$$\int d\xi'\,|\,\xi')\;\;(\xi'\,| = 1. \tag{6.68}$$

Now, note that

$$(\xi\,|\,\mathfrak{I}\hat{\Lambda}\,|\,\xi_0) = \int d\xi'\;\; q(\xi)\;\; \lambda(\xi_0)\;\; \delta(\xi_0 - \xi') = \lambda(\xi_0)\;\; q(\xi). \tag{6.69}$$

Hence, Equation (6.67) yields

$$(\xi \mid \tilde{P}(s) \mid \xi_0) = \frac{1}{s+\lambda(\xi_0)}\left[\delta(\xi-\xi_0)+\lambda(\xi_0)\frac{q(\xi)}{s+\lambda(\xi)}+\frac{\lambda(\xi_0)}{s+\lambda(\xi)}q(\xi)\int d\xi' \frac{\lambda(\xi')q(\xi')}{s+\lambda(\xi')}+\ldots\ldots\right]. \tag{6.70}$$

The remaining terms can again be grouped together as a geometric series that can be summed back, thus yielding*

$$(\xi \mid \tilde{P}(s) \mid \xi_0) = \frac{1}{s+\lambda(\xi_0)}\left\{\delta(\xi-\xi_0)+\lambda(\xi_0)\frac{q(\xi)}{s+\lambda(\xi)}\left[1-\int d\xi' \frac{\lambda(\xi')q(\xi')}{s+\lambda(\xi')}\right]^{-1}\right\}. \tag{6.71}$$

Evidently, when $\lambda(\xi')$ equals λ, independent of ξ', $q(\xi)$ becomes $p(\xi)$ and the above equation reduces to

$$(\xi \mid \tilde{P}(s) \mid \xi_0) = \frac{1}{s+\lambda}\left[\delta(\xi-\xi_0)+\frac{\lambda}{s}p(\xi)\right], \tag{6.72}$$

which, as expected, yields the Laplace transform of the probability matrix for the Kubo–Anderson process [cf. Equation (6.49)]. The efficacy of the kangaroo process can be assessed by examining the correlation function.

$$C(t) = \iint d\xi_0\, d\xi \;\; p(\xi_0) \;\; \xi_0 \;\; \xi(\xi \mid \hat{P} \mid t) \mid \xi_0). \tag{6.73}$$

Clearly, the Laplace transform of $C(t)$ can be written from Equation (6.71) as

$$\tilde{C}(s) = \int d\xi_0 \frac{p(\xi_0)\xi_0^2}{s+\lambda(\xi_0)} + \left(\int d\xi_0 \frac{\lambda(\xi_0)\xi_0}{s+\lambda(\xi_0)}\right)\left(\int d\xi \;\; \frac{q(\xi)\xi}{s+\lambda(\xi\;)}\right)\bullet\left[1-\int d\xi' \frac{\lambda(\xi')q(\xi')}{s+\lambda(\xi')}\right]^{-1}. \tag{6.74}$$

The above equation assumes a particularly simple structure if we take $\lambda(\xi)$ to be symmetric, i.e., $\lambda(\xi) = \lambda(-\xi)$ within its range of integration from $-\infty$ to ∞. In that case, the second term vanishes while the first leads to

$$\tilde{C}(s) = \int_{\infty}^{\infty} d\xi \frac{p(\xi)\;\xi^2}{s+\lambda(\xi)}. \tag{6.75}$$

Transforming back to the time domain,

$$C(t) = \int_{-\infty}^{\infty} d\xi\,\xi^2 \;\; p(\xi) \;\; e^{-\lambda(\xi)\;t}. \tag{6.76}$$

---

* See Exercise 3.

A variety of modelings of $\lambda(\xi)$ can be effected to obtain various forms of the correlation function *C(t)* which is now a continuous superposition of exponentially decaying functions of time. Such non-exponential forms are found in nature as in stretched exponential decay $\sim \exp(-t^{\alpha})$, $0 < \alpha < 1$ in glassy materials or in a power law decay $\sim (t)^{-\beta}, \beta > 0$, as in long time tails in hydrodynamic flows. In any case, although the kangaroo process is derived from jump diffusion following a stationary Markov process, the associated correlation function is very much un-diffusion-like. Indeed, it typifies anomalous diffusion—the subject of Chapter 7.

Exercises

1. Prove Equation (6.35).
2. Validate Equations (6.42) and (6.43).
3. Prove Equation (6.71).

References

Anderson, P. W. 1954. An introduction to probability theory and its applications. *J. Phys. Soc. (Japan)*. 9, 316.

Balakrishnan, V. 1979. *Pramana J. Phys.* 13, 337.

Blume, M. 1968. Stochastic theory of line shape: Generalization of the Kubo–Anderson model. *Phys. Rev.* 174, 351.

Brissaud, A. and U. Frisch. 1974. Solving linear stochastic differential equations. *J. Math. Phys.* 15, 524.

Brissaud, A. and U. Frisch. 1971. *J. Quant. Spectrosc. Rad. Transfer* 11, 1767.

Dattagupta, S. and M. Blume. 1974. Stochastic theory of line shape I. Nonsecular effects in the strong-collision model. *Phys. Rev.* B10, 4540.

Dattagupta, S. 1987. *Relaxation Phenomena in Condensed Matter Physics.* Orlando: Academic Press (new edition published by Levant, Kolkata, 2011).

Feller, W. 1972. *An Introduction to Probability Theory and Its Applications*, Vols. 1 and 2. New Delhi: Wiley Eastern.

Fixman, M and K. Rider. 1969. The *J*-diffusion and the *M*-diffusion models. *J. Chem. Phys.* 51, 2425.

Flynn, C.P. and A.M. Stoneham. 1970. *Phys. Rev.* B1, 3966.

Gordon, R. G. 1966. On the rotational diffusion of molecules. *J. Chem. Phys.* 44, 1830.

Kubo, R. 1954. Note on the stochastic theory of resonance absorption. *J. Phys. Soc. (Japan)*. 9, 935.

Rautian, S. G. and I. I. Sobelman. 1967. Effect of collisions on the Doppler broadening of spectral lines. *Sov. Phys. Usp.* 9, 701.

Rautian, S. G. and I. I. Sobelman. 1966. Effect of collisions on the Doppler broadening of spectral lines. *Usp. Fiz. Nauk.* 90, 209.

7

Random Walk and Anomalous Diffusion

7.1 Introduction to Continuous Time Random Walk (CTRW)

The primary notion of a CTRW is to break a time interval *(0,t)* into subintervals defined by the instants $t_1, t_2, \ldots t_n$, at which points in time a system is assumed to be subject to certain fluctuations or collisions that mimic the random events caused by the surrounding heat bath. The underlying stochastic process is then a chain of collisions or an ongoing renewal of a stochastic sequence. The basic idea is illustrated in Figure 7.1.

Eventually, the epochs $t_1, t_2, \ldots t_n$ themselves will be assumed to be randomly distributed. That is indeed the genesis of the CTRW phrase coined by Montroll and Weiss (1965) to contrast the case of pure random walks in which the times between successive steps of the walk are considered equal.

With respect to Figure 7.1, consider the following sequence in which the stochastic process $x(t)$ starts with the value x_0 at time $t = 0$ and ends with the value x at time t. We want to find the conditional probability for this process. In one possible trajectory for a member of an ensemble of trajectories, the stochastic variable x has a constant value x_0 until time t_1, at which point the system suffers a collision. The latter has a certain probability of throwing the variable x into another value, selected from a possible ensemble of values, much like the velocities of a Maxwellian gaseous particle.

The corresponding transition probability can then be written as $(x_1 \,|\, \hat{\mathfrak{I}} \,|\, x_0)$, where $|\, x_1)$ is the new stochastic state, and $\mathfrak{I}$ is a transition or collision operator. The variable retains the value x_1 from times t_1 to t_2 and at t_2, another jump to the stochastic state $|\, x_2)$ occurs, and so on. Finally, at instant t_n, the system transits to the state $|x)$ and remains there until time t. We will demonstrate below that the process is stationary and Markovian when the collisions are triggered by a Poisson pulse sequence with a constant mean rate λ.

We will separately list below the consequences of zero collisions, one collision, … n collisions, etc., in the time interval 0 to t, add the results, and sum n from zero to infinity.

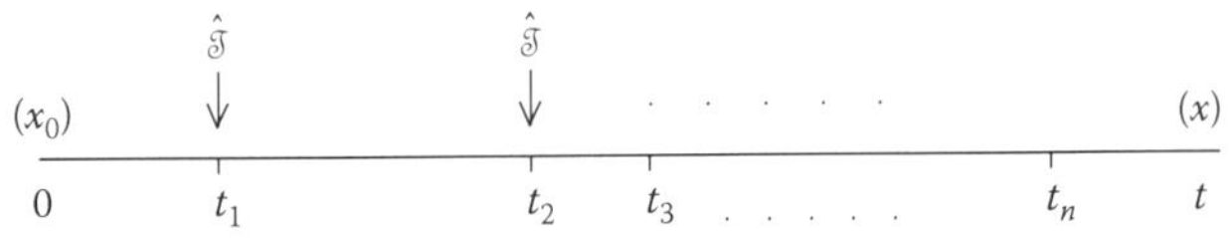

FIGURE 7.1
Renewal of a random walk at time intervals t_1, t_2, ... within the domain 0 to t. The stochastic process $x(t)$ starts with the value x_0 at time $t = 0$ and ends up with the value x at time t.

No-collision term — The conditional probability for this event is clearly $\delta(x_0 - x_1)\ e^{-\lambda t}$ because for a Poisson distribution with a mean rate λ, the probability that no collision occurs between 0 and t is $e^{-\lambda t}$. The corresponding (coarse-grained) transition operator is $1e^{-\lambda t}$.

One-collision term — The probability that a collision does occur between t_1 and $t_1 + dt_1$ is $\lambda dt_1\ \ e^{-\lambda t_1}$ and the probability that no further collision occurs until time t is $e^{-\lambda(t-t_1)}$. Hence the conditional probability for the one-collision event is:

$$\int_0^t dt_1\ \ e^{-\lambda(t-t_1)}\ \ (x\,|\,\lambda\hat{\mathfrak{I}}\,|\,x_0)e^{-\lambda t_1}.$$

Since a collision can occur at any time between 0 and t_1, t_1 is integrated over. The corresponding transition operator is $\int_0^t dt_1\ e^{-\lambda(t-t_1)}\ (\lambda\hat{\mathfrak{I}})e^{-\lambda t_1}$.

Two-collision term — The conditional probability for this event

$$\sum_{x_1}\int_0^{t_2} dt_1 e^{-\lambda(t-t_2)}(x\,|\,(\lambda\hat{\mathfrak{I}})\,|\,x_1)\ e^{-\lambda(t_2-t_1)}\ \ (x_1\,|\,(\lambda\hat{\mathfrak{I}})\,|\,x_0)\ e^{-\lambda t_1}$$

and the transition operator is

$$\int_0^t dt_2\int_0^{t_2} dt_1 e^{-\lambda(t-t_2)}\ \ (\lambda\hat{\mathfrak{I}})\ e^{-\lambda(t_2-t_1)}\ \ (\lambda\hat{\mathfrak{I}})\ e^{-\lambda t_1}.$$

The summation over x_1 implies that all possible intermediate values of x are considered.

***n*-Collision term** — Generalizing, the conditional probability for n collision events is

$$\sum_{x_1x_2\ldots.x_{n-1}}\ \int_0^t dt_n\int_0^{t_n} dt_{n-1}\ldots\int_0^{t_2} dt_1\ e^{-\lambda(t-t_n)}\ (x\,|\,\lambda\hat{\mathfrak{I}}\,|\,x_{n-1})\,e^{-\lambda(t_n-t_{n-1})}\ (x_{n-1}\,|\,(\lambda\hat{\mathfrak{I}})\,|\,x_{n-2})\ldots e^{-\lambda(t_2-t_1)}$$

$(x_1\,|\,(\lambda\hat{\mathfrak{I}})\,|\,x_0)e^{-\lambda t_1}$, which in operator notation yields

$$\sum_{x_1x_2\ldots.x_{n-1}}\ \int_0^t dt_n\int_0^{t_n} dt_{n-1}\ldots\int_0^{t_2} dt_1\ \ e^{-\lambda(t-t_n)}\ \ (\lambda\hat{\mathfrak{I}})e^{-\lambda(t_n-t_{n-1})}\ \ .(\lambda\hat{\mathfrak{I}})\ldots e^{-\lambda(t_2-t_1)}\ \ .(\lambda\hat{\mathfrak{I}})e^{-\lambda t_1}.$$

When all the contributions are added the conditional probability operator for the entire chain of events turns out to be

$$\hat{P}(t) = \sum_{n=0}^{\infty} \int_0^t dt_n \int_0^{t_n} dt_{n-1} \ldots \int_0^{t_2} dt_1 \; e^{-\lambda(t-t_n)}(\lambda\hat{\mathfrak{I}}).e^{-\lambda(t_n-t_{n-1})}.(\lambda\hat{\mathfrak{I}})\ldots e^{-\lambda(t_2-t_1)}.(\lambda\hat{\mathfrak{I}}) \; e^{-\lambda t_1}.$$

All the exponential functions combine to simply yield $e^{-\lambda t}$ whereas the multiple time integrals lead to a factor $t^n/n!$. Therefore,

$$\hat{P}(t) = \sum_{n=0}^{\infty} \frac{(\lambda\hat{\mathfrak{I}}t)^n}{n!} e^{-\lambda t} \; e^{\lambda t(\hat{\mathfrak{I}}-1)}. \tag{7.1}$$

Interestingly, by defining

$$W = \lambda(\hat{\mathfrak{I}} - \hat{1}), \tag{7.2}$$

as we split the W matrix earlier for discussing the telegraph process and the Kubo–Anderson process [cf. Equation (7.24)], we obtain

$$\hat{P}(t) = e^{\hat{W}t}, \tag{7.3}$$

the defining equation for the stationary Markov process (Equation (6.19).

It is interesting at this stage to note the following important property of a CTRW (which remains valid even on further generalization, as seen below): the composite conditional probability is a factored product of two terms—the n-fold transition probability $(\hat{\mathfrak{I}})^n$ and the probability $f_n(t)$ that exactly n collisions occur in time t:

$$\hat{P}(t) = \sum_{n=0}^{\infty} f_n(t)(\hat{\mathfrak{I}})^n. \tag{7.4}$$

Assuming that collisions are governed by a Poisson distribution,

$$f_n(t) - \frac{(\lambda t)^n}{n!} e^{-\lambda t}. \tag{7.5}$$

The result (7.1) and hence Equation (7.3) follow directly.

The reader may wonder why we must proceed in a roundabout way in writing such an elaborate expression as Equation (7.1) when we could have easily arrived at the final result (7.3) by employing Equation (7.4). The reason is, as we shall indicate below, the expanded form of Equation (7.1) allows for a straightforward generalization to non-Poisson pulse sequences yielding a non-Markov process.

We may remark in passing that while the Kubo–Anderson jump process falls within the above scheme of a Poisson pulse-driven stochastic process, in which the transition operator $\hat{\mathfrak{I}}$ has a specific form [cf. Equation (6.41)], the kangaroo process, on the other hand, is outside the purview of such a treatment. We return to this point in the next section when we consider a generalized CTRW that is not governed by a Poisson distribution.

7.2 Non-Markovian Diffusion in CTRW Scheme

In many physical applications, it is necessary to consider memory effects. Such effects will have to be dealt with by going beyond the Markovian distribution of the underlying stochastic process within the CTRW scheme described in Section 7.1 (Montroll and Scher 1973, Scher and Lax 1972, Scher and Montroll 1975, Lax and Scher 1977). Non-Markovian diffusion can be treated by relaxing the Poissonian assumption for the pulse sequences. Evidently, the two-point conditional probability $P(\xi_0 \,|\, \xi, t)$ no longer suffices for depicting non-Markovian processes, though it continues to remain relevant for computing two-point correlation functions. The question that we would like to pose is: given a basic transition operator $\hat{\mathfrak{I}}$ how does one construct the conditional probability operator $\hat{P}(t)$?

To answer this question it is essential to introduce the concepts of persistent probability, denoted by $\phi(t)$ and first waiting-time distribution designated by $\phi(t)$ (Feller 1966). This concept is required because $t = 0$ need not coincide with the beginning of a waiting period (Tunaley 1976). If the stochastic state $|\xi_0)$ has just come into being at $t = 0$, $\phi(t)$ defines the probability that the same state $|\xi_0)$ persists at time t. Thus $\phi(t = 0) = 1$ and $-\mathring{\phi}(t)dt$ is the probability that a transition to some other state occurs in the interval $(t,\ t + dt)$. The first waiting time distribution is given by (Feller 1966):

$$\phi_1(t) = \phi(t) \Big/ \int_0^\infty dt' \phi(t'). \tag{7.6}$$

Note that for a Poissonian distribution, $\phi(t) = e^{-\lambda t}$ and $\phi_1(t) = \lambda e^{-\lambda t}$. On the other hand, the probability of no transition occurring in the time interval from the randomly chosen origin of time up to time t is

$$\mathbf{\Theta}(t) = 1 - \int_0^t dt' \ \phi_1(t'). \tag{7.7}$$

Again, for a Poisson process, $\mathbf{\Theta}(t) = \phi(t) = e^{-\lambda t}$. The conditional probability matrix $\hat{P}(t)$ may then be developed as the series

$$(\xi \,|\, \hat{P}(t) \,|\, \xi_0) = \mathbf{\Theta}(t)\delta(\xi - \xi_0) + \sum_{n=1}^{\infty} \int_0^t dt_n \ldots \int_0^{t_2} dt_1 (-1)^{n-1} \mathbf{\Theta}(t - t_n)$$

$$\bullet \dot{\mathbf{\Theta}}(t_n - t_{n-1}) \ldots \dot{\mathbf{\Theta}}(t_2 - t_1) \;\; \mathbf{\Theta}_1(t_1) \;\; (\xi \,|\, \hat{\mathbf{T}}^n \,|\, \xi_0). \tag{7.8}$$

7.2.1 Application to Interpolation Model

We will now examine the structure of conditional probability when the primitive transition operator $\hat{\mathfrak{I}}$ is governed by the form associated with the interpolation model [cf. Equation (6.53)] (Balakrishnan 1979). In the latter $\hat{\mathfrak{I}}$ is given by Equation (6.54) and we may express

$$\hat{\mathfrak{I}}^n = \frac{d^n}{d\alpha^n} \exp(\alpha \, \hat{\mathfrak{I}}) \quad \alpha = 0, \tag{7.9}$$

and $\hat{\mathfrak{I}}_0$ has the idempotent property

$$\hat{\mathfrak{I}}_0^2 = \hat{\mathfrak{I}}_0 \tag{7.10}$$

It is easy to see that

$$\exp(\alpha \, \hat{\mathfrak{I}}) = \hat{\mathfrak{I}}_0 e^{\alpha} + (1 - \hat{\mathfrak{I}}_0) \;\; e^{\alpha(1-\gamma)}. \tag{7.11}$$

Hence,

$$\hat{\mathfrak{I}} = \hat{\mathfrak{I}}_0 + (1-\gamma)^n \;\; \cdot (1 - \hat{\mathfrak{I}}_0) \tag{7.12}$$

Before substituting for $(\hat{\mathfrak{I}})^n$ as in Equation (7.9) into Equation (7.8), it is convenient to take the Laplace transform of the latter when exploiting the convolution structure of that equation. Thus

$$(\xi \,|\, \tilde{\hat{P}}(s) \,|\, \xi_0) = \tilde{\mathbf{\Theta}}(s) \;\; \delta(\xi - \xi_0)$$

$$+ \sum_{n=1}^{\alpha} (-1)^{n-1} \tilde{\mathbf{\Theta}}(s)(s\tilde{\mathbf{\Theta}}(s) - 1)^{n-1} \;\; \tilde{\mathbf{\Theta}}_1(s) \;\; (\xi \,|\, (\hat{\mathbf{T}})^n \,|\, \xi_0) \tag{7.13}$$

We now substitute for $(\hat{\mathfrak{I}})^n$ as in Equation (7.12) and obtain

$$(\xi\,|\,\tilde{\hat{P}}(s)\,|\,\xi_0)=\left[\tilde{\phi}(s)+\frac{(1-\gamma)\tilde{\phi}(s)\tilde{\phi}_1(s)}{\gamma+(1-\gamma)s\tilde{\phi}_1(s)}\right]\delta(\xi-\xi_0)+\frac{\tilde{\phi}_1(s)}{s}\left[1-\frac{(1-\gamma)s\tilde{\phi}(s)}{\gamma+(1-\gamma)s\tilde{\phi}(s)}\right]p(\xi) \tag{7.14}$$

Note, from Equations (7.6) and (7.7) that

$$\tilde{\phi}_1(s)=\frac{\tilde{\phi}(s)}{\tilde{\phi}(s=0)}, \tag{7.15}$$

and

$$\tilde{\boldsymbol{\Theta}}(s)=\frac{1}{s}\left[1-\frac{\tilde{\phi}(s)}{\tilde{\phi}(s=0)}\right]. \tag{7.16}$$

Therefore, Equation (7.14) yields

$$(\xi\,|\,\hat{P}(s)\,|\,\xi_0)=\frac{1}{s}[1-F(s)]\ \delta(\xi-\xi_0)+\frac{F(s)}{s}p(\xi), \tag{7.17}$$

where

$$F(s)=\frac{\gamma\tilde{\phi}(s)/\tilde{\phi}(s=0)}{\gamma+(1-\gamma)\ s\tilde{\phi}(s)} \tag{7.18}$$

Evidently, for a Markov process, $\tilde{\phi}(s)=(s+\lambda)^{-1}$ and we recover from Equation (7.18) our previously derived Equation (6.54) for the interpolation model. We now demonstrate the connection between the CTRW and the master equation, hitherto considered only for a Markov process, by rewriting Equation (7.17) as an integro-differential equation in the time domain.

$$\frac{1}{\gamma}\frac{\partial}{\partial t}(\xi\,|\,\hat{P}(t)\,|\,\xi_0)=\frac{p(\xi)}{\int_0^\infty dt'\tilde{\phi}(t')}-f(t)\delta(\xi-\xi_0)$$

$$-\int_0^t dt'K(t-t')\ (\xi\,|\,\hat{P}(t')\,|\,\xi_0), \tag{7.19}$$

where *f(t)* is the inverse Laplace transform of

$$\tilde{f}(s) = \left\lfloor \tilde{\phi}(s) = \tilde{\phi}(s=0)\,(1 - s\tilde{\phi}(s)) \right\rfloor_{s\tilde{\phi}(s=0)\tilde{\phi}(s)},$$

and the memory kernel *K(t)* is the inverse Laplace transform of

$$\tilde{K}(s) = \frac{\left[1 - s\tilde{\phi}(s)\right]}{\tilde{\phi}(s)}. \tag{7.20}$$

The general solution given in Equation (7.19) is still too unwieldy (in analytic form) for practical computation, but it simplifies considerably when $\gamma = 1$ that corresponds to the strong collision approximation. In that case, Equation (7.19) yields.

$$(\xi \,|\, \hat{P}(t) \,|\, \xi_0) = \phi(t)\delta(\xi - \xi_0) + \left[1 - \phi(t)\right]p(\xi). \tag{7.21}$$

This result is useful for calculating the autocorrelation function defined as

$$C_\xi(t) = \langle \xi(0)\xi(t) \rangle = \iint d\xi_0 d\xi \;\, p(\xi_0) \;\, \xi_0 \;\, \xi \;\, (\xi \,|\, \hat{P}(t) \,|\, \xi_0). \tag{7.22}$$

Substituting Equation (7.21) into Equation (7.22), we find

$$C_\xi(t) = \langle \xi^2 \rangle \;\, \varphi(t) + \left[1 - \varphi(t)\right] \, \langle \xi \rangle^2, \tag{7.23}$$

where

$$\langle \xi \rangle = \int d\xi \;\, \xi \;\, p(\xi), \;\; \langle \xi^2 \rangle = \int d\xi \xi^2 p(\xi). \tag{7.24}$$

If the average value of ξ, i.e., $\langle \xi \rangle$ happens to be zero, certain general conclusions can be drawn from Equation (7.23). Because mixing (Lebowitz and Penrose 1973) requires that

$$\lim_{t\to\infty} C_\xi(t) = \langle \xi(o)\xi(t) \rangle = \langle \xi(o) \rangle \;\, \langle \xi(t) \rangle, \tag{7.25}$$

we conclude that

$$\lim_{t\to\infty} \phi(t) = 0, \quad \text{if} \quad \langle < \xi > \rangle = 0 \tag{7.26}$$

This condition is obviously valid for a stationary Markov process when $\phi(t)$ falls off to zero as an exponential function of time. However, if $\phi(t)$ falls off more slowly than exponential function (e.g., via an inverse power law), persistent memory effects ensue. Such memory effects are evidently more severe when $\gamma \neq 1$ or when the primitive transition operator $\hat{\mathfrak{F}}$ does not have the simple form, as in the interpolation model.

7. 2.2 Anomalous Diffusion

The non-Markovian CTRW model has interesting consequences for diffusion. Recall that the mean-squared displacement governed by a stationary velocity process $\upsilon(t)$ can be written as

$$\langle x^2(t)\rangle = 2\int_0^t d\tau(t-\tau)\ \langle \upsilon(O)\upsilon(\tau)\rangle. \tag{7.27}$$

Evidently, in the Markovian limit

$$C_\vartheta(t) = <\upsilon(o)\upsilon(t)\ > \ = \ <\upsilon^2> e^{-\lambda t},$$

in which case,

$$\lim_{t\to\alpha}\ \langle x^2(t)\rangle = 2Dt,$$

where the diffusion coefficient

$$D = \int_0^\infty d\tau\langle \upsilon(0)\ \upsilon(\tau)\rangle. \tag{7.28}$$

The linear dependence of $\lim_{t\to\infty}\langle x^2(t)\rangle$ on t is, of course, the hallmark of ordinary diffusion and hence a Markov process. We may recall in passing that the same conclusion remains valid when $\upsilon(t)$ is not a jump-diffusion process, as discussed here, but is instead a Gauss–Markov process when the autocorrelation function of $\upsilon(t)$ is governed by Doob's theorem (Section 3.3.1). On the other hand, if the correlation function $C_\vartheta(t)$ acquires long-time tails, as in hydrodynamics, arising from waiting time distribution $\Phi_1(t) \sim t^{-1-\alpha}, \alpha < 1$. In that case, Equation (7.27) leads to

$$\lim_{t\to\infty}\ \langle x^2(t)\rangle = \frac{2}{(1-\alpha)} < \upsilon^2 > t^{1-\alpha}, \tag{7.29}$$

yielding sub-diffusion behavior. However, if $C_\upsilon(t)$ does not vanish asymptotically (characterizing non-ergodicity), the system exhibits super-diffusion, when $\langle x^2(t)\rangle$ grows faster than t in the long time limit.

References

Balakrishnan, V. 1979. *Pramana J. Phys.* 13, 337.

Feller, W. 1966. *Introduction to Probability Theory and Its Applications*, Vol. 2. Oxford: John Wiley & Sons.

Lax, M. and H. Scher. 1977. Dynamic bond percolation theory: A microscopic model for diffusion in dynamically disordered systems. I. Definition and one-dimensional case. *Phys. Rev. Lett.* 39, 781.

Lebowitz, J. L. and O. Penrose. 1973. Modern ergodic theory. *Physics Today.* 26(2), 23.

Montroll, E. and G. H. Weiss. 1965. Random walks on lattices. II. *J. Math. Phys.* 6, 167.

Montroll, E.W. and H. Scher. 1973. *J. Stat. Phys.* 9, 101.

Scher, H. and E. W. Montroll. 1975. Anomalous transit-time dispersion in amorphous solids. *Phys. Rev.* B, 12, 2455.

Scher, H. and M. Lax. 1972. *J. Non-Cryst. Solids* 8–10, 497.

Tunaley, J. K. E. 1976. A theory of 1/f current noise based on a random walk model. *Journal of Statistical Physics*. 15, 149.

8

Spectroscopic Structure Factor

8.1 Introductory Remarks

One important application of diffusion is to spectroscopic measurement of what is called the structure factor that is detected in the time domain or the frequency domain. The problem is best understood in the context of gas phase spectroscopy in an infrared or microwave regime. Consider an optically active atom in the surroundings of a buffer gas composed of optically neutral atoms. The density of optically active atoms is so low that we can consider them one at a time. If an active atom emits a radiation of wave vector $\vec{k}$ at a time t_1, the phase of the emitted radiation is contained in a factor:

$$\phi = \exp(-i\vec{k} \bullet \vec{r}(t_1)), \tag{8.1}$$

where $\vec{r}(t_1)$ is the instantaneous position of the emitter. The measured structure factor (for $t_1 < t_2$) given by Dattagupta (1987) is

$$S\ (\vec{k};t_1,t_2) = < e^{-i\vec{k}\ \bullet\ \vec{r}(t_1)}\ e^{-i\vec{k}\ \bullet\ \vec{r}(t_2)} >, \tag{8.2}$$

where the angular brackets <….> indicate a statistical average over the randomness of the position vectors. Now,

$$\vec{r}(t_2) = \vec{r}(t_1) + \int_{t_1}^{t_2} dt'\ \upsilon(t'), \tag{8.3}$$

and $\vec{\upsilon}(t)$ is the velocity vector. Hence

$$S\ (\vec{k};t_1,t_2) = < e^{i\int_{t_1}^{t_2} dt'(\vec{k}.\vec{\upsilon}(t'))} >. \tag{8.4}$$

If $\vec{\upsilon}(t)$ is a stationary stochastic process (see Chapters 2 through 5), time t_1 can be chosen to be the origin of time, in which case the structure factor becomes

$$S\ (\vec{k};t) = < e^{i\int_0^t dt'(\vec{k}.\vec{\upsilon}(t'))} >. \tag{8.5}$$

Interestingly, the experimentally measurable structure factor is identical to the characteristic functional of a stochastic process [cf. Equation (3.31)] if we take $\vec{k}(t)$ to be independent of *t*. By fixing the direction of observation of the emitted radiation, the relevant velocity is the component along that direction, denoted by (*t*). Hence, Equation (8.5) simplifies to:

$$S\ (k,t) = < e^{ik\int_0^t \nu(t')\ dt'} >. \tag{8.6}$$

8.2.1 Weak Collision Model: Gaussian Process

We first consider a "heavy" active atom in the midst of lighter buffer gas atoms. The active atom is kicked around randomly due to thermal fluctuations but because of the mass mismatch, the effects of collisions are weak so that the post- and pre-collision velocities differ only marginally. Consequently, the underlying stochastic process, i.e., velocity $\nu(t)$ can be characterized by a linear damping term in the equation of motion according to the Langevin equation (Chapter 4). In such a weak collision model, the resultant stochastic process is a Gaussian and hence the expressions derived earlier for the characteristics functional can be directly utilized here. Thus, from Equation (3.33),

$$S\ (k,t) = \exp[ik\int_0^t dt_1 < \upsilon(t_1) > - \frac{k^2}{2}\int_0^t dt_1 \int_0^t dt_2 < \upsilon(t_1)\ \upsilon(t_2) >]. \tag{8.7}$$

However, because $\nu(t)$ is a stationary process,

$$\begin{aligned} < \upsilon(t_1) > &= < \upsilon >, \\ < \upsilon(t_1)\ \upsilon(t_2) > &= < \upsilon(0)\ \upsilon(t_2 - t_1) >, \quad t_2 > t_1. \end{aligned} \tag{8.8}$$

Furthermore, for an optically active atom in thermal equilibrium with a buffer gas, its velocity is governed by a Maxwellian distribution. Additionally, Doob's theorem for a one-dimensional Gaussian process applies (cf. Section 3.3.1). Therefore, from Equations (3.34) and (4.28),

$$\begin{aligned} &< \upsilon > = 0, \\ &< \upsilon(0)\ \upsilon\ (t_2 - t_1) > = \frac{K_B T}{m} e^{-\lambda|t_2 - t_1|}. \end{aligned} \tag{8.9}$$

In this context, λ has the interpretation of the mean rate of collisions, i.e., λ^{-1} is the mean free time of collisions. Thus

$$S(k,t) = \exp[-k^2 < \upsilon^2 >_{eq} \int_0^t d\tau(t-\tau)e^{-\lambda\tau}] \quad = \exp\left[-k^2 \frac{K_B T}{m\lambda^2} \ (\lambda t - 1 + e^{-\lambda t})\right] \tag{8.10}$$

Alternatively, we may avoid introducing the velocity in Equation (8.3) altogether and work directly in the position space, i.e., with Equation (8.2) which yields

$$S(\vec{k},t) = < e^{-ik \ x(t)} > \ = e^{-k^2 <x^2(t)>}, \tag{8.11}$$

for a Gaussian process, in which *x(t)* follows the Langevin equation. Substitution of Equation (4.44) into Equation (8.11) directly leads to Equation (8.10). The structure factor in the frequency domain, also called the spectral line shape, is obtained from the one-sided Fourier transform of Equation (8.10):

$$S(k,\omega) = \frac{1}{\pi} \text{Re} \int_0^\infty dt \ \exp\left[-i\omega t \ - \frac{k^2 K_B T}{m\lambda^2}(\lambda t - 1 + e^{-\lambda t})\right], \tag{8.12}$$

which can be developed as a continued fraction:

$$S(k,\omega) = \frac{1}{\pi} \text{Re} \cfrac{1}{i\omega + \cfrac{k^2 K_B T / m}{i\omega + \lambda + \cfrac{2k^2 K_B T / m}{i\omega + 2\lambda + \ldots}}}, \tag{8.13}$$

that is convenient for computational purposes. For gaining further insight into the meaning of the line shape, it is useful to consider the limiting cases of very slow ($\lambda \approx 0$) and very fast ($\lambda \sim$ large) collisions.

8.2.1.1 Very Slow Collisions

In the limit of $\lambda \to 0$, Equation (8.12) reduces to

$$S(k,\omega) \simeq \frac{1}{\pi} \text{Re} \int_0^\infty dt \ \exp\left[\left(-i\omega t - \frac{k^2 K_B T}{2m\gamma^2} t^2\right)\right] = \sqrt{\frac{m}{4\pi k^2 K_B T}} \ \exp\left[\left(-\frac{m\omega^2}{4k^2 K_B T}\right)\right]. \tag{8.14}$$

Evidently, the line shape is a Gaussian centered at $\omega = 0$ and with a full width at half maxima (FWHM) given by

$$\Delta W = 4k\sqrt{\frac{KT \ln 2}{m}}. \tag{8.15}$$

Hence, the width increases with temperature leading to what is called the Doppler broadening of spectral line shapes.

8.2.1.2 Very Fast Collisions

In this regime, $\lambda t >> 1$, and Equation (8.12) leads to

$$S(k,\omega) = \frac{1}{\pi} \frac{k^2 K_B T / m\lambda}{\omega^2 + \left(\frac{k^2 K_B T}{m\lambda}\right)^2}. \tag{8.16}$$

The line shape is now a Lorentzian with a width proportional to $k^2 K_B T / m\lambda$. Therefore, for a fixed temperature, the width decreases as the collision rate λ increases, leading to the well known motional narrowing of spectral lines. For intermediate rates of collisions that are neither slow nor fast, the line shape must be computed numerically from Equation (8.13) and the results are plotted in Figure 8.1.

8.2.2 Strong Collision Model: Kubo–Anderson Process

We now consider a situation that is exactly opposite to what is to be described by the weak collision model. The case involves a light optically active atom in the company of much heavier buffer gas atoms. Each collision is so drastic, as far as the active atom is concerned, that the velocity is completely thermalized, i.e., the post-collision velocity has no memory of the pre-collision velocity. Such strong collisions can be approximately described by the Kubo–Anderson process (cf. Section 6.5) in which the collision operator $\mathfrak{J}$ has the following matrix elements:

$$(\upsilon | \hat{\mathfrak{J}} | \upsilon_0) = p(\upsilon), \tag{8.17}$$

where $p(\upsilon)$ is the equilibrium Maxwellian appropriate to the post-collision velocity υ. Manifestly, Equation (8.17) satisfies the detailed balance condition. The calculation of the structure factor in Equation (8.6) can now be set up in the CTRW scheme along the lines described in Section 7.1.

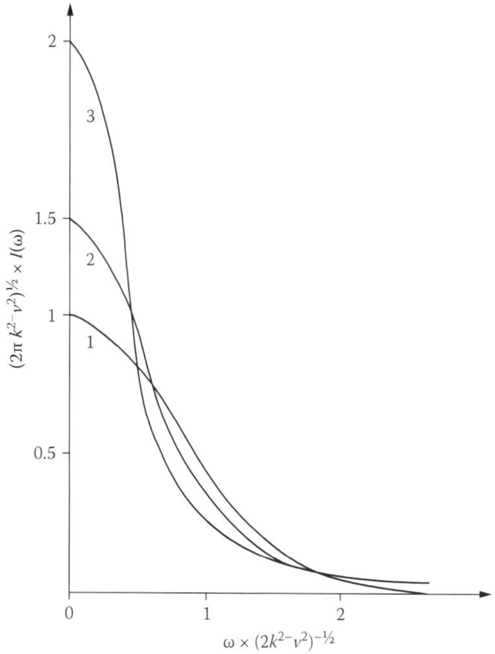

FIGURE 8.1
Collision-broadened line shape in a weak collision model for three values of collision rate λ. Curve 1: $\lambda = 0$. Curve $2: \lambda = 0.7(2k^2\bar{\upsilon}^2)^{1/2}$. Curve $3: \lambda = (4k^2\bar{\upsilon}^2)^{1/2}$. (From Dattagupta, S., *Relaxation Phenomena in Condensed Matter Physics*, Academic Press, New York, 1987. With permission.)

8.2.2.1 No-Collision Term

If the time 0 to t elapses without any collision, the probability of which is $e^{-\lambda t}$, the velocity remains constant (say υ_0) and the zeroth order structure factor can be written from Equation (8.6) as

$$S_0(k,t) = \int d\upsilon_0 p(\upsilon_0)\ e^{ik\upsilon_0 t - \lambda t}. \tag{8.18}$$

8.2.2.2 One-Collision Term

Here the velocity remains constant (i.e., υ_0) until time t_1 at which point a collision occurs that throws the system into the velocity state υ_1 that retains its value until time t. Thus, the corresponding structure factor is

$$S_1(k,t) = \iint d\upsilon_0 p(\upsilon_0)\ d\upsilon_1\ p(\upsilon_1) \quad \int_0^t \lambda dt_1\ e^{-(t-t_1)\ (\lambda - ik\upsilon_1) - t_1(\lambda - ik\upsilon_0)}$$

$$= \int_0^t S_0(k;\ t - t_1)\ S_0(k;\ t_1)\ \lambda dt_1. \tag{8.19}$$

8.2.2.3 Two-Collision Term

For two-collision events, one occurring at time t_1 and the other at time t_2 $(t_2 > t_1)$, between 0 and t, the renewal process yields for the structure factor,

$$S_1(k,t) = \iiint d\upsilon_0 p(\upsilon_0)\ d\upsilon_1 p(\upsilon_1)\ \ d\upsilon_2 p(\upsilon_2) \int_0^t \lambda dt_2 \int_0^{t_2} \lambda dt_1$$

$$\times e^{-(t-t_2)\ (\lambda - ik\upsilon_2)}\ e^{(t_2-t_1)(\lambda - ik\upsilon_1)}\ e^{-t_1(\lambda - ik\upsilon_0)} \tag{8.20}$$

$$= \int_0^t (\lambda dt_2) \int_0^{t_2} (\lambda dt_1)\ S_0(k;t-t_2)\ S_0(k;t_2-t_1)\ S_0(k;t_1)$$

and so on. Collecting all the terms together, the Laplace transform of the structure factor, upon employing the convolution theorem, can be written as

$$\tilde{S}(k,s) = \tilde{S}_0(k,s+\lambda) + \lambda[\tilde{S}_0(k,s+\lambda)]^2 + \lambda^2[\tilde{S}_0(k,s+\lambda)]^3 + \ldots$$

which, upon summing, yields a form reminiscent of the random phase approximation (RPA) of many body physics:

$$\tilde{S}(k,s) = \frac{\tilde{S}_0(k,s+\lambda)}{1-\lambda\ \tilde{S}_0(k,s+\lambda)}, \tag{8.21}$$

where [cf. Equation (8.18)]

$$\tilde{S}_0(k,s+\lambda) = \int d\upsilon\ p(\upsilon) \frac{1}{s+\lambda - ik\upsilon}. \tag{8.22}$$

As in the case of the weak collision model, it is instructive to consider the following two limiting cases.

8.2.2.4 Very Slow Collisions

For

$$\lambda \to 0,$$
$$\tilde{S}(k,s) \approx \tilde{S}_0(k,s), \tag{8.23}$$

which has the same form as Equation (8.14) when s is replaced by $-i\omega + O^+$.

8.2.2.5 Very Fast Collisions

When λ is very large, we may expand $S_0(k, s+\lambda)$ as [cf. Equation (8.22)]:

$$\tilde{S}_0(k, s+\lambda) \approx \frac{1}{(s+\lambda)}\left[1 - \frac{k^2 \Delta\overline{\upsilon^2}}{(s+\lambda)^2} + \ldots\right] \tag{8.24}$$

where

$$\upsilon^2 = \int d\nu\ p(\upsilon)\ \upsilon^2 = \frac{K_B T}{m}. \tag{8.25}$$

Therefore, Equation (8.21) can be developed as

$$\tilde{S}(k,s) = \frac{1}{[\tilde{S}_0(k, s+\lambda]^{-1} - \lambda} \approx \frac{1}{s + \dfrac{k^2 K_B T}{m(s+\lambda)}} \approx \left(s + \frac{k^2 K_B T}{m\lambda}\right)^{-1}, \tag{8.26}$$

which again, when re-expressed in terms of ω ($s = -i\omega + 0^+$), becomes identical to Equation (8.16). To derive a closed-form expression for the line shape for intermediate values of λ, we note that Equation (8.22) can be rewritten as

$$\tilde{S}_0(k, s+\lambda) = \int_0^\infty dt\ e^{-(s+\lambda)t} e^{-k^2 \frac{K_B T}{m} t^2},$$

wherein we have first re-expressed the resolvent in Equation (8.22) as an integral over a fiduciary time t, and then performed the thermal average with the aid of $p(\upsilon)$. Upon a change of variable of integration, the above can be expressed as a complementary error function:

$$\tilde{S}_0(k, s+\lambda) = \left(\frac{\pi m}{z k^2 K_B T}\right)^{1/2} \exp(z^2)\ \mathrm{erfc}(z), \tag{8.27}$$

where

$$z = \left[(s+\lambda)^2 \Big/ 2k^2 \frac{KT}{m}\right]^{1/2}. \tag{8.28}$$

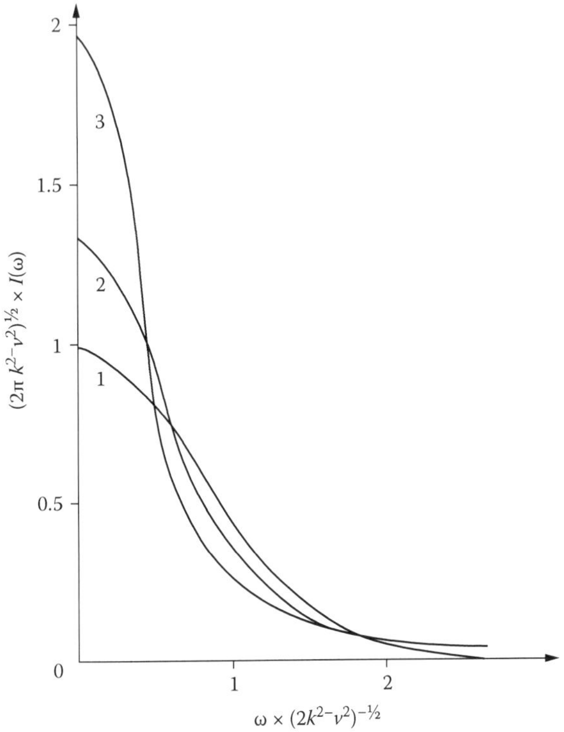

FIGURE 8.2
Collision-broadened line shape in a strong collision model for two values of collision rate λ. Curve 1: $\lambda = 0$. Curve $2: \lambda = 0.7(2k^2\bar{\upsilon}^2)^{1/2}$. (From Dattagupta, S., *Relaxation Phenomena in Condensed Matter Physics*, Academic Press, New York, 1987. With permission.)

We substitute Equation (8.27) into Equation (8.21) to generate plots of the line shape as a function of ω shown in Figure 8.2. Contrasting these plots with those of Figure 8.1 reveals that the motional narrowing and accompanying enhancement of the line intensity near $\omega = 0$ is more prominent in the strong than in the weak collision model.

8.2.3 Boltzmann–Lorentz Model

The Boltzmann–Lorentz model of collisional effects borrowed from the kinetic theory of gases is similar in spirit to the strong collision model, except that the magnitude of the velocity vector is assumed unchanged due to collisions. Only its direction is randomized (Dattagupta and Turski 1985). Further, unlike in the strong collision model in which the rate of collisions is a coarse-grained rate parameterized by the temperature and pressure of the gas, the rate is now a dynamical variable, taken to be a function of the instantaneous value of the velocity. It is then necessary at the outset to treat the velocity as a vector that is endowed by its magnitude υ and its orientation specified by the Euler angles $\vec{\Omega}$ $\{\theta,\varphi\}$ and adopt the expression for the structure factor in Equation (8.5). We may now develop our method of calculation, as in the CTRW treatment of the strong collision model in Section 8.2 as follows:

8.2.3.1 No-Collision Term

$$S_0(\vec{k},t) = \frac{1}{4\pi}\int d\upsilon \;\; p(\upsilon) \;\; \int d\Omega \;\; e^{-t(\lambda - i\upsilon(\vec{k}.\hat{u}_\Omega))}, \tag{8.29}$$

where $\hat{u}_\Omega$ is the unit vector in the direction $\vec{\upsilon}$ of the velocity. As we are ultimately interested in the Laplace transform of the structure factor, we find for the latter

$$\tilde{S}_0(\vec{k},s+\lambda) = \int d\upsilon \;\; p(\upsilon)\frac{1}{4\pi}\int d\pi \frac{1}{(s+\lambda) - i\upsilon(\vec{k}.\hat{u}_\Omega)}. \tag{8.30}$$

Choosing the direction of the $\vec{k}$ vector as the space-fixed z axis, we may write in spherical polar coordinates

$$\begin{aligned}\frac{1}{4\pi}\int d\Omega \frac{1}{(s+\lambda) - i\upsilon(\vec{k}.\hat{u}_\Omega)} &= \frac{1}{2}\int d\upsilon \;\; \sin\theta \frac{1}{(s+\lambda) - i\upsilon \; k\cos\theta} \\ &= \frac{1}{k\upsilon} \;\; \tan^{-1}\left(\frac{k\upsilon}{s+\lambda(\upsilon)}\right).\end{aligned} \tag{8.31}$$

Thus,

$$\tilde{S}_0(\vec{k},s+\lambda) = \int_0^\infty d\upsilon \;\; p(\upsilon)\frac{1}{k\upsilon} \;\; \tan^{-1}\left(\frac{k\upsilon}{s+\lambda(\upsilon)}\right). \tag{8.32}$$

8.2.3.2 One-Collision Term

$$S_1(k,t) = \int d\upsilon \;\; p(\upsilon)\frac{1}{4\pi}\int d\Omega_0 \frac{1}{4\pi}\int d\Omega \int_0^t (\lambda dt_1) \;\; e^{-(t-t_1)(\lambda - ik\upsilon \;\cos\theta)} e^{-t_1(\lambda - ik\upsilon \;\cos\theta_0)}$$

the Laplace transform of which is

$$\begin{aligned}S_1(\vec{k},s+\lambda) &= \lambda\int d\upsilon \;\; p(\upsilon)\left(\frac{1}{2}\int d\theta \frac{\sin\theta}{(s+\lambda) - i\upsilon k \;\; \cos\theta}\right)^2 \\ &= \lambda\int d\upsilon \;\; p(\upsilon) \;\; \left\{\frac{1}{k\upsilon} \;\; \left[\tan^{-1}\left(\frac{k\upsilon}{s+\lambda(\upsilon)}\right)\right]\right\}^2.\end{aligned} \tag{8.33}$$

The two and higher collision terms can be calculated along the above lines and in parallel to Section 8.2 to yield

$$S(\vec{k},s) = \tilde{S}_0(\vec{k},s+\lambda) + \tilde{S}_1(\vec{k},s+\lambda) + \tilde{S}_2(\vec{k},s+\lambda) + \ldots.$$

$$= \int_0^\infty d\upsilon \; p(\upsilon) \frac{\dfrac{1}{k\upsilon} \tan^{-1}\left(\dfrac{k\upsilon}{s+\lambda(\upsilon)}\right)}{1 - \dfrac{\lambda}{k\upsilon}\tan^{-1}\left(\dfrac{k\upsilon}{s+\lambda(\upsilon)}\right)}. \tag{8.34}$$

Borrowing ideas from classical kinetic theory, the velocity-dependent rate $\lambda(\upsilon)$ can be expressed as

$$\lambda(\upsilon) = \pi n_s b^2 \upsilon, \tag{8.35}$$

where n_s is the number of scatterers or buffer gas atoms per unit volume and b is the so-called impact parameter, i.e., the effective scattering radius. We may now introduce a dimensional parameter:

$$\alpha = n_s \frac{b^2}{k}, \tag{8.36}$$

to discuss different regimes of rapidity of collisions. Slow collisions correspond to $\alpha \approx 0$, whereas fast collisions are described by $\alpha >> 1$. For $\alpha \approx 0$,

$$\tilde{S}(\vec{k},s) = \int_0^\infty d\upsilon \; \frac{p(\upsilon)}{k\upsilon} \tan^{-1}\left(\frac{k\upsilon}{s+\pi\alpha k\upsilon}\right), \tag{8.37}$$

which upon using a Maxwellian distribution for $p(\upsilon)$ and going to the limit of $s = -i\omega$, yield for the structure

$$S(\vec{k},\omega) = \frac{1}{k^2}\left(\frac{\pi m}{2K_B\,T}\right) \exp\left(-\frac{m\omega^2}{2K_B T k^2}\right). \tag{8.38}$$

The Gaussian character (in ω) of the structure factor is exactly akin to what was found earlier for weak and strong collision models yielding Doppler broadening, but with a different prefactor reflecting the three-dimensional setting for the computation. On the other hand, when $\alpha >> 1$, i.e., there are many scatterers within the wavelength k^{-1} of the radiation and we find

$$\tilde{S}(\vec{k},s+\lambda) \approx \int_0^\infty d\upsilon \; \frac{1}{s + \dfrac{1}{3}\dfrac{k^2\upsilon^2}{3+\lambda(\upsilon)}}, \tag{8.39}$$

which can be further approximated because of a pole at $s \approx 0$, as

$$\tilde{S}(\vec{k}, s) \approx \int_0^\infty d\upsilon \ \ p(\upsilon) \frac{1}{s + \dfrac{k^2\upsilon^2}{3 + \lambda(\upsilon)}}, \tag{8.40}$$

which, upon substituting for $\lambda(\upsilon)$ from Equation (8.35) and integrating over the Maxwellian distribution $p(\upsilon)$, reduces to

$$\tilde{S}(\tilde{k}, s) = \left[s + \frac{2k}{3\pi\alpha} \left(\frac{2K_B T}{m\pi} \right)^{\frac{1}{2}} \right]. \tag{8.41}$$

The line shape, as expected, is a Lorentzian with a width Γ_L given by

$$\Gamma_L = \frac{3\pi}{\alpha} \sqrt{\pi \ln 2} \ \ \Gamma_G, \tag{8.42}$$

where Γ_G is the width appropriate for the Gaussian result of Equation (8.38). Hence, as the collisions become more frequent, the line shape changes from a Gaussian to a Lorentzian with a concomitant reduction in the width by a factor $\sim \alpha^{-1}$, in agreement with the notion of motional narrowing of spectral lines, as in the weak and strong collision models.

8.3 Cyclotron Motion in Weak Collision and Boltzmann–Lorentz Models

The issue of dissipative cyclotron motion was already dealt with in Chapters 4 and 5 in terms of the Langevin and Fokker–Planck equations. This subsection is concerned with the computation of the structure factor for such dynamics and delineating the effect of the magnetic field (Singh and Dattagupta 1996, Ghosh et al. 2001).

8.3.1 Structure Factor in the Weak Collision Model

Because the dynamics is now described by either the Langevin equation or the complementary Fokker–Planck equation, the structure factor in Equation (8.5) is provided by the characteristic functional of a Gaussian process for the three-dimensional velocity $\vec{\upsilon}$ with zero mean [cf. Equation (8.7)] in which all cumulants higher than the second vanish. Hence we can directly write

$$S(\vec{k},t) = \exp\left[-\frac{1}{2}\sum_{jl} k_j k_l \int_0^t dt_1 \int_0^t dt_2 < \upsilon_j(t_1)\upsilon_l(t_2) >\right]. \tag{8.43}$$

Stationarity further implies that

$$S(\vec{k},t) = \exp\left[-\frac{1}{2}\sum_{jl} k_j k_l \int_0^t d\tau(t-\tau) \;< \upsilon_j(0)\upsilon_l(\tau) >\right]. \tag{8.44}$$

Specializing to the geometry of Figure 8.3, we have

$$S(\vec{k},t) = \exp\left\{-\frac{1}{2}\left[\left(k_x^2 + k_y^2\right) \; S_{\perp}(t) + k_z^2 S_{\parallel}(t)\right]\right\}, \tag{8.45}$$

where,

$$S_{\parallel}(t) = \frac{2K_BT}{m\gamma^2}(\lambda t - 1 + e^{-\lambda t})$$

$$S_{\perp}(t) = \frac{2K_BT}{m\left(\lambda^2+\omega_c^2\right)^2}\left\{\lambda t\left(\lambda^2+\omega_c^2\right) - \left(\lambda^2-\omega_c^2\right) + e^{-\lambda t\left[\left(\lambda^2+\omega_c^2\right)\;\cos(\omega_c t) - 2\lambda\omega_c \sin(\omega_c t)\right]}\right\}. \tag{8.46}$$

It is clear that in the limit of zero magnetic field ($\omega_C = 0$),

$$S_{\perp}(t) = S_{\parallel}(t), \tag{8.47}$$

wherein the latter describes the structure factor for free diffusive behavior [cf. Equation (8.10)].

To analyze what influence the magnetic field has on the line shape we select, for the sake of definiteness, $k_y = k_z = 0$, and $k_x = k$. The spectral line shape is obtained from the structure factor in the frequency domain defined as in Equation (8.12). We plot the latter in Figure 8.4 in terms of the dimensionless parameters $\bar{\omega}_c \equiv \omega_c(k^2 < v^2 >)^{-1/2}$ and $\bar{\gamma}$ $(\equiv \gamma\ (k^2 < v^2 >)^{-1/2}$, but unlike the earlier figures for weak and strong collision models (in the absence of the $\bar{B}$ field) we now consider a fixed collision rate $\bar{\lambda}$ and check on the variation of the line shape as a function of the strength of the magnetic field. Interestingly, the field appears to

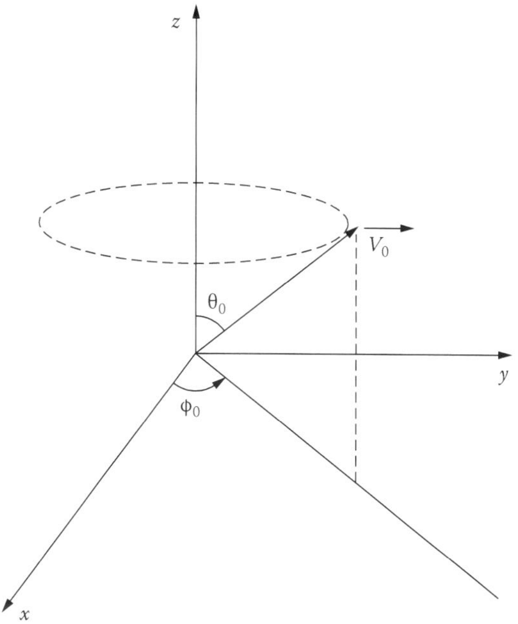

FIGURE 8.3
In the absence of collisions, the velocity vector $\vec{\upsilon}$ precesses about the direction of magnetic field $\vec{B}$ is taken to be along the i.e. axis. For an initial velocity vector $\vec{\upsilon}_0 = \upsilon \hat{u}_{\Omega 0}$, where $\hat{u}_{\Omega 0} = (\sin\theta_0 \cos\theta_0, \sin\theta_0 \sin\phi_0, \cos\theta_0)$, the velocity vector after time t is given as $\vec{\upsilon}(t) = \upsilon \hat{u}_{\pi}(t)$, where $\hat{u}_{\Omega}(t) = (\sin\theta_0 \cos\phi_0(t), \sin\theta_0 \sin\phi_0(t), \cos\theta_0)$ and $\phi(t) = \phi_0 - \omega_c t$. (From Dattagupta, S., *Relaxation Phenomena in Condensed Matter Physics*, Academic Press, New York, 1987. With permission.)

yield the same narrowing phenomenon as does an enhanced collision rate, evidently because of the constraining effect that makes the particles go into cyclotron orbits.

8.3.2 Structure Factor in the Boltzmann–Lorentz Model

In this subsection we culminate the discussion on the structure factor by considering the Boltzmann–Lorentz model that as a jump model provides a contrast to the continuous diffusion model of the weak collision type. The physics is as follows. The velocity vector of the active atom starts out as the vector $\vec{\upsilon}_0$ defined by the co-latitude angle ϕ_0 and the azimuthal angle ϕ_0, as in Figure 8.4. Under the influence of the magnetic field $\vec{B}$, assumed to be along the z axis, the velocity vector rotates in the x,y plane and acquires, for instance, for its x component the form:

$$\upsilon_x(t) = \upsilon_0 \quad \sin\phi_0 \quad \cos(\phi_0 + \omega_c t). \tag{8.48}$$

The reason we pick the x component is that it is the only relevant x component, having decided to choose the direction of observation of the

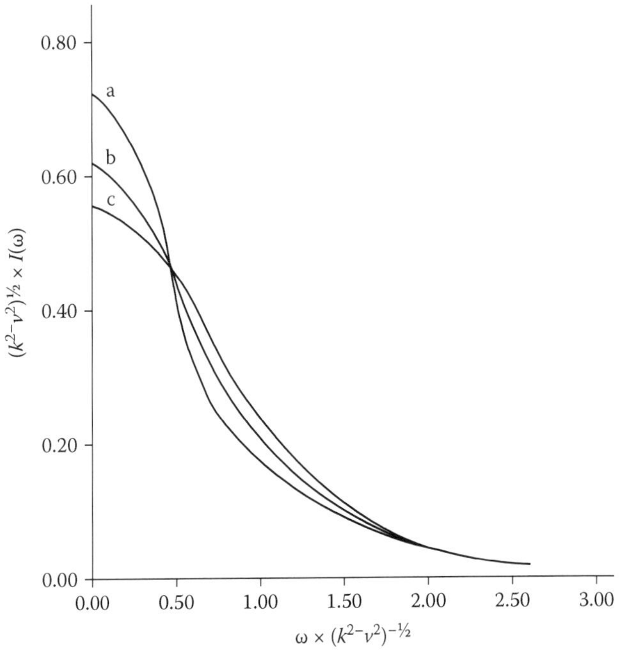

FIGURE 8.4
Collision-broadened line shape in a weak collision model in the presence of magnetic field. Curves a, b, and c correspond to $\bar{\omega}_c = 1$, $\bar{\omega}_c = 7$, and $\bar{\omega}_c = 0$ respectively ($\bar{\omega}_c = \omega_c(k^2 < \upsilon^2 >)^{-1/2}$). (From Dattagupta, S., *Relaxation Phenomena in Condensed Matter Physics*, Academic Press, New York, 1987. With permission.)

radiation to be x, i.e., $k_x = k$, and $k_y = k_z = 0$ [see discussion following Equation (8.47)]. As we set up the CTRW scheme for the Boltzmann–Lorentz model as per Section 8.2, the streaming operator or the propagator, if no collision took place during the time interval 0 to t, can be written as [cf. Equation (8.5)]:

$$S_0(k,t) = \exp\left[ik\int_0^t dt'\upsilon_x(t')\right],$$

which equals

$$S_0(k,t) = \exp\left\{\frac{ik\upsilon_0}{\omega_c}\sin\theta_0\left[\sin\ (\phi_0 + \omega_c t) - \sin\phi_0\right]\right\}. \tag{8.49}$$

The basic RPA like structure for the coarse-grained streaming operator as expressed in Equation (8.21) for the strong collision model or Equation (8.34) for the Boltzmann–Lorentz model is still applicable. Therefore, the Laplace transform of the structure factor reads

$$\tilde{S}(k,s) = \frac{\tilde{S}_0(k,s)}{1-\lambda\tilde{S}_0(k,s)}, \tag{8.50}$$

where, from Equation (8.49),

$$\tilde{S}_0(k,s) = \frac{1}{4\pi}\int d\upsilon_0 \;\; p(\upsilon_0)\int_0^{\pi} \sin\theta_0\, d\theta_0 \int_0^{2\pi} d\phi \int_0^{\infty} dt\; S_0(k,t) e^{-t(s+\gamma(\upsilon_0))}. \tag{8.51}$$

The line shape, obtained from $\tilde{S}_0(k,s)$ by going to the limit $s \to -i\omega + \overset{+}{0}$, is numerically computed and the results are shown in Figure 8.5. We once again employed the scaled parameters $\bar{\lambda}$ and $\bar{\omega}_c$ (see end of Section 8.3) and noted a narrowing of the spectral line with increasing strength of the magnetic field that appears more pronounced than that in the weak collision model.

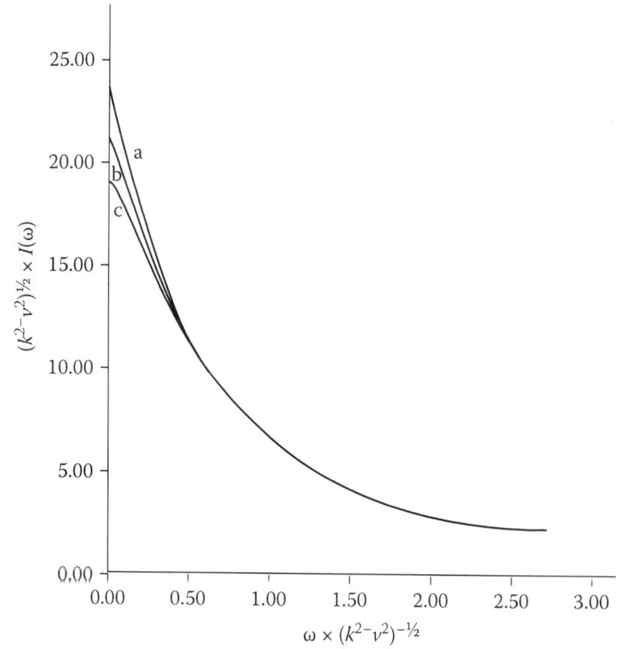

FIGURE 8.5
Collision-broadened line shape in a Boltzmann–Lorentz model in the presence of a magnetic field. Curve a corresponds to $\bar{\omega}_c = 1$, curve b to $\bar{\omega}_c = 7$, and curve c to $\bar{\omega}_c = 0$. All curves are for $n_s b^2/k = 1$. (From Dattagupta, S., *Relaxation Phenomena in Condensed Matter Physics*, Academic Press, New York, 1987. With permission.)

A similar conclusion was drawn earlier vis-á-vis the strong collision model, as in our comments following Equation (8.28). Summarizing, the magnetic field can be used as a distinct handle in modulating spectral lines in gases, in addition to temperature and pressure that control the collision rate.

8.4 Neutron Scattering from a Damped Harmonic Oscillator

Neutrons thermalized in a nuclear reactor have wavelengths comparable to the typical lattice spacing in a crystalline solid. Hence, neutron scattering is a useful complementary tool to x-ray scattering for studying lattice structures. Besides, the energy difference between the scattered and the incident neutron is comparable to typical energy scales for phonons and thus neutron scattering serves as a convenient probe for investigating vibrational modes.

If a system under study is a nonmagnetic solid, the neutron interacts with it via the Fermi contact interactions with the nuclei because of nuclear forces. Each nucleus may be considered an Einstein oscillator, the motion of which is damped because of anharmonic forces or other interactions. We would now like to focus on the scatterer as a classical, damped harmonic oscillator, the dynamics of which is governed by the Langevin equation (Chapter 4) and by the Fokker–Planck (FP) equation (Chapter 5) in the phase space.

What is experimentally measured is a structure factor in the frequency domain which is the Fourier transform of the so-called van Hove self-correlation function [cf. Equation (8.2)]:

$$G_s(\vec{k},t) = \left\langle e^{-i\vec{k}.\vec{r}(0)} e^{i\vec{k}.\vec{r}(t)} \right\rangle. \tag{8.52}$$

As discussed earlier, by controlling the direction of the $\vec{k}$-vector, Equation (8.24) can be written in terms of one-dimensional projection of the displacement, thus [cf. Equation (8.11)]

$$G_s(k,t) = \left\langle e^{-ikx(t)} \right\rangle = e^{-k^2 <x^2(t)>}, \tag{8.53}$$

where in the second equality, we employed the Gaussian property of $x(t)$. We may now directly employ the result for mean-squared displacement for the Langevin model from Equation (4.76), so that

$$G_s(k,t) = \exp\left\{-k^2 \frac{K_B T}{m\omega_0^2}\left[1 - \frac{e^{-\lambda t}}{\omega^2}\left(\omega_0^2 + 2\lambda\omega\ \sin(2\omega t) - \frac{\lambda^2}{4}\cos(2\omega t)\right)\right]\right\}, \tag{8.54}$$

where it may be recalled that [cf. Equation (6.70)]

$$\omega = \sqrt{\omega_0^2 - \frac{1}{4}\lambda^2}$$

We denote the term within the parentheses along with the prefactor $K_B T/m\omega_0^2$ as

$$D(t) = \frac{K_B T}{m\omega_0^2}\left[1 - \frac{e^{-\lambda t}}{\omega^2}\left(\omega_0^2 + \frac{1}{2}\lambda\omega\ \sin(2\omega t) - \frac{\lambda^2}{4}\ \cos(2\omega t)\right)\right]. \quad (8.55)$$

The Fourier transform of $G_s(k,t)$ then has the diffusive form

$$G_s(x,t) = \sqrt{\frac{\pi}{4D(t)}}\exp\left(-\frac{x^2}{4D(t)}\right), \quad (8.56)$$

albeit with a time-dependent diffusion coefficient $D(t)$. For a free particle, of course, Equation (8.54) reduces to the structure factor for the weak collision model of Equation (8.10). The result [Equation (8.54)] for the Einstein oscillator must be modified to treat a more realistic model of lattice vibrations such as the Debye model that requires an integration over ω_0, weighted by an appropriate density.

8.5 Restricted Diffusion over Discrete Sites

Until now our discussion about calculation of the structure factor was confined to continuous stochastic processes of the Gaussian kind described by the Langevin equation or the FP equation. We turn our attention in this subsection to spectroscopic studies of jump diffusion over discrete lattice sites, as is appropriate for solid structures (Dattagupta 1987). The underlying stochastic process is still viewed as Markovian and stationary and requires the formulation developed in Chapters 2 and 4.

8.5.1 Two-Site Case

The simplest example of confined diffusion is that of a particle moving in a one-dimensional double well (Figure 8.6). We imagine an atom such as

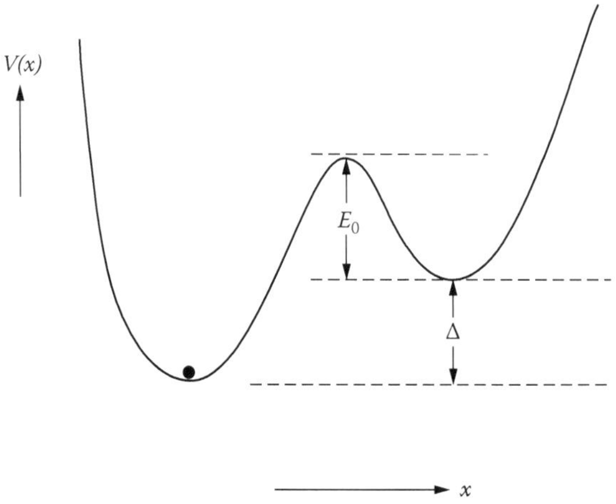

FIGURE 8.6
One-dimensional asymmetric double well, the minima of which correspond to indices + and −. The black dot represents one of two residing sites of diffusing particle. (From Dattagupta, S., *Relaxation Phenomena in Condensed Matter Physics*, Academic Press, New York, 1987. With permission.)

hydrogen that can occupy the two minima of the well denoted by plus (+) and minus (–) symbols with Boltzmann weights

$$p_+ = \frac{1}{1+e^{-\beta\Delta}}, \quad p_- = \frac{e^{-\beta\Delta}}{1+e^{-\beta\Delta}}. \tag{8.57}$$

Because of thermal fluctuations, a particle can jump from one well to the other, governed by the Arrhenius rates:

$$\lambda_{+\to-} = \lambda_0 \; e^{-\beta(E_0+\Delta)}, \quad \lambda_{-\to+} = \lambda_0 e^{-\beta E_0}. \tag{8.58}$$

Alternately,

$$\lambda_{+\;\to-} = \lambda \;\; p_-, \quad \lambda_{-\;\to+} = \lambda \;\; p_+,$$

where

$$\lambda = \lambda_0 \;\; e^{-\beta E_0}(1+e^{-\beta\Delta}). \tag{8.59}$$

The relaxation or the jump matrix $\hat{W}$ can them be formed as (see Chapter 6)

$$\hat{W} = \lambda \begin{pmatrix} -p_- & p_+ \\ p_- & -p_+ \end{pmatrix}, \tag{8.60}$$

which has the form associated with the Kubo–Anderson process:

$$W = \lambda\ (\hat{\mathfrak{I}} - 1), \tag{8.61}$$

where

$$(n\,|\,\hat{\mathfrak{I}}\,|\,m) = p_n, \quad (n\,|\,1\,|\,m) = \delta_{nm}. \tag{8.62}$$

The probability matrix is then given by Equation (6.43). Because neutrons scatter strongly from hydrogen, the most versatile technique for studying hydrogen diffusion is neutron scattering. The van Hove self-correlation function [Equation (8.52)] can be written in the present case as

$$G_s(k,t) = \sum_{n,m=\pm} p_n\ e^{-ikx_n}\,(m\,|\,\hat{P}(t)\,|\,n)\ e^{ikx_m}. \tag{8.63}$$

Substituting Equation (6.43) and after some algebra we find

$$G_s(k,t) = (1 - 4p_+p_-\sin^2(kb)) + 4p_+p_-\sin^2(kb)\ e^{-\lambda t}, \tag{8.64}$$

where b is defined in Figure (8.6) as the difference between the two minima. Finally, the Fourier transform of Equation (8.64), called the incoherent structure factor, is given by

$$S(k,\omega) = (1 - 4p_+p_-\sin^2(kb))\ \delta(\omega) + \frac{4}{\pi}\frac{\lambda}{\lambda^2+\omega^2}\,p_+p_-\sin^2(kb). \tag{8.65}$$

The line shape is then a Lorentzian with a width proportional to λ superimposed on an elastic component centered around $\omega = 0$. The Lorentzian relates to inelastic scattering in which the neutron exchanges energy with the diffusing H atom. Clearly, when the jump rate becomes extremely rapid, the inelastic component fades away and we are essentially left with the elastic line, albeit with a smaller intensity. The trigonometric expressions related to the product of the wave number k and the scattering length b are called *form factors.* The occurrence of an elastic line riding on top of an inelastic component is a generic feature of localized diffusion (Dicke 1953). Evidently, when a double well is symmetric, $\Delta = 0$, $p_+ = p_- = 1$ and the structure factor reduces to

$$S(k,\omega) = \cos^2(kb)\ \delta(\omega) + \sin^2(kb)\frac{1}{\pi}\frac{\lambda}{\lambda^2+\omega^2}. \tag{8.66}$$

All the comments that follow Equation (8.65) remain valid for Equation (8.66) as well.

We shall return to the case depicted in Figure 8.6 in which the hydrogen (black dot in one of the two sites designated by + or –) diffuses, not by jumps across the barrier, but by quantum tunneling, especially at very low temperatures (see Chapter 15).

8.5.2 Cage Diffusion

A remarkable example of restricted diffusion is the Mossbauer spectroscopy of iron (Fe) that undergoes diffusion within a cage of face centered cubic (FCC) aluminium (Al). The experiment by Vogl et al. (1976) involved electron-irradiated Al that caused "self-interstitials" consisting of Al atoms displaced outward from their face centers (Figure 8.7). The said displacement is facilitated by ^{57}Fe impurities that were pre-implanted in the lattice, resulting in Al–Fe dumbbells (Al indicated by shaded circles and Fe by dark circles, forming a dumbbell pair). The concentration of ^{57}Fe is so low that only one Fe atom at a time is within a cage. However, the Fe atom can jump within the cage due to thermal fluctuations among six possible discrete sites, thus randomly changing dumbbell partners.

The case at hand therefore does not belong to the two-site telegraph process or multi-site Kubo–Anderson process (Chapter 6). It involves uncorrelated

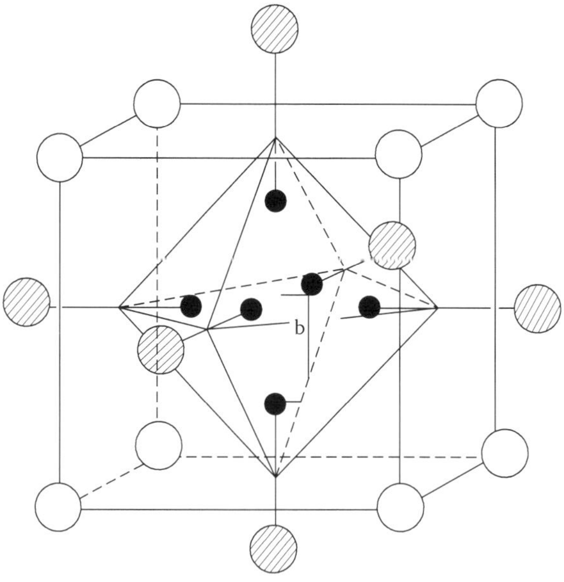

FIGURE 8.7
Equivalent configuration for mixed dumbbells that can be formed by jumps of ^{57}Co impurity atom in octahedral cage in FCC lattice of Al. Closed circles are ^{57}Co atoms. Open circles and diagonally striped circles are Al atoms. (From Dattagupta, S., *Relaxation Phenomena in Condensed Matter Physics*, Academic Press, New York, 1987. With permission.)

jumps over six possible sites and can be treated as a stationary Markov process. The corresponding jump or relaxation matrix, assuming nearest neighbor jumps only, can then be written as

$$\hat{W} = \begin{pmatrix} -4w & 0 & w & w & w & w \\ 0 & -4w & w & w & w & w \\ w & w & -4w & 0 & w & w \\ w & w & 0 & -4w & w & w \\ w & w & w & w & -4w & 0 \\ w & w & w & w & 0 & -4w \end{pmatrix}, \tag{8.67}$$

where the rows and columns are labeled by $n = \pm 1, \pm 2,$ and ± 3 and the plus and minus symbols refer to diametrically opposite sites in the cage that are not accessible from one another if only nearest neighbor jumps are permitted. Each jump rate denoted by *w is the same.* The diagonal elements of the $\hat{W}$ matrix are, of course, dictated by probability conservation:

$$\sum_m (m|\hat{W}|n) = 0, \quad \text{for any } n. \tag{8.68}$$

Though $\hat{W}$ cannot be expressed in the form of Equation (6.24) in which the $\hat{\mathfrak{I}}$ matrix is idempotent, we can still write

$$\hat{W} = \lambda\hat{\mathfrak{I}} - \lambda'(\hat{\mathfrak{I}} + \hat{\mathfrak{I}}') - \lambda_0 \hat{1}, \tag{8.69}$$

where

$$\lambda = 6w, \quad \lambda' = 2w \quad \text{and} \quad \lambda_0 = 4w. \tag{8.70}$$

Both the $\hat{\mathfrak{I}}$ and $\hat{\mathfrak{I}}$ matrices are still idempotent because their elements are

$$(m|\hat{\mathfrak{I}}|n) = \frac{1}{6}, \tag{8.71}$$

whereas the $\hat{\mathfrak{I}}'$ matrix can be expressed as

$$(m|\hat{\mathfrak{I}}|n) = \frac{1}{6}(\delta_{nm} + \delta_{n,-m}), \quad n, m = \pm 1, \ \pm 2, \ \pm 3. \tag{8.72}$$

In addition, they commute

$$[\hat{\mathfrak{I}}, \hat{\mathfrak{I}}'] = 0 \tag{8.73}$$

Using Equation (8.49) we may first write $\hat{P}(t) = \exp(\hat{W}t) = e^{-\lambda_0 t}\, e^{\lambda \hat{\mathfrak{J}} t}\, e^{-\lambda' \mathfrak{J}' t}$ which, upon further employment of the idempotency property $\hat{\mathfrak{J}}^k = \hat{\mathfrak{J}}, (\hat{\mathfrak{J}}')^k = \hat{\mathfrak{J}}',\ k \geq 1$, can be re-expressed as

$$\hat{P}(t) = \hat{\mathfrak{J}} + (\hat{1} - \hat{\mathfrak{J}}') = e^{-\lambda_0 t} + (\hat{\mathfrak{J}}' - \hat{\mathfrak{J}})\, e^{-\lambda t} \tag{8.74}$$

In writing Equation (8.74) we have further used the property that $\hat{\mathfrak{J}}\hat{\mathfrak{J}}' = \hat{\mathfrak{J}}$.

Let us now get back to a brief discussion of Mossbauer spectroscopy. It involves the emission of gamma radiation (of energy 14.3 Kev) from the excited state of ^{57}Fe that has a nuclear spin of $I_e = \frac{3}{2}$ to the ground state that has a spin of $I_e = \frac{1}{2}$. As in gas phase spectroscopy (Sections 8.1 and 8.2) and neutron scattering (Section 8.4), the instantaneous phase of the emitted radiation is contained in the factor exp(*ik.r*). If we neglect the presence of hyperfine interactions between the ^{57}Fe and its environment (distorted from cubic symmetry due to dumbbell formations), the Mossbauer line shape has the identical form of van Hove self-correlation [Equation (8.52)]. However, in this case [cf. Equation (8.74)]

$$(m|\hat{P}(t)|n) = \frac{1}{6} + \frac{1}{2}(\delta_{nm} - \delta_{n,-m})\, e^{-\lambda_0 t} + \frac{1}{6}[3(\delta_{nm} - \delta_{n,-m}) - 1]e^{-\lambda\, t}. \tag{8.75}$$

Further, the polycrystalline nature of the samples used in the experiments allows us to perform an angular average over all possible directions of the $\vec{k}$ vector, yielding (Dattagupta 1977 and 1983):

$$\begin{aligned} G_s(k,t) = {} & \frac{1}{36}\sum_{n,m} \frac{\sin k|\vec{r}_n - \vec{r}_m|}{k|\vec{r}_n - \vec{r}_m|} + \frac{1}{12}\left[6 - \sum_n \frac{\sin k|\vec{r}_n - \vec{r}_m|}{k|\vec{r}_n - \vec{r}_m|}\right] e^{-\lambda_0 t} \\ & + \frac{1}{12}\left[6 + \sum_n \frac{\sin k|\vec{r}_n - \vec{r}_m|}{k|r_n - \dot{r}_m|} - \frac{1}{3}\sum_n \frac{\sin k|\vec{r}_n - \vec{r}_m|}{k|\vec{r}_n - \vec{r}_m|}\right] e^{-\lambda\, t}. \end{aligned} \tag{8.76}$$

We denote by b the distance between the cage center and any one of the six possible dumbbell sites for the Fe atom. In addition, we take the Fourier transform of Equation (8.76) to finally write for the Mossbauer emission intensity:

$$\begin{aligned} I(\omega) = {} & \frac{1}{\pi}\, \frac{\Gamma/2}{\omega^2 + (\Gamma/2)^2} \cdot \frac{1}{6}\left[1 + \frac{\sin(2kb)}{(2kb)} + \frac{4\sin(\sqrt{2}kb)}{(\sqrt{2}kb)}\right] \\ & + \frac{1}{\pi}\, \frac{\lambda_0 + \Gamma/2}{\omega^2 + (\lambda_0 + \Gamma/2\ )^2} \cdot \frac{1}{2}\left[1 - \frac{\sin(2kb)}{(2kb)}\right] \\ & + \frac{1}{\pi}\, \frac{\lambda + \Gamma/2}{\omega^2 + (\lambda\ + \Gamma/2)^2} \cdot \frac{1}{3}\left[1 - \frac{\sin(2kb)}{(2kb)} - \frac{2\sin(\sqrt{2}kb)}{(\sqrt{2}kb)}\right]. \end{aligned} \tag{8.77}$$

If we compare Equation (8.77) with the expression for neutron structure factor in the two-site case [Equation (8.66)], we notice that the elastic component is not a delta function [first term in Equation (8.77)] but a Lorentzian centered around $\omega = 0$ because of the ubiquitous presence of a natural line width of the excited state of the Mossbauer nucleus.

The other two quasi-elastic components [second and the third terms in Equation (8.77)] are characterized by a width that depends on the fundamental jump rate λ. All three expressions are endowed by form factors [parentheses in Equation (8.77)] that depend on the relative magnitudes of cage size and the wave number k^{-1}. Separate measurements of the form factor allow the determination of cage size. Computations of the temperature dependence of the line widths of the quasi-elastic components yield the height of the Arrhenius barrier over which the Fe atom must jump. We stress once more that the occurrence of the elastic component [first term in Equation (8.77)] is characteristic of restricted diffusion probed by spectroscopic techniques.

8.6 Unbounded Jump Diffusion in Empty Lattice

We turn our discussion now from diffusion in a restricted domain to diffusion in unbounded space. First we consider the simpler situation in which a diffusive atom (such as H in neutron scattering or a radioactive atom such as ^{57}Fe in Mossbauer spectroscopy) finds itself in an interstitial site of a lattice. The atom vibrates around its mean position like a harmonic oscillator. The corresponding self-correlation function was calculated in Section 8.4. Vibrations, of course, occur on a much faster timescale such as the Debye frequency ($\sim 10^{-12}$ sec) compared to the mean time between diffusive jumps ($\sim 10^{-8}$ sec). Thus, the diffusive contribution to the self-correlation function can be decoupled from the vibrational contribution because of the timescale separation noted above. Further, because interstitial diffusion does not require the presence of vacancies (Section 8.7) the jumps can be treated as uncorrelated random events as in the random walk model (Chapter 1).

Assuming that the diffusing atom starts from the origin $\vec{r} = 0$ at time $t = 0$ and all interstitial sites (N in number) are equivalent, we may rewrite Equation (8.52) as (Dattagupta and Schroeder 1987):

$$G_s(\vec{k},t) = \frac{1}{N}\sum_m e^{i\vec{k}\bullet\vec{r}_m}(\vec{r}_m \,|\, \hat{P}(t) \,|\, \vec{0}) \,. \tag{8.78}$$

The above expression may then be viewed as a discrete Fourier transform of the probability matrix $(\vec{r}_m \mid \hat{P}(t) \mid \vec{0})$ that measures the conditional probability that the diffusing atom will reach site $\vec{r}_m$ at time t, given that it started from the origin at time $t = 0$.

The probability matrix $\hat{P}(t)$ can be developed along the lines of the continuous time random walk model of Section 7.1. Thus, diffusion is viewed as an uncorrelated sequence of a renewal process in which the walker stops and jumps at a rate governed by a Poisson distribution as implicit in a stationary Markov process. Therefore

$$\begin{aligned}(\vec{r}_m \mid \hat{P}(t) \mid \vec{0}) &= e^{-\lambda t}\, \delta(\vec{r}_m - \vec{0}) \\ &+ \int_0^t (\lambda\, dt_1)\, e^{-\lambda(t-t_1)}\, g(\vec{r}_m - \vec{0}) e^{-\lambda t_1} \\ &+ \frac{1}{N^2} \sum_{\vec{r}_1} \int_0^t \lambda\, dt_2 \int_0^{t_2} \lambda\, dt_1\, e^{-\lambda(t-t_2)}\, e^{-\lambda(t_2-t_1)} g(\vec{r}_m - \vec{r}_1)\, e^{-\lambda t_1}\, g(\vec{r}_1) \\ &+ \frac{1}{N^3} \sum_{\vec{r}_1 \vec{r}_2} \int_0^t \lambda\, dt_3 \int_0^{t_3} \lambda\, dt_2 \int_0^{t_2} \lambda\, dt_1 e^{-\lambda(t-t_3)}\, g(\vec{r}_m - \vec{r}_2 \,|\, e^{-\lambda(t_3-t_2)} g(\vec{r}_2 - \vec{r}_1)\, e^{-\lambda(t_2-t_1)}\, g(\vec{r}_1),\end{aligned} \tag{8.79}$$

where $g(\vec{r} - \vec{r}\,')$ is the elementary jump probability from the site $\vec{r}\,'$ to $\vec{r}$. Taking the combined temporal Laplace transform and spatial Fourier transform and employing the convolution theorem, we have from Equations (8.78) and (8.79),

$$G_s(\vec{k}, t) = \frac{1}{(s+\lambda)} + \frac{\lambda g(\vec{k})}{(s+\lambda)^2} + \frac{\lambda^2 (g(k))^2}{(s+\lambda)^3} + \frac{\lambda^3 (g(k))^3}{(s+\lambda)^4} \cdots = \frac{1}{s + \lambda(1 - g(\vec{k})}, \tag{8.80}$$

where

$$g(\vec{k}) = \frac{1}{N} \sum_{\vec{r}} e^{i\vec{k}.\vec{r}}\, g(\vec{r}). \tag{8.81}$$

The result in Equation (8.80) was derived by Chudley and Elliot (1961) and independently by Singwi and Sjoelander (1960). An extension of the result that includes possible fluctuations in hyperfine fields, as a Mossbauer atom jumps from one site to another, was carried out by Dattagupta (1975). Equation (8.80) predicts a quasi-elastic line shape governed by a width $2\lambda[1 - \delta(\vec{k})]$, λ^{-1} that represents the mean time between two successive jumps. We discuss below two special cases.

8.6.1 Large Jumps in Random Directions

If we assume the diffusive atom is at the center of a sphere of radius a and can jump to any arbitrary point of the sphere,

$$g(\vec{r}) = \frac{1}{4\pi a^2} \ \delta(r-a) \ . \tag{8.82}$$

The width is then given by

$$\bar{\Gamma} = 2\lambda \left[1 - \frac{\sin(ka)}{ka}\right]. \tag{8.83}$$

For Mossbauer experiments in a liquid like glycerol, $a \approx 3\mathring{A}$, whereas for ^{57}Fe, $k \approx 7.3 \ (\mathring{A})^{-1}$ and hence the factor within the square parentheses is nearly 0.96.

8.6.2 Small Jumps in Random Directions

We now have $|\vec{k}.\ \vec{r}| << 1$ which, for ^{57}Fe corresponds to a jump distance $\sim 0.14 \ \mathring{A}$. Assuming isotropic jumps, $g(\vec{r}) = g(r)$ and expanding $e^{i\vec{k}.\vec{r}}$ in a power series, we obtain from Equation (8.81):

$$g(\vec{k}) \approx 1 - \frac{1}{6} k^2 < r^2 >, \tag{8.84}$$

where

$$< r^2 > = \int r^2 \ g(\vec{r}) \, d\vec{r}. \tag{8.85}$$

The width now is

$$\bar{\Gamma} - \frac{\lambda}{3} k^2 < r^2 > = 2Dk^2, \tag{8.86}$$

the usual Einstein result (Chapter 1) wherein the diffusion coefficient is given by

$$D = \frac{1}{6} \ \lambda < r^2 >. \tag{8.87}$$

8.7 Vacancy-Assisted Correlated Diffusion in Solids

In dealing with jump diffusion processes in solids and the accompanying structure factors, we have considered uncorrelated jumps. This picture, while valid for interstitial diffusion in an empty lattice, will break down with vacancy-assisted jump diffusion in a solid, whether the process is self-diffusion (Fe in Fe) or impurity diffusion (Fe in Al). In the latter case, correlation between jumps becomes important.

We will treat the case of impurity diffusion (and not self-diffusion) which has an immediate bearing on spectroscopic measurement that focuses on the impurity atom. The reason the correlation effects are important is based on the simple fact that when an impurity (or tagged) atom jumps into a vacant site, the subsequent jumps are not independent because the impurity has more than random probability of jumping back, thus exchanging sites with the vacancy.

We make certain simplifying assumptions: (1) the elementary jumps of the vacancy are taken to be over nearest neighbor sites only; (2) both the impurity and vacancy concentrations are small so that only a single vacancy–impurity pair must be treated at a time; (3) the detailed dynamics of vacancy jumps is considered only for nearest neighbor sites of the impurity; outside the nearest neighbor shell, only an average vacancy concentration is taken into account.

This last assumption calls for a few comments. When the average vacancy concentration c_v is small, the probability of finding a vacancy in the nearest neighbor shell of the impurity atom is also small. As mentioned earlier, the important correlation in vacancy-assisted diffusion arises from immediate returns of the vacancy from the nearest neighbor shells. Vacancy paths beyond the nearest neighbor shell can make only negligible contributions. Hence our assumption of taking an average vacancy contribution outside the nearest neighbor shell and neglecting the correlated re-entry of the same vacancy yields an effective (coarse-grained) rate of re-entry that will be proportional to c_v itself.

A significant factor that distinguishes impurity diffusion from self-diffusion in a solid is the perturbation of the vacancy jump frequencies in the immediate vicinity of the impurity. This factor is incorporated in terms of the so-called five-frequency model, shown in Figure 8.8. The five corresponding jump rates carry information about the impurity–vacancy bonding and the potential barriers over which the vacancy must jump. These are shown in Figure 8.9.

We need to calculate $G_D\,(\boldsymbol{R}_{n,}t)$ which measures the conditional probability that the impurity atom is found at the site $\boldsymbol{R}_{n,}$ given that it was at the origin at time $t = 0$. With reference to Figure 8.7, we may write (Dattagupta and Schroeder 1987):

$$G_D(\boldsymbol{R}^n, t) = \sum_{\alpha\beta} (\boldsymbol{R}^n, \alpha \,|\, \underline{P}(t) \,|\, \mathbf{0}, \beta) p_\beta, \tag{8.88}$$

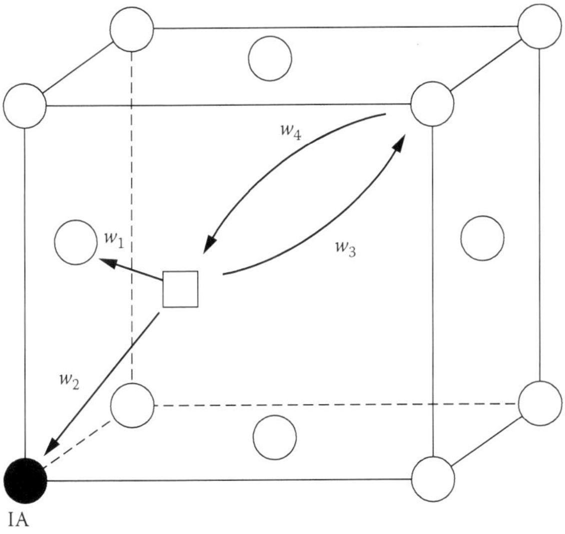

FIGURE 8.8
Five-frequency model for impurity diffusion in FCC solids. w_0 is vacancy jump frequency in pure host lattice. w_1 is vacancy jump rate in nearest neighbor shell of impurity. w_2 is exchange (of vacancy and impurity) jump rate. w_3 is dissociation rate of vacancy jump from nearest neighbor shell. w_4 is association rate of vacancy jump into nearest neighbor shell. (From Dattagupta, S., *Relaxation Phenomena in Condensed Matter Physics*, Academic Press, New York, 1987. With permission.)

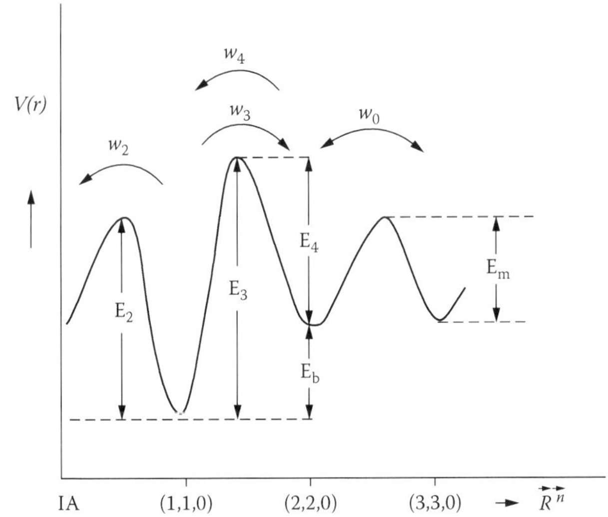

FIGURE 8.9
Energy barriers for vacancy jumps in radial direction. $E_b = E_3 - E_4$ = impurity–vacancy binding energy. $E_0 = E_m$ = migration energy in pure host lattice. (From Dattagupta, S., *Relaxation Phenomena in Condensed Matter Physics*, Academic Press, New York, 1987. With permission.)

where the matrix element $(\boldsymbol{R}^n, \alpha \,|\, \underline{P}(t) \,|\, \mathbf{0}, \beta)$ measures the conditional probability that the impurity atom is found at the site $\boldsymbol{R}^n$ while the vacancy is found at the state $|\alpha)$ at time t, given that the atom was at the origin and the vacancy was in the state $|\beta)$ at time $t = 0$. The state $|\alpha)$ or $|\beta)$ refers to one of the vacancy sites within the nearest neighbor shell of the impurity.

We shall call the latter an associated state as opposed to a nonassociated state that corresponds to any of the vacancy sites outside the nearest neighbor shell. Thus the associated states run over Z sites where Z is the so-called coordination number of the underlying lattice (for instance $Z = 12$, for an FCC lattice) whereas the nonassociated state is just one lumped state. Hence, α or β assumes $Z+1$ values. The summations in Equation (8.88) consider that vacancies are not directly probed (such as in a spectroscopy experiment like the Mossbauer effect of ^{57}Fe) and hence we must sum over all final vacancy states and average over the initial vacancy state, aided by the a priori probability of vacancy occupation defined by p_β.

As has been our wont in treating jump diffusion, we assume that the underlying stochastic process is Markovian. Thus the operator $P(t)$ obeys the master equation:

$$\frac{\partial}{\partial t}\hat{P}(t) = \hat{W}\;\hat{P}(t), \tag{8.89}$$

where the elements of the jump matrix $\hat{W}$ measure the rates of various allowed jumps for the impurity–vacancy pair. Considering the case of a simple cubic lattice, for the sake of simplicity (Figure 8.10), we can write

$$(\boldsymbol{R}^n, \alpha \,|\, \hat{W} \,|\, \boldsymbol{R}^m, \beta) = \delta_{nm}(\alpha \,|\, \hat{W}_\upsilon \,|\, \beta) + (\boldsymbol{R}^n, \alpha \,|\, \hat{W}_{MA-\upsilon} \,|\, \boldsymbol{R}^m, \beta). \tag{8.90}$$

The first term above describes the case in which the impurity atom is stationary but the vacancy jumps between the associated and nonassociated states. The matrix $\hat{W}_\upsilon$ (whose subscript implies vacancy jumps only) can be constructed as

$$\hat{W}_\upsilon = \begin{array}{r} \\ 1 \\ -1 \\ 2 \\ -2 \\ 3 \\ -3 \\ N \end{array}\left[\begin{array}{ccccccc} 1 & -1 & 2 & -2 & 3 & -3 & N \\ -(\upsilon+4w_1) & 0 & w_1 & w_1 & w_1 & w_1 & \lambda \\ 0 & -(\upsilon+4w_1) & w_1 & w_1 & w_1 & w_1 & \lambda \\ w_1 & w_1 & -(\upsilon+4w_1) & 0 & w_1 & w_1 & \lambda \\ w_1 & w_1 & 0 & -(\upsilon+4w_1) & w_1 & w_1 & \lambda \\ w_1 & w_1 & w_1 & w_1 & -(\upsilon+4w_1) & 0 & \lambda \\ w_1 & w_1 & w_1 & w_1 & 0 & -(\upsilon+4w_1) & \lambda \\ \upsilon & \upsilon & \upsilon & \upsilon & \upsilon & \upsilon & -6\lambda \end{array}\right]. \tag{8.91}$$

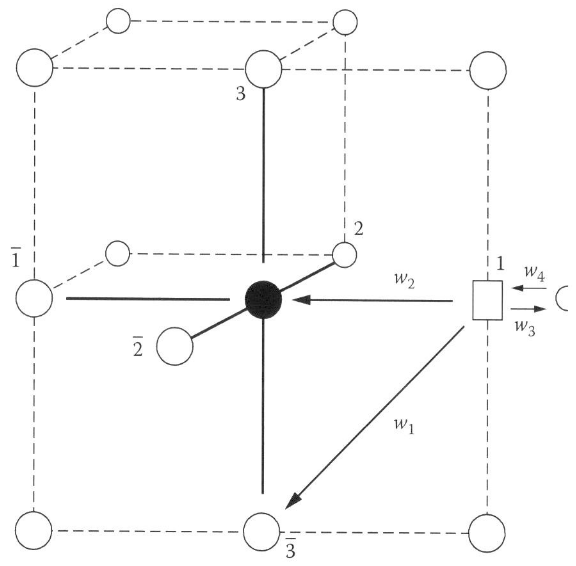

FIGURE 8.10
Enumeration of nearest neighbor sites of impurity atom (•) in a simple cubic lattice and vacancy (▫) and corresponding jump frequencies. (From Dattagupta, S., *Relaxation Phenomena in Condensed Matter Physics*, Academic Press, New York, 1987. With permission.)

The diagonal elements of $\hat{W}_{\upsilon}$ contribute to the loss term in Equation (8.90) while the off-diagonal elements contribute to the gain term. In Equation (8.91), the first six rows correspond to the associated states q (six in this case, see Figure 8.8). The seventh row labeled N represents the nonassociated state. The rate ω_1 measures the rotational jump rates of the impurity–vacancy pair. Due to the specific geometry of the simple cubic (sc) structure, these jumps are actually over the next neighbor sites. The parameter υ defines the rate of jumps from any associated state while λ designates the rate of jumps from the nonassociated state to *any* associated state. The rates υ and λ can in turn be re-expressed as elementary jump rates w_3 and w_4 (Figure 8.9). For instance, in the sc case,

$$\upsilon = 5w_3. \tag{8.92}$$

Similarly,

$$\lambda = 5w_4\ c_{\upsilon}, \tag{8.93}$$

where it may be recalled that c_{υ} is the average vacancy concentration. The a priori probabilities can be estimated from the detailed balance condition:

$$p_{\beta}(\alpha\,|\,\hat{W}_{\upsilon}\,|\,\beta) = p_{\alpha}(\beta\,|\,\hat{W}_{\upsilon}\,|\,\alpha), \tag{8.94}$$

and the probability conservation condition:

$$\sum_{\beta} p_{\beta} = 1. \tag{8.95}$$

Combining Equations (8.94) and (8.95), we find

$$\begin{aligned} p_A &= \frac{\lambda}{\upsilon + 6\lambda} = \frac{\omega_4 \ c_\upsilon}{\omega_3 + 6\,\omega_4 \ c_\upsilon}, \\ p_N &= \frac{\upsilon}{\upsilon + 6\lambda} = \frac{\omega_3}{\omega_3 + 6\,\omega_4 \ c_\upsilon}. \end{aligned} \tag{8.96}$$

p_A refers to the occupational probability of one of the associated states while p_N is the corresponding probability of the non-associated state. Note that

$$\frac{\omega_3}{\omega_4} = \frac{c_\upsilon}{c_{NN}} = \exp(-E_b / K_B T), \tag{8.97}$$

where c_{NN} is the nearest neighbor vacancy concentration and E_b is the impurity–vacancy binding energy (Figure 8.8). Employing these parameters, p_A and p_N can be re-expressed as

$$\begin{aligned} p_A &= \frac{c_{NN}}{1 + 6c_{NN}}, \\ p_N &= \frac{1}{1 + 6c_{NN}}. \end{aligned} \tag{8.98}$$

Returning to Equation (8.90) the second term represents the impurity–vacancy exchanges. Figure 8.9 demonstrates that each such exchange will place the vacancy into the opposite state. For instance, if the vacancy is in state $|1)$ before an exchange, it will be in state $|-1)$ after the exchange. Accordingly, the $W_{MA-\upsilon}$ matrix can be constructed as

$$\underline{W}_{MA-v} = \begin{array}{c} \\ 1 \\ -1 \\ 2 \\ -2 \\ 3 \\ -3 \\ N \end{array} \left[\begin{array}{ccccccc} 1 & -1 & 2 & -2 & 3 & -3 & N \\ -\omega_2\delta_{nm} & \omega_2 S_{nm}\delta_{n-r_1} & 0 & 0 & 0 & 0 & 0 \\ \omega_2 S_{nm}\delta_{n-r_1} & -\omega_2\delta_{nm} & 0 & 0 & 0 & 0 & 0 \\ 0 & 0 & -\omega_2\delta_{nm} & \omega_2 S_{nm}\delta_{n-r_2} & 0 & 0 & 0 \\ & & \omega_2 S_{nm}\delta_{n-r_2} & -\omega_2\delta_{nm} & 0 & 0 & 0 \\ 0 & 0 & 0 & 0 & -\omega_2\delta_{nm} & \omega_2 S_{nm}\delta_{n-r_3} & 0 \\ 0 & 0 & 0 & 0 & \omega_2 S_{nm}\delta_{n-r_3} & -\omega_2\delta_{nm} & 0 \\ 0 & 0 & 0 & 0 & 0 & 0 & 0 \end{array} \right]$$

$$S_{nm} = 1, \quad \text{if } \boldsymbol{R}^n \text{ and } \boldsymbol{R}^m \text{ are nearest neighbor sites but is zero otherwise.} \tag{8.99}$$

The factor S_{nm} ensures that only nearest neighbor exchanges can take place. Thus $S_{nm} = 1$, if R^n and R^m are the nearest neighbor sites, but is zero otherwise.

Additionally, $\mathbf{r}_\alpha$ is the vector distance between the impurity atom and the vacancy in the $|a)$ state. Thus the transition matrix element $(\boldsymbol{R}_n, \alpha | \hat{W} | \boldsymbol{R}_m, \beta)$ builds on the fact that jumps depend only on the difference vector $\boldsymbol{R}^n - \boldsymbol{R}^m$. From Equation (8.89) we find

$$\frac{\partial}{\partial t}(\boldsymbol{R}^n, \alpha | \hat{\boldsymbol{P}}(t) | 0, \beta) = \sum_m \sum_{\alpha'} (\boldsymbol{R}^n, \alpha | \hat{W} | \boldsymbol{R}^m, \alpha')(\boldsymbol{R}^m, \alpha' | \hat{\boldsymbol{P}}\ (t) | \boldsymbol{0}, \beta). \quad (8.100)$$

We may recall that what we need for structure factor calculation is the Fourier transform of $(\boldsymbol{R}^n, \alpha | \hat{\boldsymbol{P}}(t) | \boldsymbol{0}, \beta)$:

$$\sum_n \exp\ (ik.\boldsymbol{R}^n)\ (\boldsymbol{R}^n, \alpha | \hat{\boldsymbol{P}}(t) | \boldsymbol{0}, \beta) \equiv (\alpha | \hat{\boldsymbol{P}}_k(t) | \beta). \quad (8.101)$$

Using Equation (8.101),

$$\frac{\partial}{\partial t}(\alpha | \hat{\boldsymbol{P}}_k(t) | \beta) = \sum_{\alpha'} (\alpha | \hat{W}_k | \alpha')\ (\alpha' | \hat{\boldsymbol{P}}_k(t) | \beta). \quad (8.102)$$

Equivalently, in matrix notation

$$\frac{\partial}{\partial t}\hat{\boldsymbol{P}}_k\ (t) = \hat{W}_k\, \hat{\boldsymbol{P}}_k(t). \quad (8.103)$$

It is interesting to note that our problem is now reduced to dealing with the linear vector space of vacancy states only as the states of the tagged (or impurity) atom have been averaged over. This is reflected in the fact that $\hat{W}_k$, and hence $\hat{\boldsymbol{P}}_k$, are matrices among the vacancy states only. Further, note from Equation (8.101) that

$$(\alpha | \hat{\boldsymbol{P}}_k(t = 0) | \beta) = \sum_n \exp\ (ik \cdot \boldsymbol{R}^n)\ \delta_{n0}\delta_{\alpha\beta} = \delta_{\alpha\beta}. \quad (8.104)$$

and therefore,

$$\hat{\boldsymbol{P}}_k(t = 0) = \underline{1}. \quad (8.105)$$

Equation (8.104), in conjunction with the initial condition in Equation (8.105) yields for the Laplace transform of $\underline{\underline{P}}_k(s)$:

$$\tilde{\hat{\boldsymbol{P}}}_k(s) = (s.\underline{1} - \hat{W}_k)^{-1}. \tag{8.106}$$

Finally, the structure factor can be calculated as

$$S\ (k,\omega) = \mathrm{Re}\frac{1}{\pi\hbar}\sum_{\alpha,\beta}(\alpha\,|\,(s\ \underline{1} - \hat{W}_k)^{-1}\,|\,\beta)p_\beta, \tag{8.107}$$

where

$$s = i\ (\omega - \omega_0) + \frac{\Gamma_0}{2\hbar}, \tag{8.108}$$

Γ_0 is the natural line width in the context of spectroscopic measurements.

8.7.1 Analytical Results in a Simple Cubic (SC) Case

Using Equation (8.100) we find

$$(\alpha\,|\,\hat{W}_k\,|\,b) = (\alpha\,|\,\hat{W}_\upsilon\,|\,\beta) + (\alpha\,|\,\hat{W}_{MA-\upsilon,k}\,|\,\beta), \tag{8.109}$$

where the matrix $\hat{W}_{MA-\upsilon,k}$ is given from Equation (8.99) by

$$\hat{W}_{MA-\upsilon,k} = \begin{array}{c} 0 \\ 1 \\ -1 \\ 2 \\ -2 \\ 3 \\ -3 \\ N \end{array}\begin{bmatrix} 1 & -1 & 2 & -2 & 3 & -3 & N \\ -w_2 & w_2e^{ik.r_1} & 0 & 0 & 0 & 0 & 0 \\ w_2e^{ik.r_1} & -w_2 & 0 & 0 & 0 & 0 & 0 \\ 0 & 0 & -w_2 & w_2e^{ik.r_2} & 0 & 0 & 0 \\ 0 & 0 & w_2e^{ik.r_2} & -w_2 & 0 & 0 & 0 \\ 0 & 0 & 0 & 0 & -\omega_2 & \omega_2e^{ik.r_3} & 0 \\ 0 & 0 & 0 & 0 & w_2e^{ik.r_3} & -w_2 & 0 \\ 0 & 0 & 0 & 0 & 0 & 0 & 0 \end{bmatrix} \tag{8.110}$$

The complete $\hat{W}_k$ matrix is given by the combination of Equations (8.91) and (8.110). Evidently, the calculation scheme involves inverting the matrix $(s\ \underline{1} - \hat{W}_k)$, the dimension of which is 7×7 in the sc case and carrying out summations over α and β. For the sc case, this can be achieved analytically by exploiting certain features of the $\underline{\underline{W}}_k$ matrix. We quote here the final result and provide details in the appendix at the end of this chapter.

$$S(k,\omega)=\frac{1}{\pi\hbar}\text{Re}\left\{\left[1-\frac{\upsilon-\omega_1}{s-\gamma}-\frac{\omega_1(s-\gamma)+(\upsilon-\omega_1)(\lambda-\omega_1)}{s+\upsilon+6\lambda}\bar{G}^0\right]^{-1}\right.$$
$$\left.\times\left[<G^0>+p_N\frac{\lambda-\omega_1}{s+\upsilon+6\lambda}\bar{G}^0\right]\right\} \tag{8.111}$$

where

$$\bar{G}^0 = 2\sum_{i=1}^{3}\frac{s+\upsilon+4\omega_1+\omega_2\ [1+\cos\ (k\cdot r_i)]}{D_i}+\frac{1}{s-\gamma},$$

$$<G^0> = 2p_A\sum_{i=1}^{3}\frac{s+\upsilon+4\omega_1+\omega_2\ [1+\cos\ (k\cdot r_i)]}{D_i}+p_N\frac{1}{s-\gamma},$$

$$D_i=(s+\gamma')^2-[\omega_1^2+\omega_2^2-2\omega_1\ \omega_2\cos\ (k\cdot r_i)],$$

$$\gamma=-7\lambda-\upsilon+\omega_1, \tag{8.112}$$

$$\gamma'=\omega_2+5\omega_1+\upsilon.$$

Various limiting cases of Equation (8.111) provide insights into the mathematical details, as follows.

1. $\omega_2=0$: **No Impurity–Vacancy Exchange** — Since the tagged atom is taken to be totally immobile, the vacancy is expected to play no role whatsoever in the structure factor calculation. This is checked by observing that

$$D_1=D_2=D_3=(s+\upsilon+4\omega_1)\ (s+\upsilon+6\omega_1), \tag{8.113}$$

and hence,

$$\bar{G}^0=\frac{6}{s+\upsilon+6\omega_1}+\frac{1}{s-\gamma},\quad <G^0> = \frac{6p_A}{s+\upsilon+6\omega_1}+\frac{p_N}{s-\gamma}$$

Substituting the expressions for $\bar{G}^0$ and $< G^0 >$ in Equation (8.113), and after some algebra,

$$S(k,\omega) = \frac{1}{\pi\hbar}\,\mathrm{Re}\left(\frac{1}{s}\right) = \frac{1}{\pi\hbar}\frac{\Gamma_0/2\hbar}{(\omega-\omega_0)^2 + (\Gamma_0/2\hbar)^2}, \tag{8.114}$$

which is just the unperturbed structure factor governed by the natural line width Γ_0.

2. $\omega_2 = \omega_3 = 0$: **Cage Motion** — The condition $\omega_3 = 0$ implies that $\upsilon = 0$ and once the vacancy is in the nearest neighbor shell of the impurity atom, it cannot escape. Additionally, $w_1 = 0$ means that the vacancy forms a mixed dumbbell with the impurity. Thus the only allowed motion is the one in which the dumbbell partners exchange sites at a rate w_2. We now have:

$$p_A = \frac{1}{6},\quad p_N = 0.$$

Equation (8.111) then yields

$$S(k,w) = \frac{1}{\pi\hbar}\ \mathrm{Re} < G^0 >, \tag{8.115}$$

where from Equation (8.112),

$$< G^0 > = \frac{s + w_2 + \frac{1}{3} w_2 \sum_{i=1}^{3} \cos(k\cdot r_i)}{s(s+2w_2)}. \tag{8.116}$$

The structure factor may then be written in the form

$$\begin{aligned}\mathbf{S}(k,\omega) = {} & \frac{1}{2\pi\hbar}\frac{\Gamma_0/2\hbar}{(\omega-\omega_0)^2 + (\Gamma_0/2\hbar)^2} \times \left[1 + \frac{1}{3}\sum_{i=1}^{3}\cos(k\cdot r_i)\right] \\ & + \frac{\mathbf{1}}{2\pi\hbar}\frac{(2\omega_2 + \Gamma_0/2\hbar)}{(\omega-\omega_0)^2 + (2\omega_2 + \Gamma_0/2\hbar)^2} \times \left[1 - \frac{1}{3}\sum_{i=1}^{3}\cos(k\cdot r_i)\right].\end{aligned} \tag{8.117}$$

The first term corresponds to an unbroadened elastic component while the second represents a quasi-elastic line broadened by the factor $(2\omega_2 + \Gamma_0/2\hbar)$.

The occurrence of the quasi-elastic component is the signature of cage diffusion as indicated earlier. The expressions inside the square brackets provide the form factors.

3. $\omega_3 = 0, \omega_1 >> \omega_2$: **Uncorrelated Diffusion** — The condition $\omega_3 = 0$ and hence $\nu = 0$ indicates that the vacancy is trapped by the impurity. Further, the latter always finds a vacancy to jump into. The condition $\omega_1 >> \omega_2$ implies that a vacancy is uniformly distributed among all the nearest neighbor sites of the impurity. The situation is now akin to uncorrelated diffusion in an empty lattice (treated earlier) in which the impurity can jump into any of the nearest neighbor sites randomly without correlation to the previous jumps. Once again we have

$$p_A = \frac{1}{6}, \ p_N = 0.$$

which, from Equation (8.113) leads to

$$< G^0 > = \frac{1}{3} \sum_{i=1}^{3} \frac{s + 4\omega_1 + \omega_2 \ [1 + \cos \ (k, r_i)]}{D_i}. \tag{8.118}$$

Furthermore, Equation (8.110) implies

$$\bar{G}^0 = 6. < G^0 > + \frac{1}{s - \gamma}. \tag{8.119}$$

Employing $\nu = p_N = 0$ and Equation (8.119), we obtain from Equation (8.109),

$$S(k, \omega) = \frac{1}{\pi\hbar} \ \text{Re} \left\{ \left[1 - \frac{\omega_1}{s - \gamma} - \omega_1 \left(6 < G^0 > + \frac{1}{s - \gamma} \right) \right]^{-1} < G^0 > \right\} \tag{8.120}$$

that further simplifies to

$$S(k, \omega) = \frac{1}{\pi\hbar} \ \text{Re} \ [(< G^0 >^{-1} - 6\omega_1)^{-1}]. \tag{8.121}$$

We investigate now the $\omega_1 >> \omega_2$ limit. It is convenient to first decompose $<G^0>$ in Equation (8.118) as a sum of two terms, thus

$$< G^0 > \ = \ \frac{1}{3} \sum_{i=1}^{3} \left[\frac{C_i^-}{s + A_i^-} + \frac{C_i^+}{s + A_i^+} \right], \tag{8.122}$$

where

$$A_i^{\pm} = (\omega_2 + 5\omega_1) \pm [\omega_1^2 + \omega_2^2 - 2\omega_1\omega_2 \cos(k.r_i)]^{1/2},$$
$$C_i^{\pm} = \frac{1}{2}\left[1 \pm \frac{\omega_1 + \omega_2 \cos(k.r_i)}{\left[\omega_1^2 + \omega_2^2 - 2\omega_1\omega_2 \cos(k.r_i)\right]^{1/2}}\right]. \tag{8.123}$$

When $\omega_1 >> \omega_2$, it is evident that

$$\begin{aligned} A_i^{+} &\simeq 6\omega_1 + \omega_2 \ [1 - \cos(k \cdot r_i)], \\ A_i^{-} &\simeq 4\omega_1 + \omega_2 \ [1 + \cos(k \cdot r_i)], \\ C_i^{+} &\simeq 1, \\ C_i^{-} &\simeq 0. \end{aligned} \tag{8.124}$$

Equation (8.121) then leads to

$$< G^0 > \simeq \frac{1}{3} \sum_{i=1}^{3} \{s + 6\omega_1 + \omega_2[1 - \cos(k \cdot r_i)]\}^{-1}.$$

Therefore,

$$< G^0 > \simeq \frac{1}{s + 6\omega_1} \left[1 - \frac{1}{3} \frac{\omega_2}{s + 6\omega_1} \sum_{i=1}^{3} [1 - \cos(k \cdot r_i)]\right].$$

Finally, we find for the structure factor, from Equation (8.119)

$$S(k, \omega) = \frac{1}{\pi\hbar} \ \mathrm{Re} \left[s + \frac{1}{3}\omega_2 \sum_{i=1}^{3} [1 - \cos(k \cdot r_i)]^{-1}\right]$$

which matches with the Chudley–Elliot (1961) result for sc lattices mentioned earlier in Section 8.6.

While the analytical formulae derived above for the sc case provide insight into various limiting cases, it is pertinent that one will have to resort to numerical computations for other lattice structures such as body-centered cubic (BCC) and FCC crystals that necessitate inversion of large matrices.

Appendix

We provide below the mathematical steps leading to Equation (8.111) in the text. Since our basic task is to invert the matrix $(s.\underline{1} - \hat{W}_k)$, we split the matrix $\hat{W}_k$ into four terms and exploit certain symmetry properties of these pieces. We may write from Equation (8.110):

$$\hat{W}_k = \hat{W}_k^0 + w_1\hat{W}_1 + (\nu - w_1)\ \hat{W}_2 + (\lambda - w_1)\ \hat{W}_3, \tag{A.1}$$

where

$$(\alpha \mid \hat{W}_1 \mid \beta) = 1 \ \text{(independent of } \alpha \text{ and } \beta), \tag{A.2}$$

$$(\alpha \mid \hat{W}_2 \mid \beta) = \delta_{\alpha\ N} \ \text{(independent of } \beta), \tag{A.3}$$

$$(\alpha \mid \hat{W}_3 \mid \beta) = \delta_{\beta\ N} \ \text{(independent of } \alpha), \tag{A.4}$$

$$\hat{W}_k^0 = \begin{array}{c} \\ 1 \\ -1 \\ 2 \\ -2 \\ 3 \\ -3 \\ N \end{array} \begin{bmatrix} 1 & -1 & 2 & -2 & 3 & -3 & N \\ -\gamma' & (\omega_2 e^{i\delta_1} - \omega_1) & 0 & 0 & 0 & 0 & 0 \\ (\omega_2 e^{-i\delta_1} - \omega_1) & -\gamma' & 0 & 0 & 0 & 0 & 0 \\ 0 & 0 & -\gamma' & (\omega_2 e^{i\delta_2} - \omega_1) & 0 & 0 & 0 \\ & & (\omega_2 e^{-i\delta_2} - \omega_1) & -\gamma' & 0 & 0 & 0 \\ 0 & 0 & 0 & 0 & -\gamma' & (\omega_2 e^{i\delta_3} - \omega_1) & 0 \\ 0 & 0 & 0 & 0 & (\omega_2 e^{-i\delta_3} - \omega_1) & -\gamma' & 0 \\ 0 & 0 & 0 & 0 & 0 & 0 & \gamma \end{bmatrix}. \tag{A.5}$$

In Equation (A.5)

$$\delta_i = k.r_i, \quad i = 1, 2, 3, \tag{A.6}$$

and γ and γ' are defined in Equation (8.112). Using the operator identity:

$$\frac{1}{\hat{A}} = \frac{1}{\hat{B}} + \frac{1}{\hat{B}}(\hat{B} - \hat{A})\frac{1}{\hat{A}} = \frac{1}{\hat{B}} + \frac{1}{\hat{A}}(\hat{B} - \hat{A})\frac{1}{\hat{B}}, \tag{A.7}$$

we may write from Equations (2.21) and (A.1):

$$\tilde{P}_k(s) = \hat{P}' + \hat{P}'\ [w_1\hat{W}_1 + (\nu - w_1)\ \hat{W}_2]\ \tilde{\hat{P}}_k(s), \tag{A.8}$$

where

$$\hat{P}' \equiv [s - \hat{W}_k^0 - (\lambda - w_1)\ \hat{W}_3\,]^{-1}. \tag{A.9}$$

Note that we have not indicated the k and s dependence of $\hat{P}'$ for the sake of brevity. Equation (A.8) allows us to write

$$\begin{aligned}
&\sum_{\alpha,\beta}(\alpha\,|\,\hat{\tilde{P}}_k(s)\,|\,\beta)p_\beta \\
&= \sum_{\alpha,\beta}(\alpha\,|\,\hat{P}'\,|\,\beta)p_\beta + \sum_{\alpha,\beta\ \alpha',\beta'}(\alpha\,|\,\hat{P}'\,|\,\alpha') \\
&\times(\alpha'\,|[w_1\ \hat{W}_1 + (\nu - w_1)\ \hat{W}_2]\,|\,\beta') \\
&'\times(\beta'\,|[\hat{\tilde{P}}_k(s)\,|\,\beta)p_\beta,
\end{aligned}$$

where in the second term on the right we used the completeness relation:

$$\sum_{\alpha}|\,\alpha)(\alpha\,| = 1. \tag{A.10}$$

Employing next the properties of the matrices $\hat{W}_1$ and $\hat{W}_2$ as given in Equations (A.2) and (A.3), we obtain

$$\begin{aligned}
\sum_{\alpha,\beta}(\alpha\,|\,\hat{\tilde{P}}_k(s)\,|\,\beta)p_\beta &= \sum_{\alpha,\ \beta}(\alpha\,|\,\hat{P}'\,|\,\beta)p_\beta + \sum_{\alpha,\alpha'}(\alpha\,|\,\hat{P}'\,|\,\alpha')[w_1 + (\nu - w_1)\delta_{\alpha'N}\,] \\
&+ \sum_{\beta,\beta'}(\beta'\,|\,\hat{\tilde{P}}_k(s)\,|\,\beta)p_\beta,
\end{aligned}$$

and therefore,

$$\begin{aligned}
&\sum_{\alpha,\beta}(\alpha\,|\,\hat{\tilde{P}}_k(s)\,|\,\beta)p_\beta \\
&= \left\{\left[1 - \left[w_1\sum_{\alpha,\alpha'}(\alpha\,|\,\hat{\tilde{P}}'\,|\,\alpha') + (\nu - w_1)\sum_{\alpha}(\alpha\,|\,\hat{P}'\,|\,N)\right]\right]\right\}^{-1}\sum_{\alpha,\beta}(\alpha\,|\,\hat{P}\,|\,\beta)p_\beta
\end{aligned} \tag{A.11}$$

Thus the matrix of $\hat{P}$ suffices for the computation of Equation (8.107). Using again the identity in Equation (A.7), we may write $\hat{P}'$ as

$$\hat{P}' = \hat{P}^0 + \hat{P}'[\lambda - w_1)\ \hat{W}_3]\ \hat{P}^0, \tag{A.12}$$

where [cf. Equation (A.9)],

$$\hat{P}^0 \equiv \left(s - \hat{W}_k^0\right)^{-1} \tag{A.13}$$

Our aim now is to calculate the terms involving $\hat{P}'$ in Equation (A.11) in terms of the matrix elements of $\hat{P}^0$. Note that

$$\sum_{\alpha}(\alpha|\hat{P}^0|N) = \sum_{\alpha}(\alpha|\hat{P}^0|N) + \sum_{\alpha,\alpha'}(\alpha|\hat{P}'|\alpha')\ (\lambda - w_1)\ (N|\hat{P}^0|N), \tag{A.14}$$

where we used Equations (A.4) and (A.12). Next,

$$\sum_{\alpha,\alpha'}(\alpha|\hat{P}'|\alpha') = \sum_{\alpha,\alpha'}(\alpha|\hat{P}^0|\alpha') + \sum_{\alpha,\beta,\alpha'}(\alpha|\hat{P}'|\beta)\ (\lambda - w_1)(N|\hat{P}^0|\alpha')$$

and hence,

$$\sum_{\alpha,\alpha'}(\alpha|\hat{P}'|\alpha') = \left[1 - \sum_{\beta}(\lambda - w_1)\ (N|\hat{P}^0|\beta)\right] - 1 \sum_{\alpha,\alpha'}(\alpha|\hat{P}^0|\alpha'). \tag{A.15}$$

This result may now be inserted in Equation (A.14). Finally,

$$\sum_{\alpha,\beta}(\alpha|\hat{P}'|\beta)p_\beta = \sum_{\alpha,\beta}(\alpha|\hat{P}^0|\beta)p_\beta + \sum_{\alpha,\alpha'}(\alpha|\hat{P}'|\alpha')\ (\lambda - w_1) \times \sum_{\beta}(N|\hat{P}^0|\beta)p_\beta, \tag{A.16}$$

where the matrix of $\hat{P}'$ is given in Equation (A.15). Finally $\hat{P}^0$ is given by Equation (A.13) where the matrix $\hat{W}_k^0$ is enumerated in Equation (A.5). Using the latter, we evaluate below the required terms involving $\hat{P}^0$. First,

$$(N|\hat{P}^0|N) = (s - \gamma)^{-1} \tag{A.17}$$

Second,

$$\sum_{\alpha}(N|\hat{P}_0|\alpha) = \sum_{\alpha}(\alpha|\hat{P}_0|N) = (N|\hat{P}_0|N) = (s - \gamma)^{-1}. \tag{A.18}$$

Third,

$$\sum_{\alpha,\alpha'} (\alpha|\hat{P}_0|\alpha') = \bar{G}^0. \tag{A.19}$$

and

$$\sum_{\alpha,\beta} (\alpha|\hat{P}_0|\beta)p\beta = <G^0>. \tag{A.20}$$

where $\bar{G}^0$ and $<G^0>$ have been defined in Equation (8.12). In computing the left sides of Equations (A.19) and (A.20), we used the fact that $\hat{W}_k^0$ can be split into block matrices [cf. Equation (A.5)], the largest of dimension 2 × 2 only. Collecting all the terms together, we obtain [cf. Equation (A.15)]:

$$\sum_{\alpha,\alpha'} (\alpha|\hat{P}'|\alpha') = \left[1 - \frac{(\lambda - w_1)}{s-\gamma}\right]^{-1} \bar{G}^0 = \frac{(s-\gamma)}{s+\nu+6\lambda}\bar{G}^0]. \tag{A.21}$$

Equation (A.14) then yields

$$\sum_{\alpha} (\alpha'|\hat{P}'|N) = (s-\gamma)^{-1} \left[1 + \frac{(\lambda - w_1)\ (s-\gamma)}{(s+\nu+6\lambda)}\ \bar{G}^0\right]. \tag{A.22}$$

Finally, from Equation (A.16),

$$\sum_{\alpha} (\alpha'|\hat{P}'|\beta)p_\beta) = <G^0> + \frac{(\lambda - w_1)}{(s+\nu+6\lambda)}\bar{G}^0_{P_N}, \tag{A.23}$$

where we employed Equation (A.21). Substituting Equations (A.21) through (A.23) in Equation (A.11), we arrive at

$$\sum_{\alpha,\beta} (\alpha|\hat{P}_k(s)|\beta)p_\beta = \left(\left[1 - \frac{\nu - w_1}{s-\gamma} - \frac{w_1(s-\gamma)+\nu - w_1)(\lambda - w_1)}{s+\nu+6\lambda}\bar{G}^0\right]^{-1}\right.$$

$$\left. \times\left[<G^0> + p_N\frac{\lambda - w_1}{s+\nu+6\lambda}\bar{G}^0\right]\right). \tag{A.24}$$

Combining Equation (A.24) with Equation (8.107) yields Equation (8.111) as shown in the text.

References

Chaturvedi, S. 1983. In *Stochastic Processes: Formalism and Applications*, Vol. 184. Berlin: Springer.

Chudley, C.T. and R.J. Elliot. 1961. Neutron scattering from a liquid on a jump diffusion model. *Proc. Phys. Soc. Lond.* 77, 353.

Darwin, G.C. 1930. The diamagnetism of the free electron. *Proc. Cambridge Philos. Soc.* 27, 86.

Dattagupta, S. 1975. Effect of nuclear motion on Mössbauer relaxation spectra. *Phys. Rev.* B12, 47–57.

Dattagupta, S. 1987. *Relaxation Phenomena in Condensed Matter Physics*. New York: Academic Press.

Dattagupta, S. and K. Schroeder. 1987. Mössbauer spectrum for diffusing atoms including fluctuating hyperfine interactions. *Phys. Rev.* B35, 1525.

Dattagupta, S. and J. Singh. 1997. Landau diamagnetism in a dissipative and confined system. *Phys. Rev. Lett.* 79, 961.

Dattagupta, S. and J. Singh. 1996. Stochastic motion of charged particle in a magnetic field II. Quantum Brownian treatment. *Pramana* 47, 211.

Dattagupta, S. and L. A. Turski. 1985. Boltzmann–Lorentz model of collisional broadening of spectra. *Phys. Rev.* A32, 1439.

Dicke, R. H. 1953. The effect of collisions upon the Doppler width of spectral lines. *Phys. Rev.* 89, 472.

Ghosh, R., S. Dattagupta, and J. Singh. 2001. Magneto-optic drift of ions. *Phys. Rev.* A64, 063403.

Singwi, K. S. and A. Sjoelander. 1960. Resonance absorption of nuclear gamma rays and the dynamics of atomic motions. *Phys. Rev.* 120, 1093.

Vogl, G., W. Mansel, and P.H. Dederichs. 1976. Unusual dynamical properties of self interstitials trapped at co-impurities in Al. *Phys Rev. Lett.* 36, 1497.

9

Rotational Diffusion of Molecules

9.1 Introduction

A molecule is like a three-dimensional rigid body with its constituent atoms glued together by chemical bonds. The classical dynamic nature of a molecule is characterized by the translational motion of its center of mass and rotation around the center of mass, parameterized by the three Euler angles $\{\phi, \theta, \psi\}$ (Goldstein (1964). The rotation involves a symmetry axis or a set of symmetry axes. The symmetry axis of a linear molecule, e.g., CO, is the line joining the two atoms, in this case carbon and oxygen. For a spherical molecule, any direction in space can act as a symmetry axis.

When a molecule is part of a large medium, e.g., a fluid, its center of motion performs translational Brownian motion or diffusion—the subject of our detailed analysis including the spectroscopic observation of the structure factor. Concomitant with translational diffusion, the symmetry axes of a molecule perform rotational diffusion via small angular jumps that occur somewhat faster than translational diffusion. Thus the rotational contribution to the structure factor can be disentangled from the translational part, and can be studied experimentally by dielectric, infrared, neutron, ultrasonic, or Raman spectroscopy (Dattagupta 1987). Rotational diffusion is the subject of this chapter. We omit the even faster vibrational relaxation of the intramolecular normal modes that can also be investigated by spectroscopic tools.

The rotational diffusion equation is a straightforward rewrite of the angular portion of the three-dimensional generalization of Equation (5.13) that reads (Berne and Pecora 1976, Favro 1960, Gordon 1968, Ivanov 1964, McClung 1969, 1977, Rothschild 1984, Steele 1972):

$$\frac{\partial}{\partial t} P(\Omega, t) = d\, \nabla_{\Omega}^{2} P(\Omega, t), \tag{9.1}$$

wherein we omitted the instances of anisotropic diffusion, in which case d is a tensor. For a spherical molecule of radius a, the rotational diffusion constant d is simply a scaled-up version of the Stokes formula, Equation (1.45):

$$d = \frac{1}{a^2} \frac{KT}{6\pi a\eta}, \tag{9.2}$$

in which η is the isotropic viscosity of the fluid through which the molecule wades. The scale-up occurs through the factor $^{-2}$ ($a << 1$) that justifies the time scale separation of the rotational and translational structure factors as noted in the paragraph above. The solution of Equation (9.1) is most conveniently written in terms of Wigner (rotational) matrices (Favro 1960):

$$(\Omega\,|\,\hat{P}(t)\,|\,\Omega_0) = \sum_{i,m,n} \frac{(2l+1)}{8\pi^2} D_{mn}^{(l)*}(\Omega_0) D_{mn}^{(l)} \exp^{[-dl(l+1)t]}. \tag{9.3}$$

The left side of Equation (9.3) measures the conditional probability that the orientation of the molecular symmetry axis is Ω given than it was Ω_0 at time $t = 0$, captured by the initial condition:

$$(\Omega\,|\,\hat{P}(t=0)\,|\,\Omega_0) = \delta(\Omega - \Omega_0) \tag{9.4}$$

For most molecules only two Eulerian angles, the co-latitude θ and the azimuthal angle ϕ, suffice for describing the initial dynamics, and hence Equation (9.3) is replaced by

$$(\Omega\,|\,\hat{P}(t)\,|\,\Omega_0) = (\theta,\phi\,|\,\hat{P}(t)\,|\,\theta_0,\phi_0) = \sum_{l=0}^{\infty} \sum_{m=-l}^{+l} Y_{lm}^{*}(\theta_0,\phi_0) Y_{lm}(\theta\ \ \phi) e^{-dl(l+1)t}. \tag{9.5}$$

Correspondingly, Equation (9.4) reads

$$(\Omega\,|\,\hat{P}(t)\,|\,\Omega_0) = \delta(\cos\theta - \cos\theta_0)\ \ \delta(\phi - \phi_0). \tag{9.6}$$

For the sake of completeness, we write out the rotational diffusion Equation (9.1) by expanding the Laplacian in spherical polar coordinates:

$$\frac{\partial}{\partial t} P(\theta,\phi;t) = d\left\{\frac{1}{\sin^2\theta}\left[\sin\theta\frac{\partial}{\partial\theta}(\sin\theta)\frac{\partial}{\partial\theta} + \frac{\partial^2}{\partial\phi^2}\right]\right\} P(\theta,\phi\ \ ;t).$$

A further simplification ensues for a linear molecule such as CO, the orientation of which can be designated by one angle, i.e., the co-latitude θ, in which case

$$\frac{\partial}{\partial t} P(\theta,t) = d\frac{1}{\sin\theta}\ \frac{\partial}{\partial\theta}\left(\sin\theta\frac{\partial P(\theta,t)}{\partial\theta}\right). \tag{9.7}$$

The solution of Equation (9.7) is evidently

$$(\theta\,|\,\hat{P}(t)\,|\,\theta_0) = \sum_{l=0}^{\infty} P_l\ (\cos\theta)\ P_l\ (\cos\theta_0)\ e^{-dl(l+1)t}, \tag{9.8}$$

with

$$(\theta\,|\,\hat{P}(0)\,|\,\theta_0) = \delta(\cos\theta - \cos\theta_0), \tag{9.9}$$

P_l is the Legendre polynomial of order l. The fact that the isotropy of rotational diffusion is built in can be checked easily from Equation (9.5) by going to the $t \to \infty$ limit. In that case only the $l = 0$ term survives in Equation (9.5) and hence

$$\lim_{t\to\infty} (\theta,\phi\,|\,\hat{P}(t)\,|\,\theta_0,\phi_0) = \left[Y_{00}(\theta,\phi\right]^2 = \frac{1}{4\pi}. \tag{9.10}$$

The factor $(4\pi)^{-1}$ is indeed the a priori probability of any of the possible orientations to occur among a random sampling of isotropic rotation. Based on this background to an elementary discourse on rotational diffusion, we now focus on the quantities of physical interest, one of which is the rotational correlation function defined by

$$C^{(l)}(t) = 4\pi \sum_{m=-l}^{+l} < Y_{lm}(\Omega_0)\ \ Y^*_{lm}(\Omega(t)) >. \tag{9.11}$$

The rudimentary expansion of the angular brackets in Equation (9.11), indicating an average over the underlying diffusion process, is

$$C^{(l)}(t) = 4\pi \sum_{m=-l}^{+l} \int p(\Omega_0)\ \ Y^*_{lm}(\Omega)\ (\Omega\,|\,\hat{P}(t)\,|\,\Omega_0)\ \ Y_{lm}(\Omega_0)\,d\Omega_0, \tag{9.12}$$

$p(\Omega_0)$ is the a priori probability for the orientation to be Ω_0 which, for isotropic diffusion, is [cf. Equation (9.10)]:

$$p(\Omega_0) - \frac{1}{4\pi}. \tag{9.13}$$

Substituting the solution from Equation (9.5) into Equation (9.12), yields

$$C^{(l)}(t) = \sum_{m=-l}^{l} \int d\Omega_0\ \ Y^*_{lm}(\Omega)\ Y_{lm}(\Omega_0) \sum_{l'm'} Y_{l'm'}(\Omega)\ \ Y^*_{l'm'}(\Omega_0)\ \ e^{-dl(l+1)t},$$

which simply reduces to

$$C^{(l)}(t) = e^{-dl(l+1)t}, \tag{9.14}$$

upon using the orthonormal property of spherical harmonics:

$$\int d\Omega \;\; Y_{lm}(\Omega) \;\; Y^{*}_{l'm'}(\Omega) = \delta_{ll'}\delta_{mm'}.$$

The rotational correlation is characterized by an exponential relaxation in time t and correspondingly yields Debye relaxation in the frequency ω space which is a Lorentzian centered around $\omega = 0$, with a width proportional to $dl(l + 1)$. The attribution of Debye's name to this specific type of relaxation is based on Debye's pioneering work (1945) on the dielectric spectroscopy of polar molecules subjected to oscillatory electric fields.

The exponential relaxation with a relaxation time $[dl(l+1)]^{-1}$ is, of course, a gross simplification of the situation. First, the starting Equation (9.1) ignores a possible drift term. More fundamentally, Equation (9.1) derived from Equation (5.13) is the high-friction limit of a Fokker-Planck equation in the phase space (Brown, 1959, 1963). Therefore, the above discussion neglects the inertial motion of a molecule described by the angular momentum $\vec{L}$ (analogous to linear momentum $\vec{P}$ for translational motion). The point is the molecule between rotational jumps is expected to freely rotate like a rigid rotor that must also find a role in the correlation function. We shall incorporate such inertial dynamics in the calculation of rotational correlation function with the aid of extended diffusion models first discussed by Gordon (1966) and later by Dattagupta and Sood (1979) among others.

9.2 Extended Diffusion Models

To maintain simplicity, we consider the example of a linear rotor such as the CO molecule that has only one symmetry mode (that could be infrared- or Raman-active). In that case, azimuthal symmetry implies that only the $m = 0$ term contributes to the correlation function in Equation (9.11). Further, choosing the initial angle at zero, without loss of generality, we can write from Equation (9.11):

$$C^{(l)}(t) = \langle P_l(\cos\theta(t))\rangle, \tag{9.15}$$

where $\cos\theta(t)$ is the scalar product of two unit vectors:

$$\cos\theta(t) = \hat{u}(o)\cdot\hat{u}(t), \tag{9.16}$$

where $\hat{u}(t)$ is the direction of the relevant symmetry axis of molecular vibrations. For infrared spectroscopy, the symmetry axis is the electric dipole moment, a tensor of rank 1, hence, only $l = 1$ is relevant, and

$$C^{(l)}(t) = <\cos(\theta(t))>. \tag{9.17}$$

On the other hand, for Raman spectroscopy, the relevant physical quantity is the polarizability, a tensor of rank 2 and hence the object of interest is

$$C^{(l)}(t) = \left\langle \frac{1}{2}(3\cos^2\theta(t) - 1) \right\rangle. \tag{9.18}$$

Note that irrespective of the value of l, the Legendre polynomial can be expressed as a specific component of the Wigner rotation matrix elements that in turn are specific projections of a rotation operator (Messiah 1965). Thus, generally,

$$C^{(l)}(t) = \left\langle D_{00}^{(l)}(t) \right\rangle = \left\langle < l, m = 0 \,|\, e^{i(\vec{L}\bullet\vec{\Omega}(t))} \,|\, l, m = 0 > \right\rangle, \tag{9.19}$$

where $\vec{L}$ is the ordinary angular momentum vector, the components of which are the generators of rotation in Euclidean space. For instance, for $l = 1$, $\vec{L}$ *is a* 3×3 matrix (that should not be confused with a quantum mechanical operator). In fact, Equation (9.19) bears an uncanny resemblance to the structure factor for translational motion as in Equation (8.2). Thus $\vec{L}$ now plays the role of the momentum $\vec{p}(= \hbar\vec{k})$ (of the photon field) and $\vec{\Omega}(t)$ is equivalent to the position $\vec{r}(t)$. Thus the possibility of inclusion of inertial motion is embedded in the averaged time development or the rotation operator:

$$\hat{R}(t) = \langle \exp\ i(\vec{L}\ .\ \vec{\Omega}(t)) \rangle. \tag{9.20}$$

The idea of extended diffusion models is to employ the CTRW picture. A molecule is envisaged to rotate freely until the rotations are interrupted by randomly occurring collisional events that are Poisson-distributed (Chapter 7). Hence the underlying stochastic process is viewed as stationary and Markovian. The analogy with the phase space translational dynamics can be pushed further by formally writing the Euler angle $\vec{\Omega}(t)$ as the time integral of the angular velocity vector $\vec{\omega}(t)$:

$$\vec{\Omega}(t) = \int_0^t dt'\ \vec{\omega}(t'). \tag{9.21}$$

Hence, an alternative form of Equation (9.19) is

$$\hat{R}(t) = \left\langle \exp\, i \int_{o}^{t} dt'(\vec{L} \bullet \vec{\omega}(t')) \right\rangle. \tag{9.22}$$

9.3 M Diffusion Model

The formulation of the M diffusion model is based on Figure 9.1. At time $t = 0$, the linear molecule aligned with the space-fixed z axis starts rotating in a plane normal to the XY plane indicated by the parallelogram in the figure (Gordon 1966). The torques due to the surrounding molecules on the linear rotor of interest are assumed to act only about the z axis, during collisions. Thus, each collision, which is assumed instantaneous as in the impact approximation of classical kinetic theory of gases, is viewed to cause a sudden, impulsive change in the plane of rotation, i.e., the azimuthal angle ϕ measured around the z axis. A grossly simplifying feature of the M diffusion model is that the components of the angular velocity along x and y directions are assumed to stay constant. Thus we may write for $\vec{\omega}(t)$

$$\vec{\omega}(t) = \hat{i}\omega_x + \hat{j}\ \omega_y + \hat{k}\sum_i \phi_i \delta(t - t_i). \tag{9.23}$$

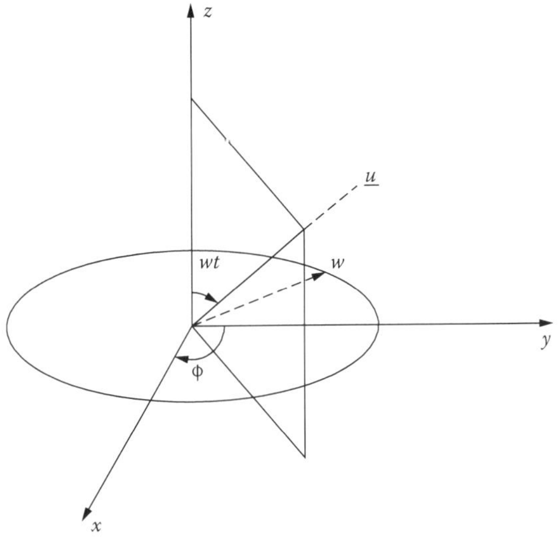

FIGURE 9.1
Geometry of rotation of a linear molecule (underscore indicates vector).

In the above, t_i represents the instants at which the Poisson process-driven collisions are taken to occur, ϕ_is designate the possible set of orientations of the rotational plane. The presence of the delta functions in time ensures that the collisions are instantaneous. Consequently,

$$\vec{L}\,.\;\vec{\omega}(t) = (L_x\omega_x + L_y\omega_y) + L_z \sum_i \phi_i\;\delta(t - t_i). \tag{9.24}$$

Eventually, of course, we will have to integrate over ω_x and ω_y in relation to the underlying Maxwellian distribution. We may now develop the same CTRW scheme developed earlier as follows.

9.3.1 No-Collision Term

$$C_0^{(l)}(t) = \int d\omega_x p(\omega_x)\; d\omega_y p(\omega_y)\; e^{-\lambda t} < l, m = 0 \,|\, e^{i(L_x\omega_x + L_y\omega_y)t} \,|\, l, m = 0 >, \tag{9.25}$$

where

$$p(\omega_i) = \left(\frac{\mathcal{G}}{2\pi KT}\right)^{1/2} \exp\,\left(-\frac{\mathcal{G}}{2KT}\omega_i^2\right),\quad i = x, y, \tag{9.26}$$

$\mathcal{G}$ is the moment of inertia of the linear molecule and λ is the mean rate of collision. Referring to Figure 9.1, it is evident that $\omega_x L_x + \omega_y L_y$ is the projection of the vector $\vec{L}$ in the direction of $\vec{\omega}$. in the xy plane and hence

$$e^{i\,(\omega_x L_x + \omega_y L_y)t} = e^{i(\pi/2 - \alpha)L_z}\;\; e^{iL_y\omega\;t} e^{-i(\pi/2-\alpha)L_z}, \tag{9.27}$$

where

$$\omega = \sqrt{\omega_x^2 + \omega_y^2}, \tag{9.28}$$

and

$$\alpha - \tan^{-1}(\omega_y / \omega_x). \tag{9.29}$$

Basically, the right side of Equation (9.27) effects a rotation of the y axis by an angle $(\pi/2 - \alpha)$ about the z axis (Figure 9.1). Hence,

$$\left\langle l, m = 0 \,|\, e^{i(\omega_x L_x + \omega_y L_y)t} \,|\, l, m = 0 \right\rangle = \; < l, m = 0 \,|\, e^{i\omega L_y t} \,|\, l, m = 0 > \; = g_0^{(l)}(t). \tag{9.30}$$

Substituting in Equation (9.25),

$$C_0^{(l)}(t) = \int_0^{\infty} d\omega\ p(\omega)\ e^{-\lambda t}\ \ g_0^{(l)}(t), \tag{9.31}$$

where

$$p(\omega) = \left(\frac{\mathcal{G}}{\pi KT}\right)\ \exp\left(-\frac{\mathcal{G}\omega^2}{2KT}\right)\ \ \omega\ d\omega. \tag{9.32}$$

9.3.2 One-Collision Term

Here, the rotor freely rotates in a hatched plane, away from the z axis, keeping the magnitude ω and the azimuthal angle ϕ fixed until time t_1. At time t_1, a collision instantaneously throws the angle ϕ to any of the possible values of ϕ within the domain 0 to 2π, described by an average (over ϕ) of a transition operator:

$$(\mathfrak{F})_{av} = \frac{1}{2\pi}\int_0^{2\pi} e^{iL_z\varphi}\,d\phi. \tag{9.33}$$

The rotor then freely rotates once again until time t. Thus the one-collision contribution to the correlation function can be written as

$$\begin{aligned} C_1^{(l)}(t) = \iint d\omega_x\, d\omega_y p\ \ (\omega_x)\ \ p(\omega_y)\int_0^t \lambda\, dt_1\ \ e^{-\lambda(t-t_1)} \\ < l, m = 0 \,|\, e^{i(\omega_x L_x + \omega_y L_y)(t-t_1)} . (\mathfrak{F})_{av}\ \ e^{-\lambda t_1} e^{i(\omega_x L_x + \omega_y L_y)t_1} l, m = 0 > . \end{aligned} \tag{9.34}$$

Now, from Equation (9.33)

$$< l\ m\ \ |(\hat{\mathfrak{F}})_{av}\,|lm' >= \delta_{mm'}\frac{1}{2\pi}\int_0^{2\pi} d\varphi\ e^{lm\varphi} = \delta_{mm'}\delta_{mo}. \tag{9.35}$$

Therefore, Equation (9.34) yields

$$C_1^{(l)}(t) = \iint d\omega_x\, d\omega_y p(\omega_x) p(\omega_y)\cdot \int_0^t (\lambda\, dt_1)\ e^{-\lambda(t-t_1)}.\ g_0^{(l)}(t-t_1)\ \ e^{-\lambda t_1} g_0^{(l)}(t_1), \tag{9.36}$$

where $g_0^{(l)}(t-t_1)\ e^{-\lambda t_1} g_0^{(l)}(t_1)$ is defined in Equation (9.31). A further use of the prescription employed in Equation (9.27) leads to

$$C_1^{(l)}(t) = \int_0^\infty d\omega\ \ p(\omega) \int_0^t (\lambda dt_1)\ e^{-\lambda(t-t_1)}\ \ g_0^{(l)}(t-t_1)\cdot e^{-\lambda t_1} g_0^{(l)}(t_1).$$

The CTRW scheme is familiar one. It was employed earlier in the strong collision and Boltzmann–Lorentz models. We therefore write successive collision terms as convolution of the no-collision term weighted by exponential damping of the Poisson–Markov process. The Laplace transform of the correlation function is then

$$\tilde{C}^{(l)}(s) = \int d\omega\ \ p(\omega)\ \left[\tilde{g}_0^{(l)}(s) + \lambda\ \ (\tilde{g}_0^{(l)}(s))^2 + \lambda^2\ \ (\tilde{g}_0^{(l)}(s))^3 + \ldots\right]$$

$$= \int_0^\infty d\omega\ \ p(\omega)\cdot \frac{\tilde{g}_0^{(l)}(s)}{1-\lambda\ \tilde{g}_0^{(l)}(s)}, \tag{9.37}$$

where, from Equation (9.30),

$$\tilde{g}_0^{(l)}(s) = \int_0^\infty dt\ \ e^{-(s+\lambda)t}. < l, m=0 \,|\, e^{i\omega L_y t}\ |\, l, m=0 >$$

$$= < l, m=0 \,|\, [(s+\lambda) - i\omega L_y)^{-1}] l, m=0 >. \tag{9.38}$$

Because the magnitude of $\vec{\omega}$ is not altered by collisions, the average over ω hangs outside the propagator as in the Boltzmann–Lorentz model (cf. Section 8.2.3). The situation is different in the J diffusion model discussed below. The Wigner rotation matrix $\exp(i\omega L_y t)$ (depicting rotation about the y axis by an angle ωt) is a standard matrix, the specific element of which can be easily read from tables, thus yielding a closed-form expression for $\tilde{g}_0^{(l)}(s)$. We shall return to the computation of infrared and Raman line shapes later.

9.4 J Diffusion Model

Recall that for a linear molecule the set $\{\omega, \phi\}$ completely specifies the motion; ω is the magnitude of the angular velocity and ϕ the angle that defines the projection of $\vec{\omega}$ in the plane normal to the anisotropy axis of intermolecular torques. In the M diffusion model, only ϕ is assumed to be randomized in collisions whereas ω is considered unchanged. This somewhat artificial premise is sought to be corrected in the J diffusion model in which ω is also assumed to be randomized in a collision (Gordon 1966).

The J diffusion model is similar in spirit to the strong collision model of collision broadening in that the post-collision value of ω has no relation to the pre-collision value. Therefore, detailed balance requires that

$$(\omega \,|\, \hat{\mathfrak{I}} \,|\, \omega_0) = p(\omega). \tag{9.39}$$

We may now set up the successive collision terms as follows:

9.4.1 No-Collision Term

The contribution from no collision in time 0 to t is identical to that in the M diffusion model of Equation (9.31):

$$\tilde{C}_0^{(l)}(t) = \int_0^\infty d\omega \;\; p(\omega) \;\; e^{-\lambda t} g_0^{(l)}(\omega, t), \tag{9.40}$$

where $g_0^{(l)}(\omega, t)$ is given by Equation (9.30).

9.4.2 One-Collision Term

We assume that at time $t = 0$, the angular velocity has a magnitude ω_0 and an azimuthal angle ϕ. The a priori probability of that is $\frac{1}{2\pi} p(\omega)$. At time t_1 the collision completely randomizes ϕ and throws the angular speed to ω with a probability $p(\omega)$. The system then freely rotates until time t without further collisional changes. Hence,

$$C_1^{(l)}(t) = \iint d\omega_o p(\omega_0) \; d\omega \;\; p(\omega) \int_0^t \lambda \;\; dt_1 e^{-\lambda(t-t_1)} g_0^{(l)}(\omega_1 t - t_1) \bullet e^{-\lambda t_1} g_0^{(l)}(\omega_0 t - t_1). \tag{9.41}$$

The Laplace transform of the above is given by

$$\tilde{C}_1^{(l)}(s) = \int d\omega_0 \;\; p(\omega_0) \;\; \tilde{g}_0^{(l)}(\omega_0, s) \cdot \lambda \int d\omega \;\; p(\omega) \;\; \tilde{g}_0^{(l)}(\omega \;\, , s) = \lambda \left[\tilde{C}_0^{(l)}(s) \right]^2. \tag{9.42}$$

The pattern is familiar by now and therefore, summing up all the contributions, we find

$$\tilde{C}^{(l)}(s) = \frac{\tilde{C}_0^{(l)}(s)}{1 - \lambda \; \tilde{C}_0^{(l)}(s)}, \tag{9.43}$$

where [cf. Equation (9.41)]:

$$\tilde{C}_0^{(l)}(s) = \int_0^\infty d\omega \ p(\omega) < l, m = 0 | \frac{1}{s + \lambda - i\omega L_y} | l, m = 0 > . \tag{9.44}$$

9.5 Interpolation Model

In our analysis of jump diffusion models in Chapter 6, we considered an interpolation model for the collision (or transition) operator that in the present context interpolates well between the M and J diffusion results (Dattagupta and Sood 1979). Thus, Equation (9.39) is modified as (cf. Section 6.7):

$$(\omega | \hat{\mathfrak{J}} | \omega_0) = r \ p(\omega) + (1 - r) \ \delta(\omega - \omega_0), \quad 0 \le r \le 1, \tag{9.45}$$

where the interpolation parameter r equals 1 for J diffusion and zero for M diffusion. Although the interpolation is linear in the sense of Equation (9.45), the composite correlation function does not lead to the complex line shape patterns as shown below. We construct in the following collision terms on the basis of the generic CTRW methodology.

9.5.1 No-Collision Term

There is no difference between the present case and the M and J diffusion results since no collision has taken effect. Thus, $C_0^{(l)}(t)$ is as given in Equation (9.40) which, in the Laplace-transformed space, is given by

$$\tilde{C}_0^{(l)}(s) = \int_0^\infty d\omega \ p(\omega) \ \tilde{g}_0^{(l)}(s), \tag{9.46}$$

where $\tilde{g}_0^{(l)}(s)$ is expressed in Equation (9.38). Following the CTRW idea, the successive collision terms are discussed below.

9.5.2 One-Collision Term

Comparing with Equation (9.42) and recognizing the critical presence of the collision operator $\hat{\mathfrak{J}}$ at every collision, we have

$$C_1^{(l)}(s) = \iint d\omega_0 \ p(\omega_0) \ \tilde{g}_0^{(l)}(\omega_0, s) \cdot \lambda(\omega | \hat{\mathfrak{J}} | \omega_0) . \tilde{g}_0^{(l)}(\omega_0, s), \tag{9.47}$$

where the matrix element $(\omega|\hat{\mathfrak{J}}|\omega_0)$ measures the (instantaneous) probability of a collision-induced jump from the angular speedy ω_0 to ω. It is convenient at this stage to introduce an operator $\hat{\Omega}$ whose matrix elements are

$$(\omega|\hat{\Omega}|\omega_0) = \omega_0 \int (\omega - \omega_0). \tag{9.48}$$

Consequently,

$$(\omega|\hat{G}_0^{(l)}(\hat{\Omega}, s)|\omega_0) = \tilde{g}_0^{(l)}(\omega_0, s)\ \delta(\omega - \omega_0), \tag{9.49}$$

$$\hat{G}_0^{(l)}\ (\hat{\Omega}, s) = < l, m = 0|(s + \lambda - i\Omega L_y)^{-1}|l, m = 0 > \tag{9.50}$$

Hence, the one-collision term in Equation (9.47) can be written in a compact matrix form as

$$\tilde{C}_l(s) = \iint d\omega_0 d\omega\ \ p(\omega_0)\ (\omega|\hat{G}_0 \cdot \lambda\ \ \hat{\mathfrak{J}} \cdot G_0|\omega_0) \tag{9.51}$$

where the arguments of $\hat{G}_0$ have been suppressed for the sake of brevity of notation. It is evident that Equation (9.51) yields Equation (9.47) if we take note of the closure relation:

$$\int d\omega\ \ |\omega)\ (\omega| = 1. \tag{9.52}$$

9.5.3 Two-Collision Term

The two- (and higher) collision terms are self-evident. For two collisions, we have

$$\tilde{C}_l(s) = \iint d\omega_0 d\omega\ \ p(\omega_0)\ (\omega|\hat{G}_0.\lambda\ \ \hat{\mathfrak{J}} \cdot \hat{G}_0 \hat{\mathfrak{J}} \cdot \hat{G}_0|\omega_0). \tag{9.53}$$

Collecting all the collision terms and summing a geometric progression we find

$$\tilde{C}_l(s) = \iint d\omega_0 d\omega\ \ p(\omega_0)\left(\omega|\hat{G}_0.\frac{1}{1 - \lambda\hat{\mathfrak{J}}\hat{G}_0}|\omega_0\right). \tag{9.54}$$

Substituting for the form of the collision operator $\hat{\mathfrak{I}}$ appropriate to the interpolation model [cf. Equation (9.45)], we rewrite Equation (9.54) as

$$\tilde{C}_l(s) = \iint d\omega_0 d\omega \; p(\omega_0) \left(\omega \left| \frac{1}{\hat{G}_0^{-1} - \lambda(1-r) - (r\hat{\mathfrak{I}}_1)} \right| \omega_0 \right), \tag{9.55}$$

where

$$\Upsilon = r\lambda, \tag{9.56}$$

and

$$(\omega \,|\, \hat{\mathfrak{I}}_1 \,|\, \omega_0) = p(\omega) \tag{9.57}$$

In order to further simplify Equation (9.55), we introduce the notation

$$\hat{G}^{-1} = \hat{G}_0^{-1} - \lambda(1-r). \tag{9.58}$$

Thus,

$$\tilde{C}_l(s) = \iint d\omega_0 d\omega \; p(\omega_0) \left(\omega \left| \hat{G} \cdot \frac{1}{1-\gamma \;\; \hat{\mathfrak{I}}_1 \hat{G}} \right| \omega_0 \right). \tag{9.59}$$

Denoting

$$\hat{U} = \hat{G} \frac{1}{1-\gamma \;\; \hat{\mathfrak{I}}_1 \hat{G}}, \tag{9.60}$$

we note that $\hat{U}$ satisfies the integral equation

$$\hat{U} = \hat{G} + \gamma \hat{G} \; \hat{\mathfrak{I}}_1 \; \hat{U}. \tag{9.61}$$

Therefore,

$$\begin{aligned}
\tilde{C}_l &= \iint d\omega_0 d\omega \; p(\omega_0) \; (\omega \,|\, \hat{U} \,|\, \omega_0) \\
&= \iint d\omega_0 d\omega \; p(\omega_0) \; (\omega \,|\, \hat{G} \,|\, \omega_0) \\
&\quad + \gamma \iint d\omega_0 d\omega \; p(\omega_0) \iint \; (\omega \,|\, \hat{G} \,|\, \omega_1)(\omega_1 \,|\, \hat{\mathfrak{I}} \,|\, \omega_2 \; (\omega_2 \,|\, \hat{U} \,|\, \omega_0),
\end{aligned}$$

where we employed Equation (9.52). Using Equation (9.57), the right side can be further manipulated:

$$\tilde{C}_l(s) = <\hat{U}> = <\hat{G}> + \gamma <\hat{G}><\hat{U}>, \tag{9.62}$$

where

$$\langle \ldots \rangle \equiv \iint d\omega_0 \; d\omega \; p(\omega_0) \quad (\omega|\ldots|\omega_0). \tag{9.63}$$

Hence,

$$\tilde{C}_l(s) = <\hat{U}> = \frac{<\hat{G}>}{1-\gamma<\hat{G}>}. \tag{9.64}$$

Clearly [cf. Equations (9.58) and (9.48)],

$$\langle \hat{G} \rangle = \int d\omega \; p(\omega) \frac{\tilde{g}_0^{(l)}(\omega, s)}{1-\lambda(1-r)\tilde{g}_0^{(l)}(\omega, s)}. \tag{9.65}$$

It is instructive to check how the M and J diffusion results ensue from Equation (9.65). In the limit $r = 0, \gamma$ vanishes, and from Equation (9.64):

$$\tilde{C}_l(s) = <\hat{G}> = \int d\omega \quad p(\omega) \frac{\tilde{g}_0^{(l)}(\omega, s)}{1-\lambda\tilde{g}_0^{(l)}(\omega, s)}, \tag{9.66}$$

the M diffusion expression of Equation (9.35). On the other hand, for $r = 0, \gamma$ equals λ, and the denominator within the integrand of Equation (9.65) reduces to unity. Thus

$$\tilde{C}_l(s) = \frac{\int d\omega \; p(\omega) \; \tilde{g}_0^{(l)}(\omega, s)}{1-\lambda \int d\omega \; p(\omega)\tilde{g}_0^{(l)}(\omega, s)}. \tag{9.67}$$

This result is in conformity with the J diffusion result of Equation (9.43).

9.6 Applications to Infrared and Raman Rotational Spectroscopy

As mentioned earlier, the infrared case corresponds to l = 1. Hence, from Equation (9.38), it is a matter of simply reading out the relevant matrix element of the $l = 1$ rotational operator, which is now a 3×3 matrix. The answer is (Dattagupta 1987):

$$\tilde{g}_0^{(l)}(\omega, s) = \int_0^\infty dt\ e^{-(s+\lambda)t} < l, m = 0 | e^{i\omega L_y t} | l, m = 0 >$$

$$= \int_0^\infty dt\ e^{-(s+\lambda)t}\ \cos(\omega t) = \frac{1}{(s+\lambda) + \dfrac{\omega^2}{(s+\lambda)}}\ . \tag{9.68}$$

In that case

$$< \hat{G} > = \int d\omega\ p(\omega)\ \frac{(s+\lambda)}{(s+\lambda)\ (s+\gamma) + \omega^2}, \tag{9.69}$$

which, when plugged into Equation (9.64), yields the infrared line shape. The latter is plotted in Figure 9.2 by taking the real part of $\tilde{C}_1(s)$, for $\lambda = 0.3$ and for three values of $r : r = 0$ (M diffusion), $r = 1$ (J diffusion), and $r = 0.5$ (interpolation). Figure 9.3 shows the line shape for $\lambda = 2.5$ for the same three values of r. The selected values of the collision rate λ correspond respectively to gas-phase-like rotation and the contrasting case of highly hindered rotation.

It is also interesting to work out the line shape in the time domain by finding the inverse Laplace transform of $\tilde{C}_1(s)$. The time–space versions of Figure 9.2 and Figure 9.3 are shown in Figure 9.4 and Figure 9.5, respectively. The fact that the correlation function does not decay to zero asymptotically for $r = 0$ as the time goes to infinity is because of the restrictive nature of the M diffusion model that does not allow the angular speed to change at collisions.

One other point is worth remarking. Since the dipole correlation is the statistical average of the linear power of the cosine of an angle, the correlation function dips below zero, especially for slow collisions. This property is obliterated when the collisions are rapid and do not let the angle to deviate far from its initial value of zero. A final point to note is that for short times, when collisions have not taken hold, the M, J, and interpolation results are similar, just as they are for small values of the collision rate at all times.

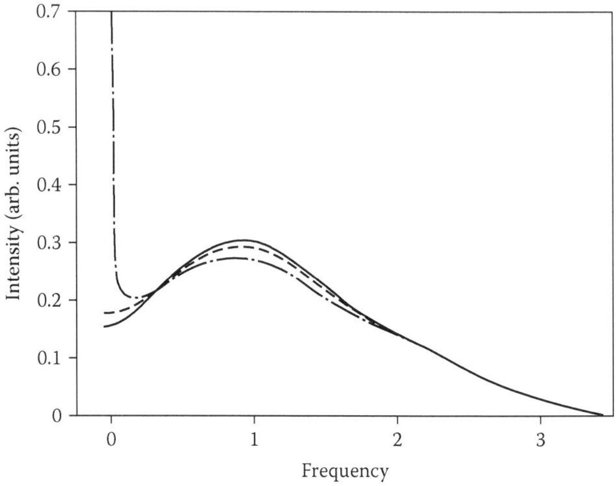

FIGURE 9.2
Infrared line shapes for $\lambda = 0.3$ and interpolation parameter $r = 0.0$ (M diffusion) (– · –), 0.5 (– – –), and 1.0 (J diffusion) (—). Frequency is in reduced units of $(K_BT/\mathfrak{I})^{1/2}$; $\mathfrak{I}$ is moment of inertia.

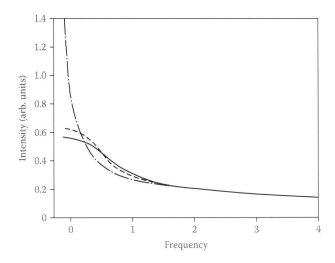

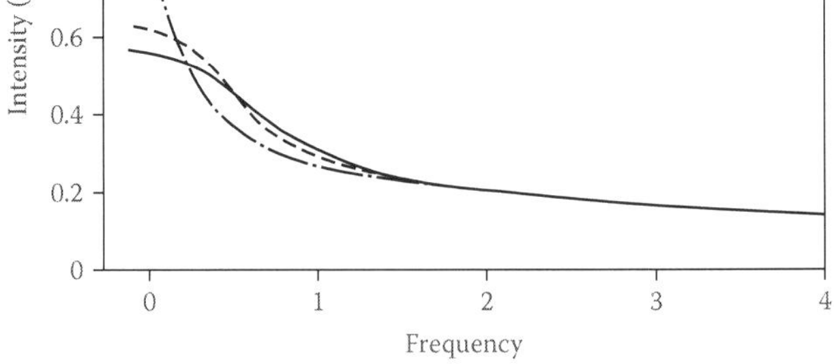

FIGURE 9.3
Infrared line shapes for $\lambda = 2.5$ and $r = 0.0$ (M diffusion) (– · –), 0.5 (– – –), and 1.0 (J diffusion) (—). Frequency is in reduced units of $(K_BT/\mathfrak{I})^{1/2}$.

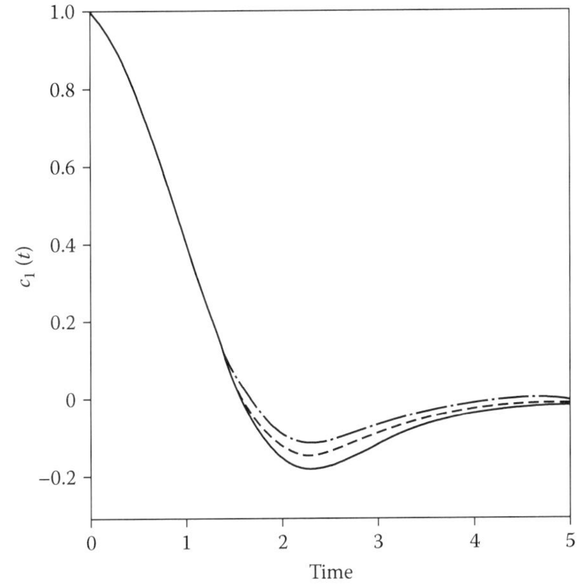

FIGURE 9.4
Dipole correlation function for same values of parameters as in Figure 9.2.

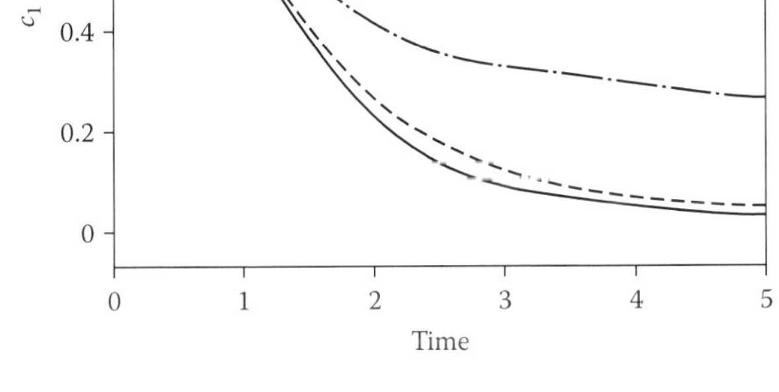

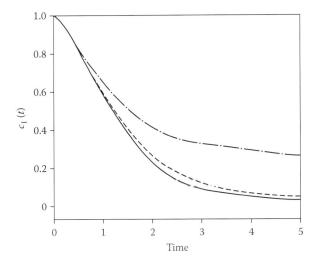

FIGURE 9.5
Dipole correlation function for same values of parameters as in Figure 9.3.

Turning now to the computation of the Raman line shape, we need the matrix elements of the $l = 2$ rotation operator (a 5 × 5 matrix) that yield

$$\tilde{g}_0^{(z)}(\omega, s) = \frac{(s+\lambda)+\omega^2/(s+\lambda)}{(s+\lambda)^2+4\omega^2} . \tag{9.70}$$

From Equation (9.68)

$$<\hat{G}> = \int d\omega \;\; p(\omega)\frac{(s+\lambda)^2+\omega^2}{(s+\lambda)^2(s+\gamma)+\omega^2(4s+3\lambda+\gamma)}. \tag{9.71}$$

Using the above expression, the Raman line shape is given by Equation (9.64). The line profiles are shown in Figure 9.6 and Figure 9.7, respectively for λ = 1.25 and λ = 2.5 for the same r values as in the infrared cases of Figures 9.4 and 9.5. It is evident that the difference between the M, J, and interpolation models becomes more discernible for larger values of λ in the Raman case than in infrared spectra. However, the feature that the J diffusion results lie closer to those of the interpolation model for $r = 0.5$ although 50% of the collisions are now M-like is common for Raman and infrared spectra. This is proof of our earlier assertion that the line shape for the interpolation model is not just a linear interpolation for J and M diffusion line shapes.

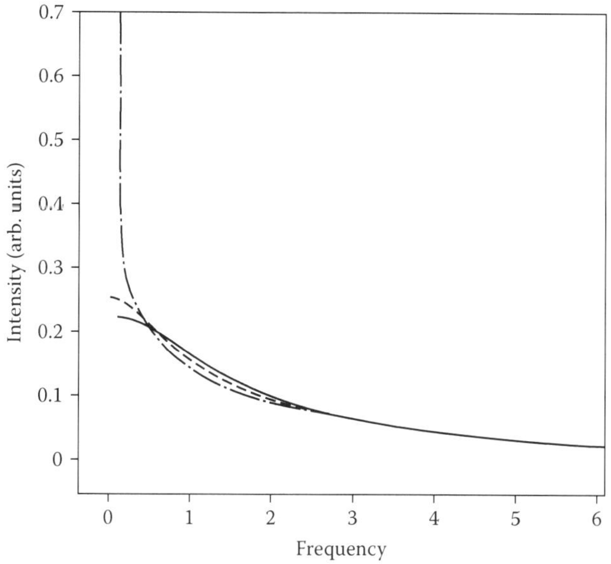

FIGURE 9.6
Raman line shapes for $\lambda = 1.25$ and $r = 0.0$ (M diffusion) (– · –), 0.5 (– – –), and 1.0 (J diffusion) (—). Frequency is in reduced units of $(K_BT/\mathfrak{I})^{1/2}$.

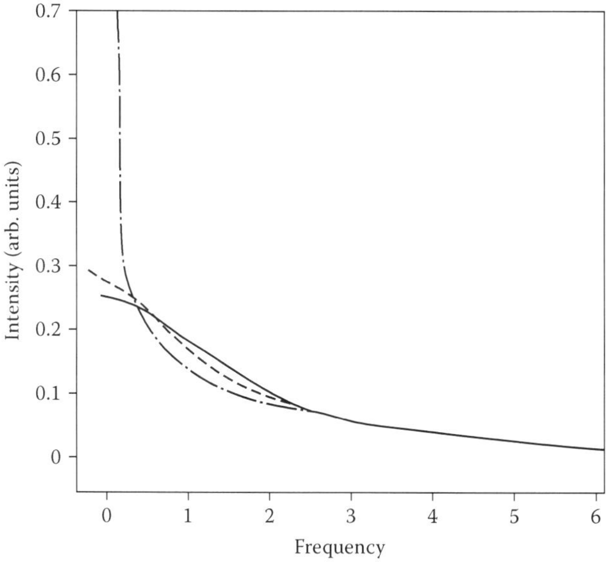

FIGURE 9.7
Raman line shapes for $\lambda = 2.5$ and $r = 0.0$ (M diffusion) (– · –), 0.5 (– – –), and 1.0 (J diffusion) (—). Frequency is in reduced units of $(K_B T/\mathfrak{I})^{1/2}$.

References

Berne, B.J. and R. Pecora. 1976. *Dynamic Light Scattering*. New York: John Wiley & Sons.

Brown Jr., W. F. 1963. Thermal fluctuations of magnetic nanoparticles. *Phys. Rev.* 130, 1677.

Brown, Jr. W.F. 1959. *J. Appl. Phys.* 30, 130.

Dattagupta, S. 1987. *Relaxation Phenomena in Condensed Matter Physics*. New York: Academic Press.

Dattagupta, S. and A.K. Sood. 1979. *Pramana J. Phys.* 13, 423.

Debye, P. 1945. *Polar Molecules*. New York: Dover.

Favro, L. D. 1960. Theory of the rotational Brownian motion of a free rigid body. *Phys. Rev.* 119(1), 53.

Goldstein, H. 1964. *Classical Mechanics*. Reading, MA: Addison Wesley.

Gordon, R. G. 1968. Correlation functions for molecular motion. *Adv. Magn. Reson*. 3, 1.

Gordon, R.G. 1966. *J. Chem. Phys*. 44, 1830.

Ivanov, E.N. 1963. Zh, Eksp. Teor. Fig. 45, 1509 [1964. *Sov. Phys. J*.18, 1041.]

McClung, R. E. D. 1977. Relaxation and interaction processes. *Adv. Mol. Relax. Proc.* 10, 88.

McClung, R.E.D. 1969. *J. Chem. Phys*. 51, 3842.

Messiah, A. 1965. *Quantum Mechanics*, Vol. II. Amsterdam: North-Holland, chap. 13.

Rothschild, W.G. 1984. *Dynamics of Molecular Liquids*. New York: Wiley (Interscience).

Steele, W. A. 1976. Infrared and Raman spectra. *Adv. Chem. Phys*. 34, 1.

10

Order Parameter Diffusion

10.1 Cahn–Hilliard Equation

We may recall from our treatment of the Einstein model of Brownian motion that the concentration $C(\vec{r},t)$ of say, sugar molecules, as solutes in a solvent like water follows the diffusion equation.

$$\frac{d}{dt}C(\vec{r},t) = D\nabla^2 C(\vec{r},t), \tag{10.1}$$

where $C(\vec{r},t)$ is the number of sugar molecules per unit volume within a region $\vec{r}$ and $\vec{r}+d\vec{r}$ and D is the diffusion constant. We now raise the question: How is Equation (10.1) modified when we have to deal with diffusion of not just a single species (e.g., sugar particles) but two different molecular species such as oil and water that interact? One important example of such interactions is the situation in which water molecules like each other and oil molecules like each other while water and oil dislike each other.

When the temperature of the oil–water mixture is significantly high, the attractive interaction between like species is overpowered by entropic fluctuations so that we have a homogeneous mixture. However, at temperatures below what is called the critical or consolute point T_c, interaction wins over entropy and phase separation occurs. The situation is depicted in Figure 10.1. Above the miscibility gap, we see a single phase of oil–water mixture. Below T_c, we have two coexisting phases of oil-rich and water-rich regions. The question then is how does the concentration difference evolve over time:

$$C(\vec{r},t) = C_A(\vec{r},t) - C_B(\vec{r},t),$$

where C_A is the concentration of A (oil), and C_B is the concentration of B (water)? Because the intrinsic number of individual A and B molecules does not change over time, the quantity $C(\vec{r},t)$ is a conserved parameter. The conservation property is then necessarily linked with an equation of continuity:

$$\frac{d}{dt}C(\vec{r},t) = -\nabla \bullet \vec{j}(r,t). \tag{10.2}$$

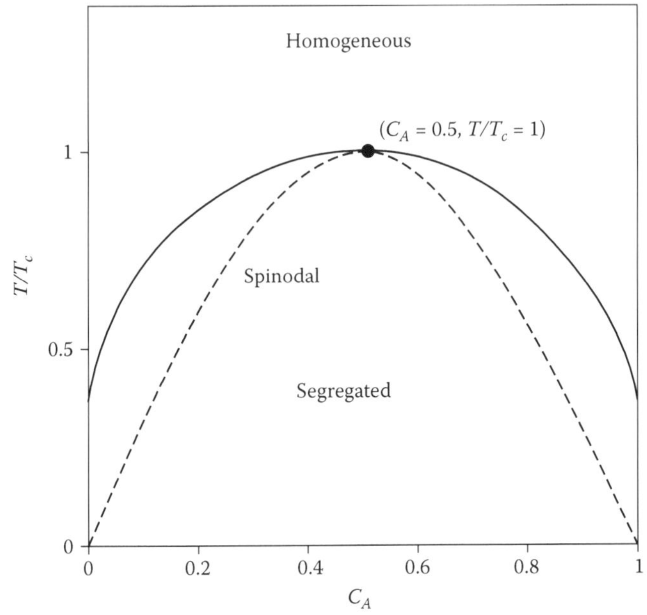

FIGURE 10.1
Phase diagram of binary mixture in $(C_A, T/T_c)$ plane. $C_A = 0.5, T/T_c = 1$ is a consolute point. The solid line represents the coexistence curve, below which the system segregates into A- and B-rich regions. The dashed line represents the spinodal.

where $\vec{j}(\vec{r},t)$ is the current density associated with the random motion or diffusion of one species past the other.

The phenomenon of phase separation described above is an example of a transition in which a system enters from one phase to another such that macroscopic properties change drastically under the influence of change of temperature or pressure or some external factor. A magnet is a common example of a phase transition when the system moves from a paramagnetic phase in which the macroscopic magnetization is zero into a ferromagnetic phase in which the magnetization $\vec{M}(\vec{r},t)$ acquires a nonzero value spontaneously, that is, in the absence of an external magnetic field. Consequently, magnetization is called an order parameter which is zero in the disordered phase governed by the entropy and nonzero in the ordered phase dominated by the interaction.

Such phase transitions are described conveniently by the Landau theory based on free energy and written as an expansion in terms of the order parameter—in this case magnetization. The crucial ingredient of this free energy expansion is symmetry because of which the free energy must have only even powers of $\vec{M}(\vec{r},t)$. The reason is that macroscopic

$\vec{M}(\vec{r},t)$ is the thermodynamic average of a microscopic spin operator that changes sign under time reversal but the free energy must have time reversal invariance.

The concentration difference $C(\vec{r},t)$ depicting phase separation also qualifies as on order parameter but unlike magnetization is not dictated to by time reversal invariance. Thus the Landau expansion of the free energy can contain terms that are odd powers of the order parameter. Furthermore, because the kinetics of $M(\vec{r},t)$ at the microscopic level involves spontaneous spin flips, it is inherently nonconserved whereas concentration fluctuations are necessarily driven by conserved kinetics and hence the continuity equation (10.2).

While the current density $\vec{j}(\vec{r},t)$ is governed by the generalized Fick's law (Chaikin and Lubensky 2004):

$$\vec{j}(\vec{r},t) = -DC(\vec{r},t)\nabla\mu(\vec{r},t), \tag{10.3}$$

$\mu(\vec{r},t)$ is the chemical potential that triggers phase separation (like the external magnetic field in the example of magnetism) and now follows the thermodynamic relation:

$$\mu(\vec{r},t) = \frac{\delta F\{C(\vec{r},t)\}}{\delta C(\vec{r},t)} \tag{10.4}$$

designating a functional derivative of the free energy with respect to the concentration. Combining Equations (10.2), (10.3), and (10.4), we arrive at the generalized diffusion equation:

$$\frac{d}{dt}C(\vec{r},t) = D\nabla^2\left[\frac{\delta F\{C(\vec{r},t)\}}{\delta C(\vec{r},t)}\right]. \tag{10.5}$$

When does Equation (10.5) reduce to the ordinary diffusion equation (10.1)? Recall that for an ideal system the free energy is quadratic in C (the so-called Gaussian model) and Equation (10.5) yields Equation (10.1). Generally, however, the free energy structure allows for additional nonlinear terms [in $C(\vec{r})$] arising from the interaction in the generalized Cahn–Hilliard (CH) diffusion equation found in the literature on phase separation, leading to rich structures as solutions, as discussed in the next section (Cahn and Hilliard 1958, 1959).

Before embarking on a detailed discussion of Equation (10.5), the point that $C(\vec{r},t)$ is a conserved order parameter becomes clear if we examine its Fourier transform:

$$C(\vec{k},t) = \int d\vec{r} e^{i\vec{k}.\vec{r}} C(\vec{r},t). \tag{10.6}$$

The presence of the Laplacian in front of the right side of Equation (10.5) immediately implies that

$$\dot{C}(\vec{k},t) = Dk^2 \int d\vec{r} e^{i\vec{k}.\vec{r}} \frac{\delta F\{C(\vec{r},t)\}}{\delta\{C(r,t)\}}, \tag{10.7}$$

and therefore,

$$\dot{C}(\vec{k}=0,t) = \int d\vec{r} C(\vec{r},t), \tag{10.8}$$

vanishes identically, indicating that the volume integral of the order parameter density remains a constant. The second point of looking beyond Equation (10.5) emerges when we add a noise term:

$$\frac{\partial}{\partial t} C(\vec{r},t) = \vec{\nabla} \bullet \left[D\vec{\nabla} \frac{\delta\ F\{C\}}{\delta C(\vec{r},t)} + \eta(\vec{r},t) \right], \tag{10.9}$$

where the vector noise $\vec{\eta}$ satisfies the usual fluctuation–desperation relation:

$$\overline{\vec{\eta}(\vec{r},t)} = 0$$

$$\overline{\eta_i(\vec{r}\,',t')\ \eta_j(\vec{r}\,'',t'')} = 2DK_BT\ \delta_{ij}\ \delta(\vec{r}\,'-\vec{r}\,'')\ \delta(t'-t''). \tag{10.10}$$

Note that the noise is inserted inside the square parentheses so that the conservation property of the order parameter is maintained even in the presence of noise.

With the Landau free energy form incorporated in Equation (10.9), the latter acquires the structure of a generalized Langevin equation endowed with nonlinear and stochastic properties. It also falls under the classification of model B of order parameter kinetics (Hohenberg and Halperin 1977). The generalized diffusion equation (10.9) has a variety of applications to condensed phases, pattern formations, biological morphogenesis, and so on (Dattagupta and Puri 2004).

One remark about the underlying structure of Equation (10.9) is in order. As discussed in Chapter 4, the Langevin equation is usually reserved for stochastic equations of motion of dynamical variables but the concentration $C(\vec{r},t)$ is not a dynamical variable. In contrast, the diffusion or CH equation (10.5) in the present instance is akin to a Fokker–Planck (FP) equation for probability sans the drift term (see Chapter 5).

While the concentration $C(\vec{r},t)$ in a coarse-grained sense is indeed like the probability of finding sugar-rich regions in a given volume and therefore must obey what we earlier called physical diffusion (Chapter 1), it also follows stochastic diffusion in the sense of random walks of sugar particles in a discretized version of the continuum. Therefore the CH equation is an important extension of Einstein's idea of Brownian motion. On the other hand, because the CH equation relies on thermodynamics, it is based on coarse graining of microscopic variables leading to a macroscopic phenomenological theory.

The addition of the noise term $\eta(\vec{r},t)$ to Equation (10.9) is simply an admission of the inherent timescale separation that leaves certain fluctuations arising from the heat bath unaveraged. This is precisely the stratagem under which Langevin equations are formed by adding noise to systematic equations, such as describing the damped motion of a harmonic oscillator or resistive Ohmic flow of electric currents in a metallic wire. Conversely, as we will show in the next section, thermal noise is essentially irrelevant in describing long-term structures based on Equation (10.9).

10.2 Pattern Formation

As mentioned above, the CH equation applies to the phase separation of an AB mixture in which A-A and B-B interactions are attractive $(E^{AA}, E^{BB} < 0)$ and A-B interaction is repulsive $(E^{AB} > 0)$. With reference to the phase diagram in Figure 10.1, we consider the dynamical evolution resulting from a quench from above the coexistence curve (homogeneous or disordered phase) to below the coexistence curve (segregated or ordered phase) (Dattagupta and Puri 2004).

Phase separation is the adopted phrase to describe how the initially homogeneous phase separates into regions of A-rich and B-rich phases. Experimentally, one distinguishes between shallow quenches (a bit below the coexistence curve) and deep quenches (far below the coexistence curve). For the former, in the region spanned by the solid coexistence line and the dashed spinodal line in Figure (10.1), the homogeneous system decomposes by nucleation and growth of droplets (of the majority phase in the midst of the minority phase). On the other hand, for deep quenches in the region much below the spinodal lines, the homogeneous system is spontaneously unstable and decomposes into A-rich and B-rich regions by a process called spinodal decomposition. The latter phenomenon is extremely important in the metallurgy of AB alloys (Cahn 1961, 1962, Cahn and Cook 1970, Langer et al. 1975).

To apply the generalized Langevin equation (9.9) to the phenomenon of pattern formation during the phase separation of a binary mixture, we write an expression for free energy *F*. Following Landau, the Helmholtz free energy is a function of the order parameter $C(\vec{r})$ (Chaikin and Lubensky 2004):

$$F\{C(\vec{r})\} = \int d\vec{r}\left[f(C(\vec{r}) + \frac{1}{2}\tilde{K}(\vec{\nabla}C(\vec{r}))^2 \right], \tag{10.11}$$

where f has the standard double-well structure:

$$f = -\frac{\bar{a}}{2}(T_c - T)C^2(\vec{r}) + \bar{b}C^4(\vec{r}), \tag{10.12}$$

$\bar{a}$ and $\bar{b}$ are constants $(\bar{a}, \bar{b} > 0), T_c$ is the critical temperature (also called the consolute point), and $\tilde{K}$ measures the surface tension due to inhomogeneities in the order parameter, and hence the energy of domain walls separating A-rich and B-rich phases.

Note that f is taken to be symmetric in C in that odd powers in C are ignored. While such symmetry considerations are not generic for binary liquid mixtures (unlike magnets), we adopt the simpler case wherein E^{AA} is taken to be equal to E^{BB} (see Section 10.3 below). We may then write Equation (10.9) as

$$\dot{C}(\vec{r},t) = \vec{\nabla} \circ \{D\vec{\nabla}[-\bar{a}(T_c - T)C(\vec{r}) + \bar{b}C^3(\vec{r}) - \tilde{K}\nabla^2 C(\vec{r})] + \eta(\vec{r},t)\}. \tag{10.13}$$

Recall that the Gaussian white noise satisfies Equation (10.9).

Rescaling the variables (for $T < T_c$) as

$$C' = \frac{C}{C_0,}, \quad C_0 = \sqrt{\frac{\bar{a}(T_c - T)}{\bar{b}}},$$

$$t' = \frac{D\bar{a}^2(T_c - T)^2}{\tilde{K}} t, \quad \vec{r}' = \sqrt{\frac{a(T_c - T)}{\tilde{K}}}\, \vec{r}$$

$$\vec{\eta}' = \frac{\sqrt{\bar{b}\tilde{K}}}{D\bar{a}^2(T_c - T)^2} \vec{\eta}, \tag{10.14}$$

and dropping the primes for the sake of brevity, we obtain the dimensionless equation:

$$\dot{C}(\vec{r},t) = \vec{\nabla} \circ \{\vec{\nabla}[-C + C^3 - \nabla^2 C] + \vec{\eta}(r,t)\}, \tag{10.15}$$

where

$$\overline{\vec{\eta}(r,t)} = 0,$$

$$\overline{\vec{\eta}_i(\vec{r}',t')\eta_j(\vec{r}'',t'')} = 2D_O\ \delta_{ij}\ \delta(\vec{r}'-\vec{r}'')\delta(t'-t''),$$

$$D_0 = \frac{K_B T \bar{b}[\bar{a}(T_c - T_1]^{(d-4)/2}}{\tilde{K}^{d/2}}, \tag{10.16}$$

d is the dimension of the underlying space. Figure 10.2 illustrates a typical evolution pattern obtained from simulating Equation (10.15) with a random initial condition. The composition of the mixture is taken as 50% A and 50% B, i.e., the average value of the order parameter is assumed zero. As mentioned earlier, thermal noise is found to be asymptotically immaterial and hence D_0 is set equal to zero.

The phase separating system depicted in Figure 10.2 is typified by a characteristic length scale $L(t) \sim t^{1/3}$ in $d \geq 2$ first derived by Lifshitz and Slyozov (LS; 1961). They were interested in the problem in which droplets of one component grow independently in a homogeneous background,

FIGURE 10.2
Simulation of Cahn–Hilliard equation (with $T < T_c$) from a disordered initial condition. A-rich phases are marked in black; B-rich phases are unmarked. Thermal noise is ignored. The Euler discretization mesh sizes are $\Delta t = 0.01$ and lattice size is 256^2. The label in each picture indicates dimensionless time after quench.

i.e., one of the components is present in a much smaller fraction than the other, corresponding to an off-critical quench. The same power law of growth applies also to spinodal decomposition wherein almost equal fractions of the two components are present, as demonstrated by Huse (1986). The Huse argument is that the chemical potential on the surface of a domain of size L is $\sim\sigma/L$, where σ is the surface tension. The concentration current is given by the diffusion constant D times the gradient of the chemical potential that scales as $D\sigma/L^2$. Hence, the domain growth is dictated by $dL/dt \sim D\sigma/L^2$ or $L(t) \sim (D\sigma t)^{1/3}$.

10.3 Jump Diffusion in Ising Model: Relation with CH Equation

We may recall from Chapter 1 how we managed by going to the continuum limit to derive the ordinary diffusion equation from a discrete random walk model. Although that treatment was given in one dimension, the generalization to higher dimensions is self-evident. The question that we ask in this section is whether the Cahn–Hilliard equation can be derived also from a discrete jump model of the random walk-type (Dattagupta and Puri 2004). It should of course be borne in mind that the CH equation deals with an interacting system characterized by a free energy that is nonlinear in the order parameter that is also capable of describing phase ordering. Hence, we must deal with a situation involving a multitude of random walkers that interact with each other.

The simplest random walk is a two-state jump or telegraph process in which the walker is constrained to jump only among two possible sites. Thus the right jump must necessarily follow a left jump and vice versa. The problem is completely isomorphic to that of a magnetic spin one half which can jump from up to down and down to up because of thermal fluctuations. Now we consider a situation involving not just a single spin but several of them, designated N that are in interaction with each other. The most familiar example of such a many-body spin system is described by an Ising Hamiltonian that can be written as (Huang 1987, Kawasaki 1966, 1972)

$$\mathcal{H} = -\sum_{<ij>} J_{ij} S_i S_j - \sum_{i=1}^{N} h_i S_i, \tag{10.17}$$

where each spin S_i (actually the z component of the spin vector) is a two-state variable that can take the values +1 or –1 depending on whether the spin points up or down (the direction determined by the quantization axis, say z). We also inserted, for sake of generality, an inhomogeneous (or site-dependent) magnetic field h_i and assumed nearest neighbor interactions. That explains the angular brackets below the summation sign.

It is interesting to note that the Ising spin problem is completely isomorphic to that of an AB mixture if the presence of an A atom at a site is viewed as the up state of the spin in which $S = 1$ and correspondingly the presence of a B atom is viewed as the down state. How does diffusion occur if we imagine a lattice for an AB alloy in which each site is occupied by either an A or B atom?

In reality, such a physical system is always endowed with defects such as vacancies. Therefore if an A atom finds a vacancy at the nearest neighbor site, it can jump into that site, leaving a hole into which a nearby B atom can jump. While the presence of a vacancy as a via medium is absolutely essential for an AB exchange, when the vacancy concentration is low, as is the case for a relatively undefected alloy system, the influence of the vacancy can be neglected and diffusion can be viewed as random AB exchanges. In spin language, such exchanges can be mediated by a flip-flop term in a Hamiltonian $\mathcal{H}_{ex}(t)$ that must be added to Equation (10.17):

$$\mathcal{H}_{ex}(t) = \sum_{<ij>} C_{ij}(t)(S_i^+ S_j^- + S_i^- S_j^+), \tag{10.18}$$

where S_i^- *flips* the spin at the ith site from up to down while S_i^+ *flops* the spin from down to up. $C_{ij}(t)$ represents random stochastic interactions arising from the heat bath. At this stage, we must clarify the meaning of the exchange term J_{ij} and the magnetic field h_i in the Ising model in the context of an AB alloy. Because $\frac{1}{2}(1 \pm < S_i >)$ is the probability that the ith site is occupied by an A(B) atom, the microscopic interaction can be expressed as

$$\mathcal{H}_{A-B} = \sum_{<ij>} \left\{ \frac{1}{2}(1+S_i)\, E_{ij}^{AA} \frac{1}{2}(1+S_j) + \frac{1}{2}(1-S_j) E_{ij}^{BB} \frac{1}{2}(1+S_0) \right.$$
$$\left. + \frac{1}{2}(1+S_i) E_{ij}^{AB} \frac{1}{2}(1-S_j) + \frac{1}{2}(1-S_i) E_{ij}^{AB} \frac{1}{2}(1+S_j) \right\}, \tag{10.19}$$

which reduces to Equation (10.18) if we identify

$$J_{ij} = \frac{1}{4}\left(2E_{ij}^{AB} - E_{ij}^{AA} - E_{ij}^{BB}\right), \tag{10.20}$$

$$h_i = \frac{1}{2}\left(E_{ij}^{BB} - E_{ij}^{AA}\right), \tag{10.21}$$

and ignore a constant spin-independent energy term. If $J_{ij} > 0$, which for the spin problem leads to a paramagnetic-to-ferromagnetic transition, is tantamount to saying that A likes A and B likes B much more than A likes B. Hence $J_{ij} > 0$ would yield phase separation or spinodal decomposition at temperatures below T_c, governed by the strength of the interaction. Incidentally, if we assume, for simplicity that $E_{ij}^{AA} = E_{ij}^{BB}$, then

$$J_{ij} = \frac{1}{2}\left(E_{ij}^{AB} - E_{ij}^{AA}\right), \tag{10.22}$$

and

$$h_i = 0. \tag{10.23}$$

Such a zero magnetic field Ising model has full-time reversal symmetry as can be easily verified by flipping both the S_i and S_j spins leaving the Hamiltonian invariant. This explains our remark in Section 10.2 concerning the free energy for the AB system in which we neglected odd powers in the order parameter.

We are now set to introduce time-dependent effects into the Ising Hamiltonian in Equation (10.17) with $h_i = 0$ to mimic jump exchanges of A and B atoms in a binary system. What we need is an N site generalization of the two-state (but single-site) jump process described by the Chapman–Kolmogorov master equation (see Chapters 2 and 6) because the underlying stochasticity is still governed by a stationary Markov process. Such a model is called the spin exchange Kawasaki model that follows the master equation (Kawasaki 1966, 1972):

$$\begin{aligned}\frac{d}{dt}P(\{m_i\},t) = &-\sum_{j=1}^{N}\sum_{k\in l_j} W(m_1,...m_j,m_k..m_N \mid m_1,..m_k,m_j..m_N)P(\{m_i\},t)\\ &+\sum_{j=1}^{N}\sum_{k\in l_j} W(m_1,...m_k,m_j..m_N \mid m_1,..m_j,m_k..m_N)P(\{m_i\}',t).\end{aligned} \tag{10.24}$$

In writing Equation (10.24), we tacitly recognize that the Ising model is essentially classical. It does not contain any non-commuting spin operators; hence, we may replace S_i by its eigenvalue. In Equation (10.24), the stochastic process is imagined to involve an exchange of m_j (or m_k) at site j with m_k (or m_j) at site $k \in L_j$ where L_j denotes a neighboring site of j. In the second term on the right side, $\{m_i\}'$ designates a new configuration that emerges form $\{m_i\}$ by interchanging $m_j \leftrightarrow m_k$.

As in the case of discrete jump processes (Chapter 6), the relaxation rate W can be prescribed from detailed balance considerations implying that if a transition occurs from the configuration $\{m_i\}$ to $\{m_i\}'$, the relaxation matrix must be proportional to the Boltzmann factor involving the energy difference between the two configurations. Thus:

$$W(m_1,...m_j,m_k,...m_N \mid m_1,..m_k,m_j..m_N)$$
$$=\frac{\lambda_K}{2}\left\{1-\tanh\left[\frac{\beta}{2}(m_j-m_k)\left(\sum_{n\in L_j'} J_{jn}m_n-\sum_{n\in L_k'} J_{kn}m_n\right)\right]\right\}, \tag{10.25}$$

where the subscript K on the jump rate lambda is in deference to Kawasaki. The primes on the summations on the right side of Equation (10.25) indicate constraints as $n \neq k$ (for $n \in L_j'$) and $n \neq j$ (*for* $n \in L_k'$). By exploiting the fact that $\frac{1}{2}(m_j - m_k)$ equals 0 or ±1, the factor $(m_j - m_k)$ from the argument of the tanh function can be brought forward as a prefactor and thus:

$$W(m_1,...m_j,m_k,...m_N \mid m_1,..m_k,m_j..m_N)$$
$$=\frac{\lambda_K}{2}\left\{1-\frac{(m_j-m_k)}{2}\tanh\left[\frac{\beta}{2}(m_j-m_k)\left(\sum_{n\in L_j'} J_{jn}m_n-\sum_{n\in L_k'} J_{kn}m_n\right)\right]\right\}. \tag{10.26}$$

We are now ready to write the equation of motion for the order parameter $< S_i(t) >$ which is obtained by multiplying Equation (10.24) by m_i and summing over all possible configurations, yielding after some algebra:

$$2\lambda_K^{-1}\frac{d}{dt}< S_i(t) > = -Z_0 < S_i > + \sum_{k\in L_i} < S_k >$$
$$+\sum_{k\in L_i}\left\langle (1-S_iS_k)\tanh\left[\beta\left(\sum_{n\in L_j'} J_{in}S_n-\sum_{n\in L_k'} J_{kn}S_n\right)\right]\right\rangle. \tag{10.27}$$

Here Z_0 denotes the number of nearest neighbors (coordination number) of a lattice site. A similar equation can also be written for correlation functions. Because our eventual goal is to derive the CH equation based on a Landau free energy, we must rewrite Equation (10.26) in the so-called mean field approximation which essentially amounts to replacing the spin operator by its average value, i.e., the order parameter itself. Thus,

$$2\lambda_K^{-1}\frac{d}{dt}C_i(t) = -Z_0C_i + \sum_{k\in L_i} C_k + \sum_{k\in L_i}(1-C_iC_k)\tanh\left[\beta\left(\sum_{n\in L_i'} J_{in}C_n-\sum_{n\in L_k} J_{kn}C_n\right)\right]. \tag{10.28}$$

We now employ suitable coarse graining by going to the continuum limit of Equation (10.28). To this end, we define the range R of the interaction as

$$R^2 = [J(0)]^{-1} \sum_{j=1}^{N} (\vec{r}_i - \vec{r}_j)^2 J_{ij}, \tag{10.29}$$

where

$$J(0) = \sum_{j=1}^{N} J_{ij}, \tag{10.30}$$

and $\vec{r}_i$ is the position vector of the site i. We may then expand as

$$\sum_{j=1}^{N} J_{ij} C_j(t) \simeq J(0)\{C(\vec{r}_i, t) + \frac{1}{2} R^2 \nabla_i^2 C(\vec{r}_i, t)] + \text{other terms}. \tag{10.31}$$

Additionally,

$$\tanh\left[\beta \sum_{j=1}^{N} J_{ij} C_j(t)\right] \simeq \beta \sum_{j=1}^{N} J_{ij} C_j(t) - \frac{1}{3}\left(\beta \sum_{j=1}^{N} J_{ij} C_j(t)\right)^3 + \cdots$$

$$\simeq \frac{T_C}{T} C(\vec{r}_i, t) - \frac{1}{3}\left(\frac{T_C}{T}\right)^3 C(\vec{r}_i, t)^3 + \frac{1}{2}\frac{T_C}{T} R^2 \nabla_i^2 C(\vec{r}_i, t) + \cdots, \tag{10.32}$$

where we hence employed the other expansion given in Equation (10.31) and defined T_c as

$$T_C = \frac{J(0)}{K_B}. \tag{10.33}$$

Next, Equation (10.28) can be simplified by using the identity:

$$\tanh(X - Y) = \frac{\tanh X - \tanh Y}{1 - \tanh X \tanh Y}. \tag{10.34}$$

Further, since we are interested in late-stage dynamics when a system is expected to be locally equilibrated, we may replace $\tanh \beta \Sigma_{n \in L_i'} J_{in} C_n$ by the local order parameter C_i as per mean field theory. Therefore we may approximately write

$$(1 - C_i C_k) \tanh\left[\beta\left(\sum_{n \in L_i} J_{in} C_n - \tan \beta \sum_{n \in L_k} J_{kn} C_n\right)\right]$$

$$= \tanh \beta \sum_{n \in L_i} J_{in} C_n - \tanh \beta \sum_{n \in L_k^z} J_{kn} C_n, \tag{10.35}$$

wherein we employed the identity in Equation (10.33). Hence Equation (10.28) can be rewritten as

$$2\lambda_K^{-1}\frac{d}{dt}C_i(t)=\sum_{k\in L_i}(C_k-C_i)-\sum_{k\in L_i}\left\{\tanh\left(\beta\sum_{n\in L_k}J_{kn}C_n\right)-\tanh\left(\beta\sum_{n\in L_i}J_{kn}C_n\right)\right\}$$

$$=\Delta_D\left\{C_i-\tanh\left(\beta\sum_{n\in L_i}J_{in}C_n\right)\right\}, \tag{10.36}$$

where Δ_D denotes the discrete Laplacian operator. Finally, using the Taylor expansion in Equation (10.32), we arrive at

$$2\lambda_K^{-1}\frac{d}{dt}C(\vec{r},t)=-a^2\nabla^2\left[\left(\frac{T_C}{T}-1\right)C-\frac{1}{3}\left(\frac{T_C}{T}\right)^3C^3+\frac{T_C}{2T}a^2\nabla^2C\right], \tag{10.37}$$

wherein we ignored higher order terms and denoted the lattice parameter by a. We may now identify (10.37) with the phenomenological CH equation (10.5) with the local free energy given in Equation (10.11) and the diffusion constant $D=\lambda_K\beta a^2/2$. Concluding, we find that the CH equation is a reasonable coarse-grained description of the microscopic Kawasaki kinetics of the Ising model, just as the ordinary diffusion equation emerges as the continuum version of the discrete random walk.

10.4 Reaction–Diffusion Models and Spatiotemporal Patterns

We may recall from the arguments in Equations (10.7) and (10.8) that the CH equation embodies conserved kinetics of an order parameter. At a microscopic level of contact between the CH equation and coarse-grained Kawasaki kinetics of an AB system, the conservation property is self-evident, as the underlying jump processes do not alter the total numbers of A and B atoms. However, the free energy in Equation (10.11) also refers to a magnet comprising N spin ½ atoms for which the concentration C (or difference between concentrations of A and B species) is simply magnetization M. The microscopic kinetics proceeds by random flips of single spins and hence M cannot be a constant of motion. The corresponding macroscopic equation for order parameter kinetics will thus not be endowed with the prefactor nabla symbol as in Equation (10.5), which is the hallmark of conserved kinetics, but can instead be written as

$$\frac{\partial}{\partial t} C(\vec{r},t) = -\Gamma \frac{\delta F\,[C]}{\delta C}, \tag{10.38}$$

Γ *is* a phenomenological constant. The logic behind Equation (10.38), attributed to Landau and Khalatnikov (1954), Chaikin and Lubensky (2004), and Hohenberg and Halperin (1977), is that thermal equilibrium ensues when the free energy is a minimum with respect to the variation of the order parameter consistent with Equation (10.38). Further, any deviation from equilibrium should be driven locally by the functional derivative of the free energy itself. If we add noise compatible with the fluctuation–dissipation relation to Equation (10.37) as we did in Equation (10.9), we are led to the time-dependent Ginzburg–Landau (1950) equation for order parameter kinetics. Combining Equation (10.38) with (10.11) and (10.12), we obtain

$$\frac{\partial C}{\partial t}(\vec{r},t) = \frac{1}{2} K\Gamma\ \nabla^2 C(\vec{r},t) + \Gamma\,\bar{a}(T_C - T)\ C(\vec{r}) - 3\Gamma\bar{b}\ C^3(\vec{r}). \tag{10.39}$$

The first term on the right has a diffusive character, whereas the last two terms have connotations to *reaction* if we borrow terminology from chemical kinetics. The equations such as those in (10.39) are designated reaction–diffusion equations. In general, they are not restricted to describing the kinetics of any single species and may be extended to multiple interacting species. The resultant reaction–diffusion model will therefore involve coupled differential equations that are usually nonlinear, leading to a variety of applications to population dynamics, predator–prey models, oscillating chemical reactions, and others.

We now discuss one application of reaction–diffusion models to bacterial colony growth arising from the addition of nutrients to a background medium. As the bacterium feeds on the nutrient, the nutrient concentration decreases and the bacteria thrive. However, if the nutrient level drastically falls, the bacteria die and the cycle continues. Denoting the nutrient concentration as $n(\vec{r},t)$ and the bacterium concentration as $b(\vec{r},t)$, the coupled equations can be modeled as (Roy et al. 2010):

$$\frac{\partial}{\partial t} n(\vec{r},t) = D_n \nabla^2 n(\vec{r},t) - f(n,b),$$

$$\frac{\partial}{\partial t} b(\vec{r},t) = D_b \nabla^2 b(\vec{r},t) + f(n,b), \tag{10.40}$$

where D_n and D_b are the respective diffusion coefficients. The reaction terms are designated by f that in general would be a nonlinear function of n and b. Note that if diffusion were absent, the total concentration $(n + b)$ would be a constant in time. The simplest situation is a reaction in which f is proportional to the product of n and b, in which case

$$\frac{\partial}{\partial t} b(\vec{r},t) = D_b \nabla^2 b(\vec{r},t) + n(\vec{r},t)\ b(\vec{r},t). \tag{10.41}$$

A similar equations holds for $n(\vec{r},t)$. The interpretation of Equation (10.41) is clear. The bacterium concentration increases over time as the bacterium finds nutrients. Hence the linear dependence on n (with a plus sign in front). At the outset, of course, some bacterium must be present to grow along with the linear dependence on b. The corresponding term in the equation for $n(\vec{r},t)$ uses a minus sign for the obvious reason that the nutrient is eaten by the bacterium.

Real life, however, is far more complex. We consider the rod-shaped and motile bacteria called *Bacillus thuringiensis* that are known to produce proteins that act as insecticides. These bacteria are also known for their therapeutic activities against hookworms and cancer. Hence the study of *B. thuringiensis* in various environments is important for discovering new applications to agricultural and health sciences. It is therefore of great interest to examine the bacterial growth and motility under the influence of glucose—the nutrient considered to be the ubiquitous energy source of nature.

We refer to the recent experiments and modeling of colony growth patterns of a *B. thuringiensis* strain in agar-based peptone nutrient media, in the absence and the presence of glucose. Experiments were carried out on a biotic surface in a semi-solid nutrient agar media contained in plates on which bacteria-forming units were inoculated. The colonies were permitted to grow at room temperature. The optical densities (at 600 nm) of the growing bacterial cultures were computed at various times and plotted as functions of growth time (see Figure 10.3). Our simulations were based on the two-dimensional version of Equation (10.40):

$$\frac{\partial}{\partial t}\begin{pmatrix} n & (x,y;t) \\ b & (x,y;t) \end{pmatrix} = \left(\frac{\partial^2}{\partial x^2} + \frac{\partial^2}{\partial y^2}\right)\begin{pmatrix} D_n & n(x,y;t) \\ D_b & b(x,y;t) \end{pmatrix} + f(n,b)\begin{pmatrix} -1 \\ 1 \end{pmatrix} \tag{10.42}$$

It was observed experimentally that the point at which the bacterium concentration b exceeds a threshold value $b_{th,} f(n, b)$ is well-approximated by the product form depicted in the second term on the right of Equation (10.41). However, below the threshold value, nonlinear terms in b dominate. We therefore model $f(n, b)$ in a computationally facile form as

$$f(n,b) = nb\left[1 - \exp\left(-\, {}^{b}\!/_{b_{th}}\right)\right]. \tag{10.43}$$

A convenient number for b_{th} is the initial bacterium concentration b $(t = 0)$. Thus, at time $t = 0$, the term within the square parentheses in Equation (10.43) acquires the value $(1 - e^{-1})$. However, as the bacterium concentration increases beyond b_{th} [$= b(t = 0)$], nonlinear terms in b begin to dominate $f(n, b)$.

On the other hand, when b far exceeds b_{th}, the exponential becomes vanishingly small and we get back to Equation (10.41). For intermediate values of b, rich spatiotemporal structures emerge, as discussed below.

To simulate the reaction–diffusion equations in (10.42), we employ a finite element method that is easy to implement for the circular geometry of the experimental Petri dish and for visualizing cell growth through coarse graining over many bacteria. The system is divided into 500 triangular grids (for x and y values between −40 and 40).

Initially, each triangle center is assigned a nutrient concentration $n(t = 0) = 4$, a bacterium concentration $b(t = 0) = 1$, and the inside circle of radius $r_c = 0$ with a nutrient diffusion constant $D_n = 1$ in suitable units. The bacterium diffusion constant D_b varied among values to obtain best match with experiment (Figures 10.4, 10.5, 10.6, and 10.7). Finally, we integrate $b\ (x, y; t)$ and $n(x, y; t)$ over the circular dish and plot them as functions of growth time (Figure 10.8) to verify that bacterial growth saturates faster when D_b is larger.

As in models of diffusion-limited aggregation (DLA), *B. Thuringiensis* colonies grow in a fractal manner that may be parameterized by the Hausdorff dimension defined by (Mandelbrot 1977):

$$D_H = \underset{\delta \to 0}{Lim}\left[\frac{\log N(\delta)}{\log(1/\delta)}\right], \tag{10.44}$$

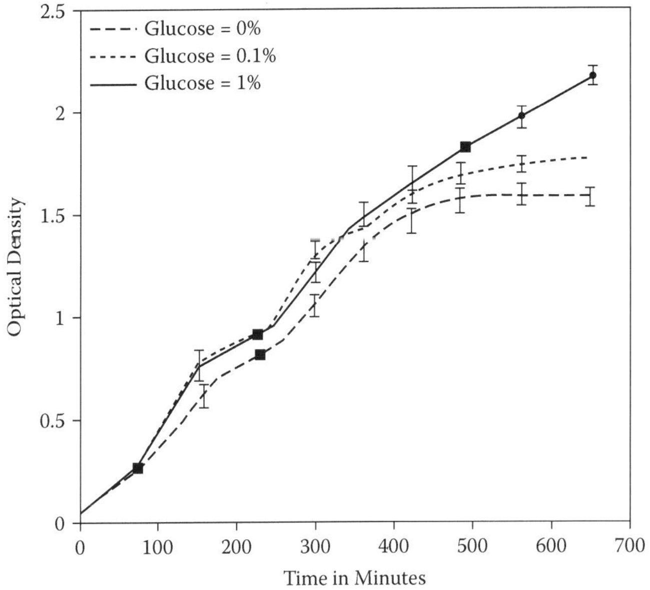

FIGURE 10.3
Optical density measurement (at 600 nm) of bacterium growth as a function of time in absence (––) and presence of 0.1% (…) and 1% (––) glucose.

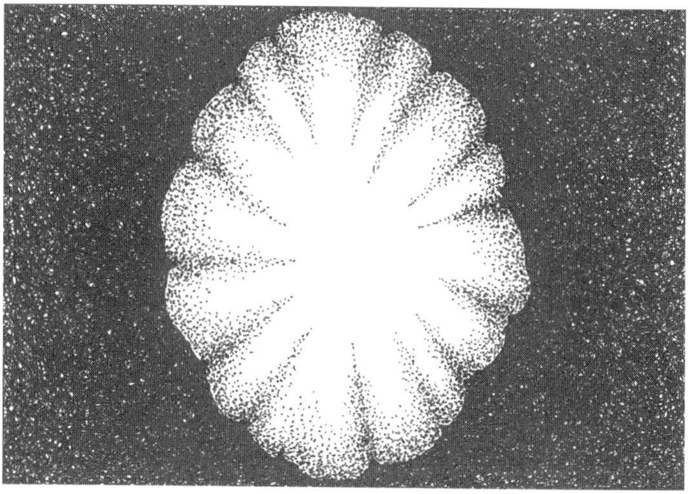

FIGURE 10.4
Original experimental images of a bacterial colony.

where δ is the length scale and N is the number of boxes necessary to cover the fractal geometry in a standard box-counting method. Taking two images as benchmarks, one from experiments (Figure 10.9) and the other from simulations (Figure 10.10), we obtain closely matching $D_{H,\exp} = 1.1969$ and $D_{H,Sim} = 1.1965$. We find that reaction–diffusion equations are satisfactory model descriptions of bacterial pattern formation caused by glucose inhibition.

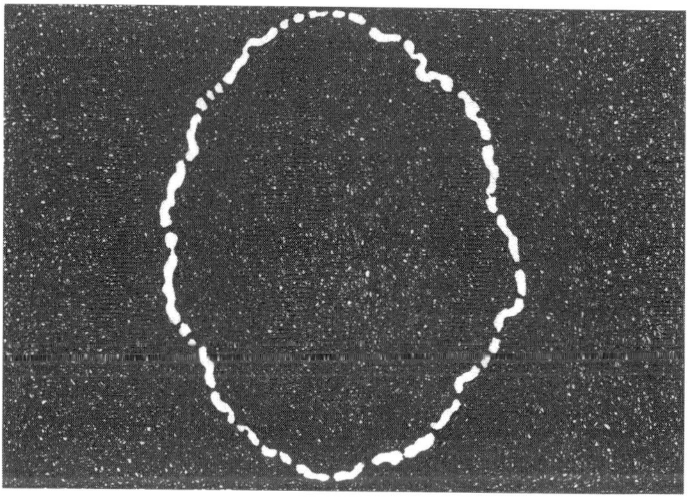

FIGURE 10.5
Edge detection (for calculating Hausdorff dimension) of experimental images.

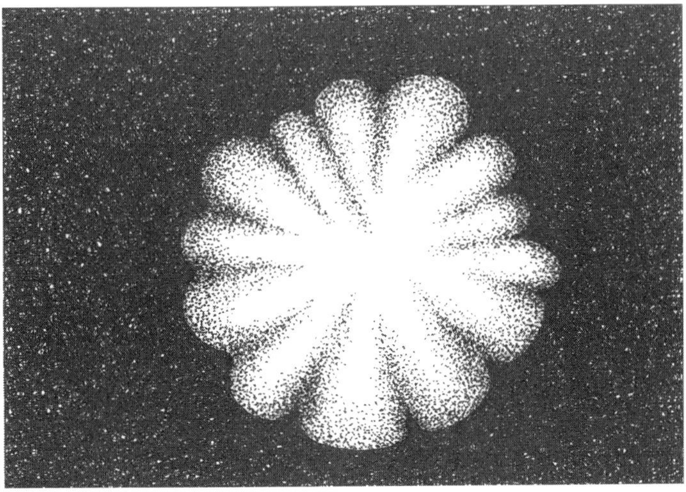

FIGURE 10.6
Simulated images of a bacterial colony.

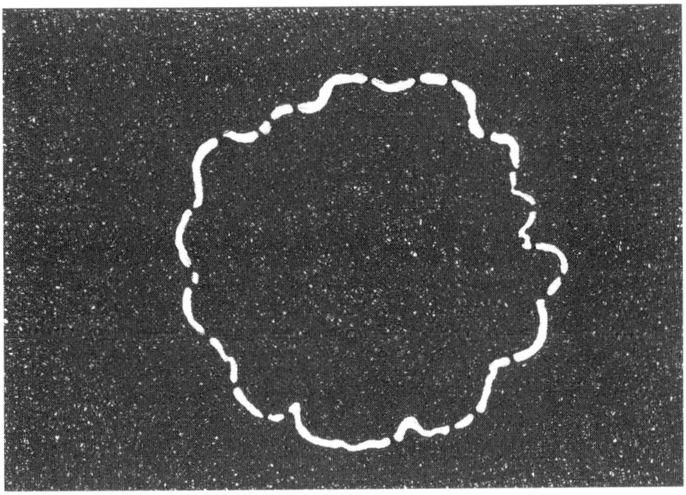

FIGURE 10.7
Edge detection of simulated images.

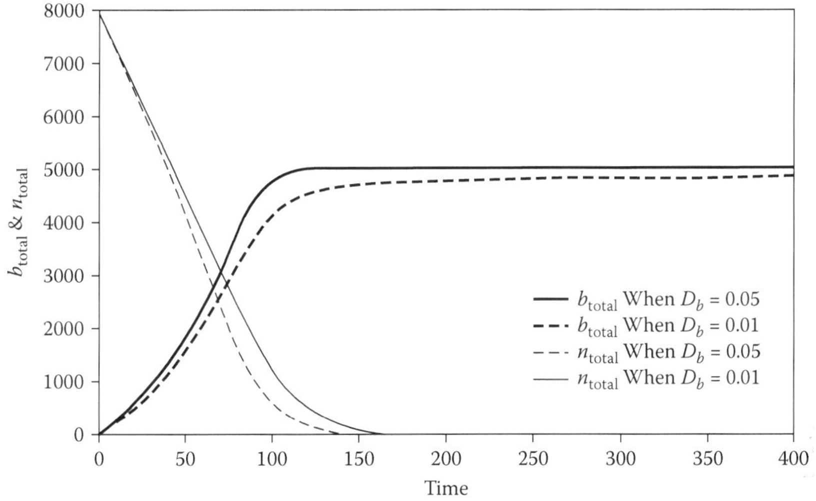

FIGURE 10.8
Total counts for cell and nutrient for 400 time steps and n (t = 0) = 4, D_n = 1, D_b = 0.01, and D_b = 0.05.

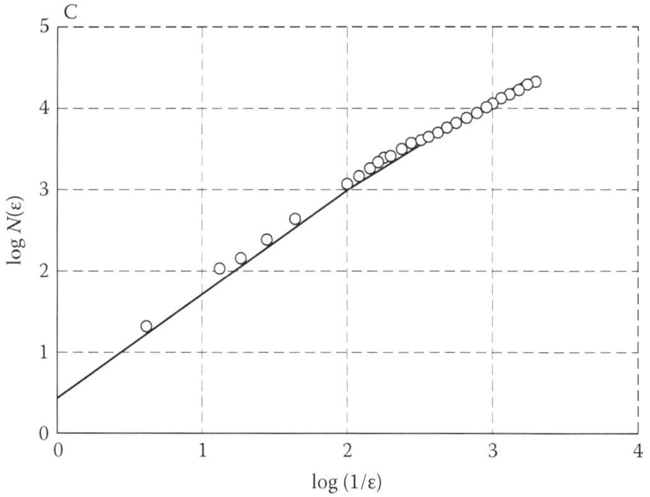

FIGURE 10.9
Plot of log $N(\in)$ versus log $(1/\in)$ and its linear fit to experimental data yielding Hausdorff dimension $D_{H,exp}$ = 1.1969.

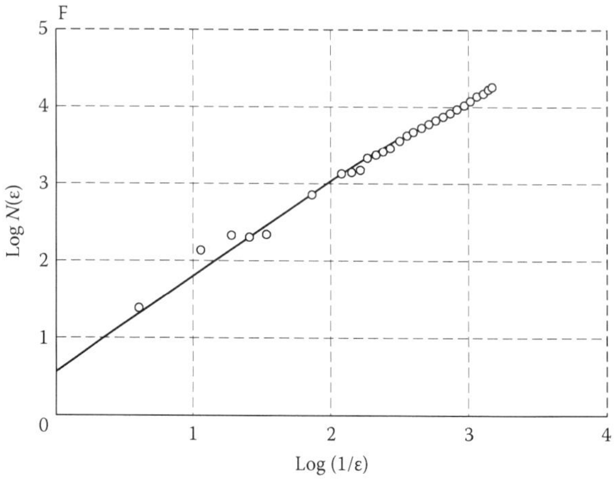

FIGURE 10.10
Plot of log $N(\in)$ versus log $(1/\in)$ and its linear fit to simulated data yielding Hausdorff dimension $D_{H,sim} = 1.1965$.

References

Cahn, J. W. 1961. On spinodal decomposition. *Acta Metall.* 9, 795.

Cahn, J. W. 1962. On spinodal decomposition in cubic crystals. *Acta Metall.* 10, 907.

Cahn, J.W. and J.E. Hilliard. 1959. *J. Chem. Phys.* 31, 668.

Chan, J. W. and J. E. Hilliard. 1958. Free energy of a nonuniform system. *J. Chem. Phys.* 28, 258.

Chaikin, P.M. and T.C. Lubensky. 2004. *Principles of Condensed Matter Physics.* Cambridge: Cambridge University Press.

Cook. H. E. 1970. Brownian motion in spinodal decomposition. *Acta Metall.* 18, 297.

Dattagupta, S. and S. Puri. 2004. *Dissipative Phenomena in Condensed Matter: Some Applications.* Heidelberg: Springer.

Ginzburg, V.L. and L.D. Landau. 1950. *Zh. Eksp. Teor, Fig. 20,* 1064.

Hohenberg, P. C. and B. I. Halperin. 1977. Theory of dynamic critical phenomena. *Rev. Mod. Phys.* 49, 435.

Kawasaki, K. 1966. Phase transitions and critical phenomena. *Phys. Rev.* 145, 224.

Kawasaki, K. 1972. *In Phase Transitions and Critical Phenomena,* Vol. II. London: Academic Press.

Landau, L. D. and I. T. Khalatnikov, 1954. On the anomalous absorption of a sound near to points of phase transition of the second kind. *Akad. Nauk SSSR* 96, 496.

Landau, L., I. Khalatnikov, and N. Meiman. 1965. In *Collected Papers of L.D. Landau.* New York: Gordon & Breach, p. 776.

Langer, J. S., M. Baron, and H. D. Miller. 1975. New computational method in the theory of spinodal decomposition. *Phys. Rev.* A, 11, 1417.

Lifshitz, I. M. and V. V. Slyozov. 1961. The kinetics of precipitation from supersaturated solid solutions. *J. Phys. Chem. Solids* 19, 35.

Mandelbrot, B. 1977. *Forms, Chance, and Dimension.* New York: Freeman.

Roy, M., P. Banerjee, T. K. Sengupta, and S. Dattagupta. 2010. Glucose induced fractal colony pattern of bacillus thuringiensis. *Jour. Theor. Biology* 265, 398.

11

Diffusion of Rapidly Driven Systems

11.1 Introduction

The mathematical treatment of the dynamics of a classical system when the latter is subjected to a time-dependent force is usually straightforward when the amplitude and the frequency of the external impetus are reasonably small. In such cases, standard perturbation treatments are available in the literature. However, when either (or both) the amplitude or (and) the frequency is (are) very large, the situation is much more complex. Ironically though, when the frequency is extremely large and the analysis may appear hopelessly complicated, a timescale separation emerges that makes the calculation much simpler.

Essentially, the system is rapidly cycled several times within its own systematic timescale(s) and hence it feels only an averaged effect of the external perturbation. As Kapitza showed in a neat argument paraphrased by Landau and Liftshitz in their treatise on classical mechanics, averaging yields a time-independent contribution to the effective potential (Kapitza 1951, Landau and Liftshitz 1976).

The latter is inversely proportional to the square of the frequency ω when the perturbation is assumed to be monochromatic characterized by a single frequency ω. While our interest in this book is to extend the analysis when the system under consideration is diffusive (i.e., Brownian), it is instructive first to reproduce Kapitza's scheme for deterministic dynamics before we add stochasticity. The argument runs as follows.

The Newtonian equation of motion for a particle of mass m moving in a one-dimensional potential $V(x)$ but subjected further to periodic forcing with amplitude f is given by

$$m\ddot{x} = -\frac{dV}{dx} + f(x)\cos(\omega t). \tag{11.1}$$

While the assumed one dimensionality of the motion does not diminish the generality of the argument, an additional space dependence of the amplitude $f(x)$ stipulated above lends further structure to the dynamics, as we will discuss below.

Because the frequency ω is taken to be much larger than all characteristic frequencies of the system, it is reasonable to assume that $x(t)$ will be a sum of two parts: a slowly varying component $X(t)$ that does not change much over times $\sim \frac{1}{\omega}$, and a rapidly varying part $\zeta(t)$ expected to be governed by the same oscillating term as is extant in the applied force:

$$x(t) = X(t) + \zeta(t). \tag{11.2}$$

Substituting in Equation (11.2) and making a Taylor series expansion of $V(x)$ and $f(x)$, we find

$$m(\ddot{X}+\ddot{\zeta}) = -\frac{d}{dX}\left(V(X)+\zeta\frac{dV}{dX}\right)+\cdots+\left(f(X)+\zeta\frac{df(X)}{dX}+\cdots\right)\cos(\omega t), \tag{11.3}$$

wherein we have explicitly retained terms of order linear in $\zeta(t)$ only, again in anticipation that ζ will be effectively small. Following that premise we may read from Equation (11.3)

$$\ddot{\zeta}(t) = \frac{f(X)}{m}\cos(\omega t), \tag{11.4}$$

which, when integrated over a timescale in which X is nearly constant, yields

$$\zeta(t) = -\frac{f(X)}{m\omega^2}\cos(\omega t). \tag{11.5}$$

Thus, as surmised above, $\zeta(t)$ oscillates in phase with the external force and is scaled down by a factor ω^{-2} and is therefore effectively small. It is evident that $\zeta(t)$ can be employed as an effective expansion parameter even if f and ω are of similar orders in magnitude (in dimensionless units). The last step is to substitute the expression for $\zeta(t)$ in Equation (11.5) into Equation (11.3) and perform an average over a complete cycle of vibration, i.e., over a time period $T(= 2\pi/\omega)$. We obtain (denoting the average by an overhead):

$$m\ddot{\bar{X}}(t) = -\frac{d}{d\bar{X}}\left(V(\bar{X})+\frac{f^2(\bar{X})}{4m\omega^2}\right). \tag{11.6}$$

Equation (11.6) is then the desired result that suggests that after the effect of the oscillatory force averages out, the averaging process leads to a modified potential given by

$$V_{eff}(\bar{X}) = V(\bar{X})+\frac{f^2(\bar{X})}{4m\omega^2}. \tag{11.7}$$

As mentioned at the outset, the correction term is inversely proportional to ω^2. Hence, in order to make this term effective, the amplitude must also be adequately large so that the ratio $\frac{f^2}{\omega^2}$ is finite. Our second point is that the assumed space dependence of the amplitude lends additional structure to the effective potential with important physical consequences as discussed below.

However, for the present, it is pertinent to point out that the main result of recalibration of the potential [cf. Equation (11.7)] remains intact even when the amplitude *f(x)* is a constant *f*. To see this, we must extend Equation (11.3) to the next higher order in ζ, thus

$$m(\ddot{X}+\ddot{\zeta}(t)) = -\frac{d}{dX}\left(V(X)+\zeta\frac{dV}{dX}+\frac{\zeta^2}{2}\frac{d^2V}{dX^2}\right)+f\cos(\omega t). \quad (11.8)$$

As before, and in partial modification of Equation (11.5), we find

$$\zeta(t) = -\frac{f}{m\omega^2}\cos(\omega t). \quad (11.9)$$

Substituting Equation (11.9) and integrating over time from 0 to T, we now obtain

$$m\ddot{\bar{X}}(t) = -\frac{d}{d\bar{X}}\left(V(\bar{X})+\frac{f^2}{4m^2\omega^4}V''(\bar{X})\right)+\cdots. \quad (11.10)$$

The effective potential in this case is given by

$$V_{eff}(\bar{X}) = V(\bar{X})+\left(\frac{f}{2m\omega^2}\right)^2 V''(\bar{X}). \quad (11.11)$$

Clearly, for the periodic forcing to have any meaningful effect on the dynamics, the potential energy must be more than quadratic in displacement.

Given this background to what transpires when a mechanical system is subjected to a rapidly oscillating high-amplitude force, our main interest in this book is naturally to extend the analysis to a diffusive system. Thus, in the remainder of this chapter, we will consider a Brownian particle moving in an arbitrary potential subjected to high-frequency forcing. It is relevant to note that our treatment has important consequences for a large number of phenomena in physics, chemistry, biology, and geology. These include stochastic resonance, Kramers' escape over barriers, molecular motors and ratchet transport, and noise-induced phase transitions, to name a few.

As has been our practice in systematically developing the idea of diffusion, we will first consider in Section 11.2 diffusion in the over-damped limit according to the Langevin equation. In that limit, the momentum variable is averaged over (Chapter 4) and we are led to restrict the treatment to the position space only.

11.2 Langevin Equation in Over-Damped Case

The Langevin model of diffusion was discussed in detail in Chapter 4. As noted there, timescale separation when friction is strong allows the momentum to be averaged away, yielding a Langevin equation for the position variable alone. That equation reads:

$$\dot{x} = -\Gamma \frac{\partial V}{\partial x} + \eta(t), \tag{11.12}$$

where $\eta(t)$ is again a random white noise with correlation given by

$$\langle \eta(t)\ \eta(t') \rangle = 2\Gamma KT\delta(t - t'). \tag{11.13}$$

We now add a periodic forcing term: $f(x)\cos(\omega t)$ to the equation of motion. In the over-damped limit of Equation (11.12), the additional term can be incorporated as (Sarkar and Dattagupta 2002)

$$\dot{x} = -\Gamma \frac{\partial V}{\partial x} + \eta(t) + \frac{f(x)}{m\gamma}\cos(\omega t). \tag{11.14}$$

Once again, by making the timescale decomposition of the coordinate variable, as in Equation (11.2), we find

$$\dot{X} + \dot{\zeta} = -\Gamma \frac{\partial V}{\partial X} + \eta(t) + \frac{f(x)}{m\gamma}\cos(\omega t), \tag{11.15}$$

where $\zeta(t)$ can now be approximated as [following the arguments below Equation (11.3)]:

$$\zeta(t) = \frac{f(x)}{m\gamma\omega}\sin(\omega t). \tag{11.16}$$

Later in this section we will take the applied force to be spatially uniform, which requires, as argued at the end of Section 11.1, the potential to be Taylor-expanded to quadratic order in ζ. Thus

$$\dot{X} \simeq -\Gamma \frac{\partial}{\partial X}\left(V(X) + \zeta \frac{\partial V}{\partial X} + \zeta^2 \frac{\partial^2 V}{\partial X^2} \right) + \eta(t). \tag{11.17}$$

We now perform a cyclic average of Equation (11.17) and assume that the noise $\eta(t)$ is unaffected by the averaging procedure. The extent to which this assumption is justified is detailed in Sarkar and Dattagupta (2002). The net result is

$$\dot{\bar{X}} = -\Gamma \frac{\partial}{\partial \bar{X}}(V(\bar{X}) + \frac{f^2}{4m^2\gamma^2\omega^2} V''(\bar{X}) + \eta(t). \tag{11.18}$$

We are thus led to a new Langevin equation:

$$\dot{\bar{X}} = -\Gamma \frac{\partial}{\partial \bar{X}} V_{eff}(\bar{X}) + \eta(t), \tag{11.19}$$

where the effective potential energy is given by

$$V_{eff}(\bar{X}) = V(\bar{X}) + \left(\frac{f}{2m\gamma\omega}\right)^2 V''(\bar{X}). \tag{11.20}$$

Comparing this result with the one in the deterministic case of Equation (11.11), it is evident that the prefactor in the correction term replaces ω^2 with $\gamma\omega$. Since γ depends on external parameters such as the temperature and pressure and on internal attributes such as density, both γ and ω can be tuned to modulate the correction term [second term in Equation (11.20)]. We discuss below two applications of such a scheme.

11.2.1 Dynamical Symmetry Breaking

Consider a particle moving in a bistable potential

$$V(x) = -\frac{1}{2}ax^2 + \frac{1}{4}x^4. \tag{11.21}$$

This potential is symmetric with two minima at $x = \pm\sqrt{a}$ and a local maximum at $x = 0$, with a corresponding energy separation that equals $\frac{a^2}{4}$. If we add the correction term as depicted in Equation (11.20), the effective potential becomes

$$V_{eff}(X) = -a\left(\frac{f}{m\gamma\omega}\right)^2 + \frac{1}{2}X^2\left[-a + \frac{3}{2}\left(\frac{f}{m\gamma\omega}\right)^2\right] + \frac{1}{4}X^4. \tag{11.22}$$

As the constant term, i.e., the first term on the right of Equation (11.22), has no effect on the dynamics [cf. Equation (11.19)], we may ignore that and focus on the rest of the right side. It is evident that when

$$\frac{3}{2}\left(\frac{f}{m\gamma\omega}\right)^2 > a, \tag{11.23}$$

the double well becomes a single well! Thus one may define a critical frequency ω_c:

$$\omega_c = \frac{f}{m\gamma}\sqrt{\frac{3}{2a}}, \tag{11.24}$$

such that when the frequency ω is scaled up beyond the value ω_c, the system transits from a single to a double well. When the well is single, the stable equilibrium occurs at $x = 0$ whereas for a double well, $x = \pm\sqrt{a}$ provides the stable equilibrium points or attractors. Because this happens by accelerating the frequency ω, keeping all other parameters fixed, such a phenomenon is referred to as dynamical symmetry breaking when the position coordinate continually changes from zero to a finite value, as in a second order phase transition.

When $\omega > \omega_c$, the particle can undergo a noise-induced jump from, say $x = -\sqrt{a - \frac{3}{2}(\frac{f}{m\gamma\omega})^2}$ to $x = \sqrt{a - \frac{3}{2}(\frac{f}{m\gamma\omega})^2}$, across a barrier, the height of which is $\frac{1}{4}[a - \frac{3}{2}(\frac{f}{m\gamma\omega})^2]^2$. Such thermal tunneling between the two attractors $\pm\sqrt{a - \frac{3}{2}(\frac{f}{m\gamma\omega})^2}$ can be studied as stochastic resonance in the presence of an additional weak signal. The interesting point is that although the original problem is time-dependent, the Kapitza scheme yields a time-independent problem. Hence, all standard techniques such as the Kramers method (Chandrasekhar 1943, Kramers 1940) and time series analysis can be applied to the present situation.

11.2.2 Decay of Metastable State

This issue was discussed in Section 5.8 in the context of the Kramers approach to the problem of escape over a barrier. The prototypical potential energy for a metastable state can be written as

$$V(x) = \frac{1}{2}ax^2 - \frac{1}{3}x^3. \tag{11.25}$$

This potential has a metastable minimum at $x = 0$ and an unstable maximum at $x = 0$, separated by a barrier of height $\frac{1}{6}a^2$. Noise-induced thermal fluctuations can trigger escape over the barrier at a rate governed by the Kramers formula discussed in Chapter 5. Our interest here is to re-examine that rate when additional frequency-dependent forcing is present. In the latter case, the effective potential is

$$V_{eff}(x) = \frac{1}{2}ax^2 - \frac{1}{3}x^3 - \frac{1}{2}\left(\frac{f}{m\gamma\omega}\right)^2 x. \tag{11.26}$$

In order to find the extrema of Equation (11.26), we consider

$$V'_{eff}(x) = ax - x^2 - \frac{1}{2}\left(\frac{f}{m\gamma\omega}\right)^2 = 0, \tag{11.27}$$

which leads to two roots at

$$x = \frac{1}{2}\left[a \pm \sqrt{a^2 - 2\left(\frac{f}{m\gamma\omega}\right)^2}\right]. \tag{11.28}$$

However, the two roots merge into one when

$$\omega = \bar{\omega} = \sqrt{2}\frac{f}{m\gamma a}. \tag{11.29}$$

Therefore, when $\omega < \bar{\omega}$, the radical in Equation (11.28) becomes imaginary (and meaningless). This implies that for such values of ω, there is no local minimum and the particle can monotonically escape to infinity.

The above discussion suggests that thermally activated decay of the metastable state can be meaningfully studied only when $\omega < \bar{\omega}$. To quantitatively verify the validity of the effective potential approach, we calculate the mean first passage time (MFPT) from the bottom of the potential well to the top of the barrier by three distinct methods: (a) numerical study of the original dynamics described by Equation (11.14) with $V(x)$ given by Equation (11.25), (b) numerics of the effective dynamics of Equation (11.19) with $V_{eff}(\bar{X})$ given by Equation (11.26), and (c) the analytical Kramers' formula of Section 5.8.

The comparison is provided in Table 11.1, in which $f = 100$ and $\Gamma KT = 0.05$. The data for various frequencies ($= 2\pi/T$) indicate satisfactory agreement (within error bars) for original and effective dynamics. The analytical prediction is also in reasonable concurrence except at the lowest frequencies, for reasons mentioned earlier: for $\omega < \bar{\omega}$, the barrier disappears.

11.3 Langevin Equation for Arbitrary Damping

In Section 11.2, we analyzed the effect of periodically rapid forcing in the over-damped limit of the Langevin equation. Here we would like to relax that condition and consider the Langevin dynamics for arbitrary damping (Bandyopadhyay et al. 2006).

Recall from Chapter 4 that the underlying scheme is based on timescale separation of a multitude of time units. The most primitive timescale is, of course, the one associated with the heat bath-induced thermal noise that is

TABLE 11.1

Mean First Passage Time Calculated by Various Methods

	Mean First Passage Time	
Perturbation Period (T)	**Original Dynamics**[a]	**Effective Dynamics**[b]
2×10^{-3}	194.0 (6.6)	197.8 (6.4)
6×10^{-3}	180.1 (4.1)	179.1 (4.7)
1.1×10^{-2}	147.6 (7.3)	147.1 (7.7)
2.2×10^{-2}	70.0 (3.5)	74.1 (3.7)
3.3×10^{-2}	27.7 (1.2)	26.8 (1.3)

Note: Figures within parentheses are estimates of error bars.

[a] From Equation (11.29). [b] From Equation (11.49).

taken to be the shortest. Then comes the momentum relaxation timescale that is roughly governed by the inverse of the friction coefficient. That is why, in the over-damped limit, the momentum is averaged out and we enter the Brownian diffusive regime of the position variable.

Next is the additional ingredient of an externally imposed timescale, i.e., that governed by the inverse of the frequency of the external perturbation. Although we want to set the timescale as the highest frequency of the problem, the natural question is whether the frequency is large enough to make the time period T smaller than the noise correlation time?

We want to stress once again that the detailed analysis in Sarkar and Dattagupta (2002) makes it amply clear that our basic premise of working within the Langevin framework will break down if we deviate from the point that the noise correlation time is indeed the shortest timescale in the dynamics. In fact, it is only under that assumption, i.e., T is small but not small enough and definitely not smaller than the noise fluctuation time, that we conclude that the Kapitza method works.

Hence, in further discussion we want to categorically reiterate the hierarchies of various timescales. The noise correlation time associated with a white noise is the shortest timescale. Then comes T, the period of external perturbation $(= 2\pi/\omega)$, followed by the momentum averaging time $(\sim 1/\gamma)$, and finally the diffusive timescale associated with the position coordinate. Within this fundamental framework, the noise has no influence on the averaging procedure, as shown in Section 11.2., in arriving at an effective potential. Our starting point is the phase space Langevin equation (Chapter 4):

$$m\ddot{x} = -\gamma \dot{x} - \frac{\partial}{\partial x} U(x) + F(t) + \eta(t). \tag{11.30}$$

As before, $\eta(t)$ is a delta-correlated white noise, white $F(t)$ is a time-dependent and space-independent external perturbation:

$$F(t) = f \cos \omega t. \tag{11.31}$$

The additional space dependence of the amplitude $f(x)$ has important consequences for physical attributes such as the effective diffusive coefficient as discussed in Dutta and Barma (2003) and Bandyopadhyay et al. (2006). For reasons elaborated above, the noise $\eta(t)$ can be ignored in our route to an effective potential, and hence will be omitted from the preliminary analysis below.

As before, we split $x(t)$ into a slowly varying $X(t)$ and a rapidly changing coordinate $\zeta(x,t)$, as

$$x(t) = X(t) + \zeta(x,t). \tag{11.32}$$

Equation (11.29) sans the noise $\eta(t)$ then reads

$$m(\ddot{X}(t) + \ddot{\zeta}(x,t) = -\gamma(\dot{X}(t) + \dot{\zeta}(x,t)) + F(X + \zeta(x,t);t) - \frac{\partial}{\partial x}(U(X + \zeta(x,t);t) \tag{11.33}$$

Upon the usual Taylor expansion, Equation (11.33) yields

$$m(\ddot{X}(t)+\ddot{\zeta}+2\dot{\zeta}\dot{\zeta}'+\ddot{\zeta}'\zeta+\zeta'\ddot{\zeta})=-\gamma(\dot{X}(t))+\dot{\zeta}+\dot{\zeta}\zeta'+\zeta\dot{\zeta}')$$

$$+F(t)+\frac{\partial}{\partial x}\left[U+\zeta(1+\zeta')U'+\frac{1}{2}\zeta^2U''\right], \quad (11.34)$$

where we stop at the second order in ζ and amplify that all space dependences are with respect to the slow variable X and hence the primes denote spatial derivatives with reference to X. Note that Equation (11.32) implies

$$\frac{\partial}{\partial X}=\frac{\partial x}{\partial X}\frac{\partial}{dx}=\frac{\partial}{\partial X}(X+\zeta+\zeta'\zeta)\frac{\partial}{\partial x}=(1+\zeta'+(\zeta')^2+\zeta\zeta'')\frac{\partial}{\partial x}. \quad (11.35)$$

Thus

$$\frac{\partial}{\partial x}=(1-\zeta'-(\zeta')^2-\zeta\zeta'')\frac{\partial}{\partial X}. \quad (11.36)$$

As before, it is evident that the fast variable ζ satisfies the equation

$$m\ddot{\zeta}+\gamma\dot{\zeta}=f\cos\omega t. \quad (11.37)$$

This equation has the solution

$$\zeta(x,t)=-\frac{1}{m\left(\omega^2+\frac{\gamma^2}{m^2}\right)}\left[f\cos\omega t-\frac{\gamma}{m\omega}f\sin\omega t\right]. \quad (11.38)$$

We substitute Equations (11.36) and (11.38) into Equation (11.34) and average over the period T during which all terms linear in ζ (ζ', ζ'', etc.) disappear (because of the appearance of $\cos\omega t$ or $\sin\omega t$). Denoting the averaged variables by overhead bars, as customary, we find:

$$m\overline{\overset{\circ\circ}{X}}(t)+\gamma\overline{\overset{\circ}{X}}(t)=U'(x)+\frac{1}{2}\overline{(\zeta')^2}U'(\bar{X})-\frac{1}{2}\frac{\partial}{\partial X}(\overline{U''\zeta^2}). \quad (11.39)$$

As mentioned earlier, for space-independent forcing, we go to the second order in ζ as indicated in Equation (11.39). To this end, we require:

$$\frac{1}{2}\bar{\zeta}^2=\frac{(f(\bar{X}))^2}{4m^2\omega^2(\omega^2+\gamma^2/m^2)}. \quad (11.40)$$

The effective potential is then

$$U_{eff}(\bar{X})=U(\bar{X})+U^{(2)}(\bar{X}), \quad (11.41)$$

where the second order correction reads

$$U^{(2)}(\bar{X}) = \frac{f^2 U''(\bar{X})}{4m^2\omega^2(\omega^2 + \gamma^2/m^2)}. \tag{11.42}$$

As stated earlier, the second order correlation term yields a coupling between the external force and the original potential.

11.4 Effective Diffusion in Periodic Potential

We apply the formalism developed in Section 11.3 to calculate the diffusion coefficient for a particle moving in a periodic potential. The problem has important applications to diffusion on crystal surfaces, intracellular transport, and ratchet motion in molecular motors, to name a few (Chowdhury 2000). The periodic potential is chosen to be of the form:

$$U(x) = -V_0\,[\sin(x) + \mu\sin(2x)]. \tag{11.43}$$

$$U'(X) = -V_0\,[\cos(x) + 2\mu\cos(2x)], \quad U''(x) = V_0\,[\sin(x) + 4\mu\sin(2x)]. \tag{11.44}$$

The effective potential is then given by

$$U_{eff}(\bar{X}) = -V_0\sin(\bar{X})\left[1 - \frac{f^2}{4m^2\omega^2(\omega^2+\gamma^2/m^2)}\right] - \mu V_0\sin(2\bar{X})\left[1 - \frac{f^2}{4m^2\omega^2(\omega^2+\gamma^2/m^2)}\right]. \tag{11.45}$$

The effective potential is inserted into the Langevin equation for the slow variable in accordance with Section 11.2, which now reads

$$m\overline{\ddot{X}} = -\gamma\overline{\dot{X}} - \frac{\partial}{\partial\bar{X}}U_{eff}(\bar{X}) + \eta(t), \tag{11.46}$$

where, for reasons stated earlier, the thermal noise $\eta(t)$ remains delta correlated, as before. Based on this equation and averaging over all possible realizations of noise and initial values of position and velocity (denoted by double brackets), the diffusion coefficient can be computed from the Einstein relation (Chapter 1):

$$D = \lim_{t\to\infty}\frac{1}{2t}\left[\langle\langle \bar{x}^2(t)\rangle\rangle - \langle\langle \bar{x}(t)\rangle\rangle^2\right]. \tag{11.47}$$

The diffusion coefficient D is an important physical attribute to analyze as it is directly related to mobility. The influence exerted by the combined presence of periodic potential and oscillating drive on D when the frequency of oscillation of the drive is very large has great significance according to *Vibrational Mechanics* (Bleckman 2000). The importance of this issue in the context of effective segregation of particle mixtures with respect to their mass and geometry has crucial implications in biology, for example, as molecular motors for transport in biological structures (Haenggi 2005), ion pumps (Doyle et al. 1998), and cell membranes (Alberts 1994) to name a few.

The problem was studied within a vibrational mechanics and Langevin equation approach by Borromeo and Marchesoni (2003 and 2007) and also by Wickenbrock et al. (2012). Borromeo and Marchesoni (2007) show that the effect of averaging over the time period of the external drive, as done above, yields a modifying prefactor that multiplies the original periodic potential. That prefactor from the vibrational mechanics mechanism (Bleckman 2000) is a Bessel function of order zero: $J_0(y)$. Interestingly, the expansion of $J_0(y)$ where $y = f/m\omega\sqrt{\omega^2 + \gamma^2/m^2}$ exactly matches our perturbation expansion [cf. Equation (11.45)] to leading order in f^2/ω^2.

With this motivating background, we plot the numerically computed diffusion coefficient D from Equation (11.47) versus the amplitude f of the external drive (Figure 11.1). Two plots are provided, one based on the original dynamics governed by Equation (11.30) with the potential given by Equation (11.43), and the other based on the effective dynamics dictated by Equation (11.46).

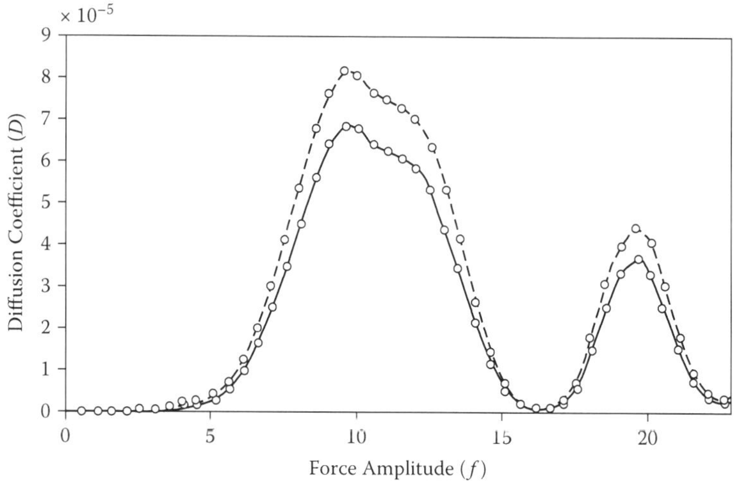

FIGURE 11.1
Diffusion coefficient D is plotted versus the forcing amplitude f. The dashed line is based on the original dynamics. The solid line depicts the effective dynamics. The chosen parameters are $m = 1$, $K_BT = 0.5$, $\omega = 3.0$, and $\gamma = 0.1$ in dimensionless units. The oscillating peaks carry the signature of the Bessel function of zeroth order.

The peaks in the diffusion coefficient D for certain values of the amplitude f are in line with the oscillatory nature of the Bessel function $J_0(y)$. The physical reason for this apparently surprising result can be understood from the following intuitive argument. In the absence of an external drive, the Brownian particle remains mostly confined to the bottoms of the wells (of the periodic potential). However, the external drive modulates the periodic potential and for $y = n\pi$ (when $J_0(y)$ vanishes) the effect of the periodic potential is nullified. The particle is then a damped free particle according to the Langevin equation, endowed with a free particle diffusion coefficient. Indeed, the AC drive forces the particle to the top of the barrier, thereby neutralizing the confining effect of the periodic potential.

References

Alberts, B. et al. 1994. *Molecular Biology of the Cell.* New York: Garland.

Bandyopadhyay, M. and S. Dattagupta. 2006a. *J. Stat. Phys.* 123, 1273.

Bandyopadhyay, M. and S. Dattagupta. 2006b. *J. Phys. Cond. Matt*, 18, 10029.

Bleckman, I.I. 2000. *Vibrational Mechanics.* Singapore: World Scientific.

Borromeo, M. and F. Marchesoni. 2007. *Phys. Rev. Lett.* 99, 150605.

Borromeo, M and F. Marchesoni. 2005. Mobility oscillations in high-frequency modulated devices. *Europhys. Lett.* 72, 362.

Chandrasekhar, S. 1943. Stochastic problems in physics and astronomy. *Rev. Mod. Phys.* 15,1.

Chowdhury, D., L. Santen, and A. Schadschneider. 2000. Statistical physics of vehicular traffic and some related systems. *Phys. Rep.* 329, 199.

Doyle, D. A., J. C. Morais, and R. A. Pfuetzner. 1998. The structure of the potassium channel: Molecular basis qfK+ conduction and selectivity. *Science* 280, 69.

Dutta, S. B. and M. Barma, 2003. Asymptotic distributions of periodically driven stochastic system. *Phys. Rev.* E67, 061111.

Haar, D., Ed. 1965. *Collected Papers of P.L. Kapitza.* Oxford: Pergamon.

Haenggi, P., F. Marchesoni, and F. Nori. 2005. Brownian motors. *Ann. Phys.* (Leipzig) 1451.

Kapitsa, P. L. 1951. Dynamic stability of pendulum with oscillating suspension point. *Zh. Teor. Fiz.* 85, 4521.

Kramers, H. A . 1940. Brownian motion in a field of force and the diffusion model of chemical reactions. *Physica (Utrecht)* 7, 284.

Landau, L.D. and E.M. Lifshitz. 1976. *Course of Theoretical Physics,* Vol. 1. New York: Pergamon.

Sarkar, S. K. and S. Dattagupta. 2002. Escape rates in the presence of high frequency perturbations. *Int. J. Mod. Phys.* B16, 1247.

Wickenbrock, A., P. C. Holz, N. A. Abdul Wahab, P. Phoonthong, D. Cubero, and F. Renzoni. 2012. Vibrational mechanics in an optical lattice: Controlling transport via potential renormalization. *Phys. Rev. Lett.* 08, 020603.

Section II

Quantum Diffusion

12

Quantum Langevin Equations

12.1 Introduction

The process of diffusion constitutes the dissipative motion of a classical particle or group of particles subjected to many-body interactions with their surroundings. When the surroundings consist of an infinitely large system that can rapidly recover from any deviation from thermal equilibrium such that its temperature remains roughly uniform, the system qualifies as a heat bath. Further, when the degrees of freedom of the heat bath are integrated out because of the implied timescale separation, the dynamics of the subsystem appear dissipative, characterized by parameters such as friction coefficient, diffusion constant, and others.

A convenient formalism to deal with this phenomenon is the set of Langevin equations as discussed in Chapter 4. The onus of randomness arising from the influence of a heat bath is thrust upon the momentum variable that directly bears the brunt of stochasticity via noisy forces. The position variable, on the other hand, is only an indirect participant in this process since it is driven by momentum.

Such an asymmetric separation of position and momentum is, however, not allowed in quantum mechanics where position and momentum are operators and cannot be measured simultaneously with precise degrees of accuracy. Another way of saying this is that quantum mechanics is characterized by wave particle duality: what is a particle is also a wave and the smaller the mass of a particle and the lower the temperature, the larger the wave length.

Thus, unlike the case of a classical particle deeply entrenched in a mechanical potential from which it cannot extricate itself to jump across a barrier, unless driven by requisite thermal fluctuations, a quantum particle can tunnel through a barrier unaided by temperature. Tunneling is, of course, a euphemism for the wave function of a particle that has a tail that extends beyond the barrier.

The mathematical implication for wave particle duality is that the kinetic energy and the potential energy are quantum operators that do not commute—the statics and dynamics are inexorably intertwined.

Consequently, irrespective of whether equilibrium is governed by the density matrix or nonequilibrium dictated by the Schrödinger time evolution operator, we cannot disentangle the kinetic energy and the potential energy. Thus, when describing the diffusive behavior of an elementary quantum particle such as a proton, neutron, or electron in many-body interactions with phonons, magnons, and other elementary excitations, we need a different formalism altogether. This in a nutshell is the genesis of quantum diffusion.

We explained earlier that classical diffusion hinges on timescale separation at various levels and utilized a hierarchy in increasing orders of impact time, mean free times of collisions, momentum relaxation times, and other factors. What happens in the quantum case? It is evident that quantum mechanics gives way to classical mechanics in the limit of the temperature T going to infinity and the Planck constant $\hbar$ going to zero. (We must note that there is nothing like a real infinity or a real zero—they are only relative concepts.) Therefore, one can introduce a quantal timescale $\tau_Q = \hbar / K_B T$ where K_B is the Boltzmann constant.

Evidently, τ_Q goes to zero in the classical limit but acquires a finite value when quantum effects are significant. It is clear then that if the time of observation is less than τ_Q, quantum effects shall dominate. There will be memory of the initial state because of the inexorable link between the momentum and the position variables. Hence, within the time τ_Q, a quantum process will have to be necessarily non-Markovian. A formal manner of expressing this within the Langevin picture (see below) will be to recognize that because the noise is an operator in quantum mechanics it cannot commute even with itself at different times.

Therefore, the auto-correlation of the noise can never be described by a delta function of the difference of the two respective times, which is the hallmark of white noise-driven classical Langevin dynamics. Such non-Markovian features will have deep consequences for time correlation functions, dissipative tunneling, quantum renewal processes, anomalous diffusion, and other processes.

A pure quantum system is characterized by unitary evolution of its Schrödinger time development operator because the underlying Hamiltonian is hermitian. A result of this unitary evolution is that the quantum wave function has a unique phase coherence—a quantum attribute that is responsible for physical properties such as interference, quantum information, and entanglement. The unitary evolution is disturbed by the dissipative influence of the surroundings, leading to dephasing, decoherence, and other effects.

One other application of quantum diffusion will be to explain coherence-to-decoherence transitions in relation to contemporary topics of quantum computation, macroscopic quantum mechanics, and Bose–Einstein condensation, among others. Based on this background, to deal with quantum diffusion we need a formalism for the quantum Langevin equation that is a generalization of its classical counterpart described in Chapter 1. However, because the position and momentum must be treated on the same footing and the heat bath must be characterized by quantum degrees of freedom, we need a derivation of the classical Langevin equation from first principles as a preliminary step before we embark on quantum generalization. Such a treatment is the subject of the next section.

12.2 Derivation of the Classical Langevin Equation

We consider the classical dynamics of a particle of momentum p moving in an arbitrary potential $U(q)$ and interacting with a number of harmonic oscillators via a linear coordinate–coordinate interaction. The corresponding Hamiltonian can be expressed as

$$\mathcal{H} = \frac{p^2}{2m} + U(q) + \sum_{j=1}^{N}\left\{\frac{p_j^2}{2m_j} + \frac{1}{2}m_j\omega_j^2\left(q_j - \frac{C_j q}{m_j\omega_j^2}\right)^2\right\}. \tag{12.1}$$

The linear nature of the coupling becomes self-evident upon completing the square, thereby attributing to C_j the epithet of the coupling constant. Of course, we acquire an additional counter term but because it is quadratic in q, it can be absorbed into $U(q)$ whose physical meaning becomes clear after a set of canonical transformations described by

$$\tilde{q}_j = \frac{m_j\omega_j^2}{C_j}q_j, \quad \tilde{p}_j = \frac{C_j}{m_j\omega_j^2}p_j. \tag{12.2}$$

By further redefining the mass m_j as

$$\tilde{m}_j = \frac{C_j^2}{m_j\omega_j^4}, \tag{12.3}$$

the canonically transformed Hamiltonian can be written as

$$\tilde{\mathcal{H}} = \frac{p^2}{2m} + U(q) + \sum_j\left\{\frac{\tilde{p}_i^{-2}}{2\tilde{m}_j} + \frac{1}{2}\tilde{m}_j\omega_j^2(\tilde{q}_j - q)^2\right\}. \tag{12.4}$$

The form of the Hamiltonian in Equation (12.4) is manifestly translationally invariant, and can be interpreted further as the energy of several j harmonic oscillators, each connected by a spring to the center of mass of the particle at q. The corresponding Hamilton equations are

$$\begin{aligned}
&\mathring{q} = p/m, \\
&\mathring{q} = -\frac{\partial U(q)}{\partial q} + \sum_j C_j\left(q_j - C_j\frac{q}{m_j\omega_j^2}\right), \\
&\mathring{q}_j = \frac{p_j}{m_j}, \\
&\mathring{p}_j = -m_j\omega_j^2\left(q_j - C_j\frac{q}{m_j\omega_j^2}\right).
\end{aligned} \tag{12.5}$$

The last two equations can be combined to yield the displaced harmonic oscillator equation:

$$\overset{\circ\circ}{q}_j = -\omega_j^2\left(q_j - \frac{C_j q}{m_j\omega_j^2}\right), \tag{12.6}$$

the solution of which has the familiar form

$$\begin{aligned}
q_j(t) = q_j(o)\cos(\omega_j t) + \frac{p_j(o)}{m_j\omega_j}\sin(\omega_j t) \\
+ \frac{C_j}{m_j\omega_j}\int_0^t dt' q(t')\sin\,[\omega_j(t-t')].
\end{aligned} \tag{12.7}$$

A partial integration leads to

$$\begin{aligned}
\left(q_j(t) - \frac{C_j q(t)}{m_j\omega_j^2}\right) = \left[q_j(0) - \frac{C_j q(0)}{m_j\omega_j^2}\right]\cos\,(\omega_j t) + \frac{p_j(0)\sin(\omega_j t)}{m_j\omega_j} \\
- \frac{C_j}{m_j\omega_j^2}\int_0^t dt'\frac{p(t')}{m}\cos[\omega_j(t-t')].
\end{aligned} \tag{12.8}$$

We now substitute Equation (12.8) into the second part of Equation (12.5) to obtain

$$\begin{aligned}
\mathring{p} = -\frac{\partial U(q)}{\partial q} - \int_0^t dt'\frac{p(t')}{m}\sum_j \frac{C_j^2}{m_j\omega_j^2}\cos[\omega_j(t-t')] \\
+ \sum_j \{C_j[q_j(o) - \frac{Cj\, q(o)}{m_j\omega_j^2}]\cos\,(\omega_j t) + \frac{C_j p_j(o)}{m_j\omega_j}\sin\,(\omega_j t)]\}.
\end{aligned} \tag{12.9}$$

Interestingly, Equation (12.9) has the structure of a generalized Langevin equation:

$$\overset{\circ}{p} = -\frac{\partial U(q)}{\partial q} - \int_0^t dt' \frac{p(t')}{m} \zeta(t-t') + \eta(t), \tag{12.10}$$

where the noise

$$\eta(t) = \sum_j \left\{ C_j \left[q_j(o) - \frac{C_j q(o)}{m_j \omega_j^2} \right] \cos\ (\omega_j t) + + \frac{C_j p_j(o)}{m_j \omega_j}\ \sin\ (\omega_j t) \right\}, \tag{12.11}$$

and the friction $\zeta(t)$ become a memory function:

$$\zeta(t) = \sum_j \frac{C_j^2}{m_j \omega_j^2} \cos(\omega_j t). \tag{12.12}$$

The significant point of the noise $\eta(t)$ in Equation (12.11) is that it is not an external implant, as in the original Langevin model of Chapter 4, but comprises the initial values of coordinates and momenta that live in the phase space of a composite system. Thus the noise is governed by the thermodynamics of the oscillator system if the latter is treated as a thermal bath. The second important point of Equation (12.10) is that the friction depends on the parameters of the surrounding harmonic system, particularly on the square of the coupling constant. The latter property is a consequence of the assumption of the harmonic nature of the surrounding system.

We now make a crucial assumption concerning irreversibility by considering an initial ensemble of states in which $p(0)$, $q(0)$ are held fixed and the harmonic oscillator variables are picked at random from a Gibbsian thermal distribution:

$$P(\{p_j(0)\},\ \{q_j(0)\}) = \frac{e^{-\beta \mathcal{H}}}{\prod_j \int .. \int dp_j(0) dq_j(0) e^{-\beta \mathcal{H}}} \tag{12.13}$$

It is evident that

$$\langle \eta(t) \rangle = \prod_j \int .. \int dp_j(0) dq_j(0)\ \eta(\{q_j\}, \{p_j\})\ P(\{p_j(0)\ q_j(0)\}) = 0. \tag{12.14}$$

Further, the auto-correlation of the noise

$$\langle \eta(t)\ \eta(t')\rangle = K_B T\ \delta(t-t'), \tag{12.15}$$

where $\beta = (K_B T)^{-1}$. Equation (12.10) then has all the attributes of a classical Langevin equation, with the fluctuation dissipation relation (12.15) inter alia emerging as a natural consequence of envisaging the harmonic oscillator system as a heat bath at a fixed temperature T such that the classical equipartition theorem applies.

12.2.1 Ohmic Dissipation

In this subsection, we ask when the Langevin equation (12.10) reduces to the memory-less ordinary Langevin equation. For an answer, we go to the limit of an infinitely large harmonic oscillator system ($N \to \infty$) wherein the summation over j is replaced by an integral, weighted by an appropriate density of a continuum of states, characterized by a spectral function $j(\omega)$:

$$\sum_j \to N \int_0^\infty j(\omega)\ d\omega. \tag{12.16}$$

Furthermore, assuming that all masses m_j are equal and so are the coupling constants (properly scaled),

$$m_j = m, \quad C_j = C/\sqrt{N}, \tag{12.17}$$

we find

$$\zeta(t) = \frac{C^2}{m}\int d\omega\ \frac{j(\omega)}{\omega^2}\cos(\omega t). \tag{12.18}$$

In addition, the spectral function $j(\omega)$ is chosen to be a quadratic function in ω with an upper cut-off D:

$$\begin{aligned} j(\omega) &= \frac{3\omega^2}{D^3}, \quad \omega < D \\ &= 0, \quad \omega > D. \end{aligned} \tag{12.19}$$

The friction coefficient then assumes the form

$$\zeta(t) = \frac{3C^2}{mD^3} \frac{\sin(Dt)}{t}. \tag{12.20}$$

If the frequency cut-off D is sufficiently large, the corresponding timescale will be on the lower end of the chart (Figure 4.2) much lower than the momentum relaxation time. In that limit,

$$\int_0^D d\omega \ \cos(\omega t) \simeq \pi \ \delta(t),$$

and

$$\zeta(t) = 2\zeta_0 \delta(t) \ ,$$

where

$$\zeta_0 = \frac{3\pi C^2}{2mD^3}. \tag{12.21}$$

This approximation allows us to recover the usual Langevin equation from Equation (12.10):

$$\mathring{p} = -\frac{\partial U(q)}{\partial q} - \gamma \ p(t) + \eta(t), \tag{12.22}$$

γ being equal to ζ_0/m, and

$$\begin{aligned} &< \eta(t) > = 0 \\ &< \eta(t') \ \eta(t'') = 2m\gamma K_B T \ \delta(t' - t''). \end{aligned} \tag{12.23}$$

It is clear that three essential steps are followed in arriving at a dissipative model, starting from deterministic evolution: (i) integrate the equations in the forward direction of time, i.e., to ascribe an arrow of time, (ii) imagine that the surroundings comprise an infinitely large number of degrees of freedom so that the states of the latter are dense, and (iii) assume the surrounding system is in perpetual thermal equilibrium characterized by a Gibbs measure so that it qualifies as a heat bath. This is how irreversibility is bestowed on a subsystem, the dynamics of which is reversible to start with.

12.3 Quantum Generalization

The derivation of the quantum Langevin equation follows similar lines as the classical case except that instead of employing Hamilton's equations we resort to Heisenberg's equations of motion for quantum operators and pay heed to noncommutativity of the position operator and its higher time derivatives. We now find from Equation (12.4) for the canonically transformed Hamiltonian (with overhead tildes dropped to achieve brevity of notation):

$$m\ddot{\hat{q}}(t)+\int_0^t dt'\zeta(t-t')\dot{\hat{q}}(t')+\frac{\partial U(\hat{q})}{\partial q}=\hat{\eta}(t), \tag{12.24}$$

where we rewrite the friction and noise by exploiting Equations (12.2) and (12.3):

$$\zeta(t)=\sum_j m_j\omega_j^2\cos(\omega_j t), \tag{12.25}$$

$$\hat{\eta}(t)=\sum_j m_j\omega_j^2\left[\hat{q}_j(0)\cos(\omega_j t)+\frac{\hat{p}_j(0)}{m_j\omega_j}\sin(\omega_j t)\right]. \tag{12.26}$$

It is tacitly understood that Equations (12.24) through (12.26) are valid for $t > 0$ because we have chosen to work with the so-called retarded solutions of the equations of motion. We should again note that the noise comprises noncommuting position and momentum operators and does not commute with itself at different times. This property is reflected in the following symmetric and antisymmetric correlation functions:

$$\left\langle\{\hat{\eta}(t),\ \hat{\eta}(t')\}\right\rangle=\frac{2}{\pi}\int_0^\infty d\omega\ \mathrm{Re}\left[\tilde{\zeta}(\omega+i\overset{+}{0})\right]\hbar\omega\ \coth\left(\frac{\hbar\beta\omega}{2}\right)\cos\left[\omega(t-t')\right], \tag{12.27}$$

$$[\hat{\eta}(t),\ \hat{\eta}(t')]=\frac{2}{\pi}\int_0^\infty d\omega\ \mathrm{Re}[\tilde{\zeta}(\omega+i0^+)]\ \hbar\omega\ \sin[\omega(t-t')]\ . \tag{12.28}$$

Here $\tilde{\zeta}(\omega+i0^+)$ is the Laplace transform of $\tilde{\zeta}(t)$. Note that in the appropriate classical limit in which $\hbar\to 0$ and $T\to\infty$, the commutator in Equation (12.28) vanishes and the symmetrized correlation function reduces to the retarded form of the Dirac delta function as expected. As in the case of the classical Langevin equation, it is pertinent to consider the limit of Ohmic dissipation when the friction ceases to be a memory function but

$\zeta(t-t') \to 2m\gamma\ \delta(t-t')$, and hence we obtain the usual Langevin equation with constant friction:

$$m\ddot{\hat{q}}(t) + m\gamma\dot{\hat{q}}(t) + \frac{\partial U(\hat{q})}{\partial q} = \hat{\eta}(t) . \tag{12.29}$$

However it is important to note that although the friction coefficient γ is independent of time, the correlation of $\hat{\eta}(t)$ is still not a delta function and is instead

$$\langle\{\hat{\eta}(t),\ \hat{\eta}(t')\}\rangle = \frac{2m\gamma}{\pi}\int_0^\infty d\omega\ \hbar\omega\ \coth\left(\frac{\hbar\beta\omega}{2}\right)\cos\left[\omega(t-t')\right], \tag{12.30}$$

$$\langle[\hat{\eta}(t),\ \hat{\eta}(t'')]\rangle = \frac{2m\gamma}{\pi}\int_0^\infty d\omega\ \hbar\omega\ \sin\left[\omega(t-t')\right] . \tag{12.31}$$

The upshot is that the quantum process is necessarily non-Markovian even in the Ohmic dissipation model. The underlying reason is the existence of the quantal timescale τ_Q, (mentioned in the introductory section above) within which quantum evolution retains strong coherence.

12.4 Free Particle

For a free quantum particle, the potential energy is absent and therefore the quantum Langevin equation (12.29) reduces to

$$\ddot{\hat{q}}(t) + \gamma\dot{\hat{q}}(t) = \frac{\hat{\eta}(t)}{m}, \tag{12.32}$$

which yields for $\dot{\hat{q}}(t)$,

$$\dot{\hat{q}}(t) = \dot{\hat{q}}(o)\ e^{-\gamma t} + \int_0^t dt' e^{-\gamma(t-t')}\ \frac{\hat{\eta}(t')}{m}. \tag{12.33}$$

A further integration over time leads to

$$\hat{q}(t) = \hat{q}(o) + \frac{\dot{\hat{q}}(o)}{\gamma}(1-e^{-\gamma t}) + \int_0^t dt' \int_0^{t'} dt''\ e^{-\gamma(t-t'')}\ \frac{\hat{\eta}(t'')}{m}. \tag{12.34}$$

Equation (12.33) yields for the momentum auto-correlation function in equilibrium:

$$\left\langle \hat{p}(t)\hat{p}(o)\right\rangle = \left(\frac{m}{2\pi}\right)\int_{-\infty}^{\infty} d\omega\ \hbar\omega \frac{\gamma}{\gamma^2+\omega^2}\ \coth\left(\frac{\hbar\beta\omega}{2}\right)\cos(\omega t), \tag{12.35}$$

where we ignored the dependence on the initial momentum $\hat{p}(o)$, which implies that the particle is decoupled from the combined density matrix for the particle and the bath. Similarly,

$$\left\langle \hat{p}^2(t)\right\rangle = \left(\frac{m\gamma}{\pi}\right)\int_0^{\infty} d\omega \frac{\hbar\omega}{\gamma^2+\omega^2}\coth\left(\frac{\hbar\beta\omega}{2}\right)\cdot\left[1+e^{-2\gamma t}-2e^{-\gamma t}\cos\omega t\right]+\left\langle \hat{p}^2(o)\right\rangle e^{-2\gamma t}. \tag{12.36}$$

Only asymptotically ($t \to \infty$) does Equation (12.36) reduce to the equilibrium result of Equation (12.35) with t set equal to zero in the latter. The three-dimensional generalization of the above results [Equations (12.35) and (12.36)] is automatic through the incorporation of a factor 3 in their right sides.

In order to gain further insights into quantum diffusive behavior, we may examine the mean-squared displacement when a particle is in equilibrium with the heat bath. From Equation (12.34), we find

$$\left\langle q^2(t)\right\rangle = \frac{\hbar}{2m\pi}\int_{-\infty}^{\infty} d\omega \left(\frac{\gamma}{\gamma^2+\omega^2}\right)\frac{\cot\hbar\dfrac{\hbar\beta\omega}{2}}{\omega}\left(1-\cos\ \omega t\right). \tag{12.37}$$

12.5 Diffusive Quantum Cyclotron Motion

The diffusive motion of an electrically charged particle such as an electron, proton, or a more massive object such as an ion under the influence of an external magnetic field was discussed earlier in Sections 4.5, 5.6, and 8.3 using classical mechanics. In this section, we will cover the corresponding treatment in the quantum domain.

The characteristic feature of quantum mechanics is that the dynamics is now governed by coherent precession of the velocity vector around the magnetic field at a rate called the cyclotron frequency $\omega_c = (|e|B)/mc$. The $|e|$ represents the magnitude of the electric charge, m is its mass, B is the strength of the magnetic field, and c is the speed of light. Diffusion parameterized by the friction coefficient causes incoherence in this motion. When $\gamma >> \omega_c$, we expect strong decoherence such that a quantum system appears classical.

In addition, the temperature enters the discussion in terms of the quantal frequency $(\beta\hbar)^{-1}$ cited in the introductory remarks in this chapter. The quantal frequency typically can be as large as $(10)^{11}$ sec^{-1} even at such low temperatures as $T \sim 1K$, i.e., quantum coherence vanishes on timescales larger than $(10)^{-11}$ sec at $T \sim 1K$. Therefore, the study of diffusive quantum cyclotron motion involves investigating the interplay of three competing timescales ω_c, γ^{-1}, and $(\beta\hbar)^{-1}$.

The formalism is now based on a Hamiltonian that is a generalization of Equation (12.1) in order to incorporate the magnetic field. If we take the field to be applied along the *z* axis, there is no Lorentz force along *z*, and hence the motion along that axis remains free. Therefore, we will have to essentially restrict our discussion to the two-dimensional *xy* plane. The underlying Hamiltonian can be written as

$$H = (1/2m)\,(\mathbf{p} - e\mathbf{A}/c)^2 + \sum_{j=1}^{N} \{(\mathbf{p}_j)^2/2m_j + m_j\,(\omega_j)^2\,(\mathbf{q}_j - \mathbf{q})^2/2\}, \quad (12.38)$$

where all operators are vectors in the *xy* plane and **A** is the vector potential. The bilinear coupling of the particle coordinate **q** and the oscillator coordinate $\mathbf{q}_j$ is the distinctive feature of models of Brownian motion considered by Ford et al. (1965, 1988), Zwanzig (1973), Caldeira and Leggett (1983), and others. Additionally, as shown by Chang and Chakravarty (1985), even when the environment is a metallic system, the fermionic bath may be represented by an effective bosonic bath, as envisaged in Equation (12.41), especially when we restrict the discussion to electron hole excitations near the Fermi surface that can be treated as Schwinger bosons.

We can now follow the same procedure as in Section 12.1 in deriving the quantum Langevin equation. Note that the magnetic field does no work on the particle and hence has no influence on the heat bath. Thus the Lorentz force simply appears as an additional term in the equation of motion. Comparing with Equation (12.22), we now have

$$m\frac{d^2}{dt^2}\mathbf{q} + m\gamma\frac{d}{dt}\mathbf{q} - \frac{e}{c}\left[\frac{d}{dt}\mathbf{q} \times B\right] = \eta(t). \quad (12.39)$$

The noise properties are two-dimensional generalizations of Equations (12.27) and (12.28) that are represented in terms of Cartesian indexes *i*, *j*, etc. Thus,

$$< \{\eta_i(t), \eta_j(t')\} >= \delta_{ij}\,\frac{2m\gamma}{\pi}\int_0^\infty d\omega\,\omega\coth\left(\frac{\beta\hbar\omega}{2}\right)\cos[\omega(t-t')], \quad (12.40)$$

and

$$<[\eta_i\,(t), \eta_j(t')]>= \delta_{ij}\,\frac{2m\gamma}{i\pi}\int_0^\infty d\omega\;\omega \quad \sin\,[\omega\;(t-t')]. \tag{12.41}$$

Although we assumed Ohmic dissipation, the quantum diffusive dynamics is still non-Markovian, reflected in the non-white noise character of the stochastic force. We now write the equations of motion separately along the x and y directions:

$$\frac{d^2}{dt^2}x(t)+\gamma\frac{d}{dt}x(t)-\omega_c\frac{d}{d}y(t)=\eta_x(t)/m,$$

$$\frac{d^2}{dt^2}y(t)+\gamma\frac{d}{dt}y(t)-\omega_c\frac{d}{d}x(t)=\eta_y(t)/m. \tag{12.42}$$

To rewrite Equation (12.42) into a more compact form, we introduce the variables:

$$\xi = x+iy, \quad \eta=\eta_x+i\eta_y, \quad \bar{\gamma}=\gamma+i\omega_c \tag{12.43}$$

We find

$$\frac{d^2}{dt^2}\xi+\bar{\gamma}\frac{d}{dt}\xi(t)=\frac{\eta(t)}{m}, \tag{12.44}$$

which has the solution

$$\xi(t)=\bar{\gamma}^{-1}\frac{d\xi}{dt}\bigg|_{t_0}\left[1-e^{-\bar{\gamma}(t-t_0)}\right]+\int_{t_0}^{t}d\tau\; e^{-\bar{\gamma}\tau}\int_{t_0}^{\tau}dt'e^{\bar{\gamma}t'}\;\frac{\eta(t')}{m}, \tag{12.45}$$

assuming the initial condition $\xi(t_0)=0$. The time derivative of the above yields the velocity operator as

$$\frac{d\xi}{dt}=\frac{d\xi}{dt}\bigg|_{t_0}\; e^{-\bar{\gamma}(t-t_0)}+e^{-\bar{\gamma}t}\int_{t_0}^{t}dt'e^{-\bar{\gamma}t'}\,\frac{\eta(t')}{m}\;. \tag{12.46}$$

We now focus attention on the correlation function:

$$C(t,t')=\left\langle\dot{\xi}(t)\;\dot{\xi}^{+}(t')\right\rangle,\; t>t', \tag{12.47}$$

where the overhead dot refers to time derivatives and the superscript indicates hermitic adjoint. Using Equation (12.46), the correlation function can be re-expressed as

$$C(t,t') = \left\langle \dot{\xi}(t_0)\ \dot{\xi}^+(t_0) \right\rangle \exp[-\bar{\gamma}(t-t_0) - \bar{\gamma}^*(t'-t_0)]$$

$$+\frac{1}{m^2} e^{-(\bar{\gamma}t+\bar{\gamma}^*t')} \int_{t_0}^{t} d\tau \int_{t_0}^{t'} d\tau'\ e^{(\bar{\gamma}\tau+\bar{\gamma}^*\tau')} \left\langle \eta(\tau)\ \eta^+(\tau') \right\rangle, \tag{12.48}$$

where we used the causality property that the initial velocity operator is uncorrelated with the noise operator, and the fact that

$$< \eta(t) > = 0. \tag{12.49}$$

As we are interested in equilibrium properties, we set t_0 to $-\infty$, tacitly recognizing that the coupling of the particle to the bath was switched on in the infinite past. Without any loss of generality, we can take $\xi(t)$ and the corresponding time derivatives as zero. Substituting for the noise correlation, we find

$$C(t,t') = \frac{1}{m\pi} \int_{-\infty}^{\infty} d\omega\ \frac{\gamma\omega}{\gamma^2 + (\omega + \omega_c)^2} \left[\coth\left(\frac{\beta\hbar\omega}{2} \right) - 1 \right] e^{i\omega(t-t')}, \tag{12.50}$$

which manifestly displays stationarity as a function of the time difference $(t - t')$. It is instructive to rewrite the above equation in terms of the longitudinal and transverse components as

$$\left\langle \upsilon_x(t)\upsilon_x(0) \right\rangle + \left\langle \upsilon_y(t)\upsilon_y(0) \right\rangle = \frac{1}{m\pi} \int_{-\infty}^{\infty} d\omega\ \frac{\gamma\omega}{\gamma^2 + (\omega + \omega_c)^2} \left[\coth\left(\frac{\beta\hbar\omega}{2} \right) - 1 \right] \cos(\omega t), \tag{12.51}$$

and

$$\left\langle [\upsilon\ (t) \times \upsilon\ (0)]_z \right\rangle = \frac{1}{m\pi} \int_{-\infty}^{\infty} d\omega\ \frac{\gamma\omega}{\gamma^2 + (\omega + \omega_c)^2} \left[\coth\left(\frac{\beta\hbar\omega}{2} \right) - 1 \right] \sin\ (\omega t). \tag{12.52}$$

Our final task in this subsection is to analyze the mean-squared displacement for which it is useful to consider the following correlation function:

$$\chi(t,t') = \left\langle \xi(t)\ \xi^+(t') \right\rangle,\ t > t'. \tag{12.53}$$

Substituting Equation (12.45) and performing certain integrals, we obtain after some algebra,

$$\chi(t,t') = \frac{\langle \dot{\xi}(t)\ \dot{\xi}^{+}(t_0)\rangle}{\gamma^2+\omega_c^2}[1+e^{(-\bar{\gamma}t-\bar{\gamma}^{*}t'+2\gamma t_0)}-e^{-\bar{\gamma}(t-t_0)}-e^{-\bar{\gamma}^{*}(t'-t_0)}]$$

$$+\frac{1}{m\pi}\int_{-\infty}^{\infty} d\omega\ \frac{\gamma\omega}{\gamma^2+(\omega+\omega_c)^2}\left[\coth\left(\frac{\beta\hbar\omega}{2}\right)-1\right]$$

$$\times\left\{\frac{1}{\omega^2}\left[1+e^{i\omega(t-t')}-e^{i\omega(t-t_0)}-e^{i\omega(t'-t_0)}\right]\right.$$

$$+\frac{1}{\gamma^2+\omega_c^2}[1+e^{(-\bar{\gamma}t-\bar{\gamma}^{*}t'+2\gamma t_0)}-e^{-\bar{\gamma}(t-t_0)}-e^{-\bar{\gamma}^{*}(t'-t_0)}]$$

$$\left.-\frac{i}{\bar{\gamma}\omega}[1-e^{-i\omega(t'-t_0)}][e^{-\bar{\gamma}(t-t_0)}-1]+\frac{i}{\bar{\gamma}^{*}\omega}[1-e^{i\omega(t-t_0)}][e^{-\bar{\gamma}^{*}(t'-t_0)}-1]\right\}. \quad (12.54)$$

We are now ready to look at the mean-squared displacement obtained from above by taking the $t = t'$ limit. However since we are also interested in studying the explicit time dependence of the mean squared-displacement, we use a different boundary condition, by setting $t_0 = 0$. We find

$$\langle x^2(t)+y^2(t)\rangle = \frac{\langle \dot{x}^2(0)+\dot{y}^2(0)\rangle}{\gamma^2+\omega_c^2}[1+e^{-2\gamma t}-2e^{-\gamma t}\cos(\omega_c\ t)]$$

$$+\frac{1}{m\pi}\int_{-\infty}^{\infty} d\omega\ \frac{\gamma\omega}{\gamma^2+(\omega+\omega_c)^2}\left[\coth\left(\frac{\beta\hbar\omega}{2}\right)-1\right]$$

$$\times\left\{\frac{4}{\omega^2}\sin^2\left(\frac{\omega t}{2}\right)+\frac{1}{\gamma^2+\omega_c^2}[1+e^{-2\gamma t}-2e^{-\gamma t}\cos(\omega_c t)]+\frac{2}{\omega(\gamma^2+\omega_c^2)}\left[2\omega_c\sin^2\left(\frac{\omega t}{2}\right)\right.\right.$$

$$\left.\left.-\gamma\sin(\omega t)+e^{-\gamma t}[\omega_c\cos(\omega+\omega_c)t-\omega_c\cos(\omega_c t)+\gamma\sin(\omega+\omega_c)t-\gamma\sin(\omega_c t)]\right]\right\}.$$

$$(12.55)$$

Once again certain insights can be derived by considering the free particle case (when the cyclotron frequency vanishes) as well as the classical case that obtains at infinite temperature.

12.6 Diffusive Landau Diamagnetism

One of the enigmatic phenomena of the condensed phase is the macroscopic response of a collection of electrons to an external magnetic field in terms of orbital dynamics. (We do not consider the response of electronic spins in the present discussion.) More than a century ago, Bohr (1911) and later van Leeuwen (1921) demonstrated that diamagnetism, as calculated from Gibbsian classical statistical mechanics, does not yield a finite answer! This is paradoxical because a metal such as bismuth (Bi) exhibits diamagnetism. The resolution of this problem brought to the fore the essential role of quantum mechanics via the work of Landau (1930), van Vleck (1932), and Peierls (1933, 1979).

Quantum mechanics is at the very core of both diamagnetism and magnetism just as it is for superfluidity and superconductivity. Before we take up the quantum treatment (and also the further effects of diffusion on diamagnetism), it is instructive to briefly explain the Bohr–van Leeuwen argument, especially to elucidate the significant influence of boundaries. Normally, for a statistical mechanical system of N particles, the ratio of the boundary particles to bulk particles vanishes as $(N)^{-1/3}$. However, for diamagnetism, the boundary does matter and that is why Peierls called diamagnetism one of the "surprises in theoretical physics" (1979). Diamagnetism is defined by the expectation value of the orbital angular momentum vector operator:

$$\mathbf{M} = (eN/2mc) <\mathbf{L}> = (eN/2mc) (\mathbf{r} \times \mathbf{p}_{kin}>. \tag{12.56}$$

In the above, $\mathbf{r}$ is the position operator and $\mathbf{p}_{kin}$ is the so-called kinematic momentum. The crucial point of electrodynamics is that the kinematic momentum is not the same as the canonical momentum $\mathbf{p}$. They differ because of the vector potential $\mathbf{A}$:

$$\mathbf{p}_{kin} = (\mathbf{p} - e\mathbf{A}/c). \tag{12.57}$$

Through $\mathbf{A}$, the magnetic field $\mathbf{B}$ appears due to the relation:

$$\mathbf{B} = \text{Curl } \mathbf{A}. \tag{12.58}$$

Peierls' argument on the essential role of the boundary is amplified by the defining equation (12.56). While the number of electrons on the boundary is indeed small, a substantial portion of the surface electrons contributes to $\mathbf{M}$ due to the large magnitude of the position vector, when the origin of the coordinate system is chosen once and for all. Bohr and van Leeuwen did not actually compute the contribution of the bulk electrons explicitly, as is amply clarified by the following argument of Peierls.

Recognizing that diamagnetism follows from the asymptotic property of the time-dependent operator $<\mathbf{L}(t)>$, as a system approaches thermal equilibrium, diamagnetism can also be calculated from the derivative with respect to $\mathbf{B}$ of the free energy, which in turn is given by the log of the partition function. Peierls and van Vleck showed that while the calculation of diamagnetism from Equation (12.56) is very sensitive to surface (or boundary) conditions, the partition function route is not! The reason is that the classical partition function involves an integration of the entire phase spaces of coordinates and momenta and hence a shift of the momentum variables inside the integral leaves the partition function independent of **B** and therefore the corresponding derivative vanishes. Thus, in estimating the diffusive effect of diamagnetism, we must carefully treat the boundary effect due to a trick attributed to Darwin (1930) as discussed below.

Landau recognized that quantum mechanics yields discrete energy levels (and the consequent degeneracy of each level), suggesting that the bulk and surface contributions to diamagnetism are different in quantum mechanics. Hence the cancellation of the two respective contributions is incomplete in quantum mechanics. Today, the boundary effects in the presence of a magnetic field play a critical role in setting up the edge currents important in the quantum Hall effect (Laughlin 1980). Our study of quantum diffusion in the context of Landau diamagnetism is motivated by this query: because classical-like results ensue for a heavily damped quantum system, should the Bohr–van Leeuwen theorem not emerge as a consequence of large damping?

With this background we calculate diamagnetism from the defining relations [cf. Equation (12.56)]:

$$M_z(t) = (|e|/2c) \langle (x\, dy/dt - y\, dx/dt) \rangle = |e|/4c \,\mathrm{Im}\, \langle \dot{\xi}\xi^{+} + \xi^{+}\dot{\xi} \rangle. \tag{12.59}$$

It is straightforward to calculate $M_z(t)$ from the result for $\xi(t)$ [cf. Equation (12.45)]. A remark is, however, in order. Because the Landau result for diamagnetism is an equilibrium attribute, we set $t_0 = 0$ and $t = \infty$, thereby relying on the built-in fluctuation–dissipation theorem. After some algebra, we find (Dattagupta and Singh 1997):

$$\lim_{t\to\infty} M_z(t) = -\frac{|e|}{2\pi mc} \int_{-\infty}^{\infty} d\omega \frac{\gamma}{\gamma^2 + (\omega - \omega_c)^2} \coth\left(\frac{\beta\hbar\omega}{2}\right). \tag{12.60}$$

This result, however, reveals a problem as can be ascertained by going to $\gamma = 0$, i.e., the dissipation-free limit, and making use of the Lorentzian form of the Dirac delta function (Dirac 1958). We find

$$M_z^0 = -\frac{|e|}{2mc} \coth\left(\frac{\beta\hbar\omega_c}{2}\right), \tag{12.61}$$

which is only a part of the Landau result (Landau 1930, van Vleck 1932) and arises in the calculation in thermal equilibrium if we ignore the role of boundary electrons. In order to include the boundary contribution, we employ a contrived parabolic potential that constrains the electrons following the scheme of Darwin. While the effect of the parabolic potential is interesting in its own right, our intent is to go ahead with the calculation for diamagnetism as a time-dependent entity from the quantum Langevin equation, then take the asymptotic limit ($t = \infty$) and finally switch the parabolic potential off. We add an extra term to the Hamiltonian that has the form of $k(x^2 + y^2)/2$ (k is the spring constant). The resulting Langevin equation now reads:

$$\frac{d^2\xi}{dt^2} + \bar{\gamma}\frac{d\xi}{dt} + \frac{k}{m}\xi = \frac{\eta(t)}{m}. \tag{12.62}$$

The solution of Equation (12.62) with $\xi(t = 0) = 0$ reads

$$\xi(t) = \frac{1}{\omega_+ - \omega_-}\left\{\dot{\xi}(0)(e^{\omega_+ t} - e^{\omega_- t}) + \left[\int_0^t d\tau(e^{\omega_+(t-\tau)} - e^{\omega_-(t-\tau)})\right]\frac{\eta(\tau)}{m}\right\}, \tag{12.63}$$

where

$$\omega_\pm = -\frac{\bar{\gamma}}{2} \pm \frac{1}{2}\sqrt{\bar{\gamma}^2 - \omega_0^2}, \quad \omega_0^2 = \frac{4k}{m}. \tag{12.64}$$

Substituting this into Equation (12.59) and letting $t = \infty$, we obtain

$$M_z = \frac{|e|\gamma}{2\pi mc}\frac{1}{|\omega_+ - \omega_-|^2} \times$$

$$\mathrm{Im}\int_{-\infty}^{\infty} d\omega\, \omega \coth\left(\frac{\beta\hbar\omega}{2}\right)\left(\frac{1}{\omega + i\omega_+^*} - \frac{1}{\omega + i\omega_-^*}\right)\left(\frac{\omega_+}{\omega - i\omega_+} - \frac{\omega_-}{\omega - i\omega_-}\right). \tag{12.65}$$

We first compute the integral over ω, before taking the Darwin limit, by closing the contour in the upper half of the complex plane and noting that the coth function has poles at $\frac{\beta\hbar\omega}{2} = in\ \pi$, n being an integer. The eventual result is

$$M_z = -\frac{|e|\theta}{mc}\,\mathrm{Im}\left\{\sum_{n=1}^{\infty}\frac{n^2\pi^2}{(n\pi + \nu_+^*)(n\pi + \nu_-^*)(n\pi - \nu_+)(n\pi - \nu_-)}\right.$$

$$\left. + \frac{1}{(\nu_+^* - \nu_-^*)}\left[\frac{(\nu_+^*)^2 \cot(\nu_+^*)}{(\nu_+^* + \nu_-)(\nu_+^* + \nu_+)} - \frac{(\nu_-^*)^2 \cot(\nu_-^*)}{(\nu_- + \nu_-^*)(\nu_-^* + \nu_+)}\right]\right\}, \tag{12.66}$$

where we have defined the dimensional parameters:

$$\theta = \frac{\beta\hbar\gamma}{2}, \quad \nu_\pm = \frac{\beta\hbar\omega_\pm}{2}. \tag{12.67}$$

Now, we carefully take the $\omega_0 \to 0$ limit by first observing that for $\omega_0 = 0$,

$$\nu_+ \simeq \frac{\nu_0^2}{4}\frac{(\theta - i\nu_c)}{\theta^2 + \nu_c^2}, \quad \nu_- \simeq -(\theta + i\nu_c),$$
$$\nu_c = \frac{\beta\,\omega_c}{2}, \quad \nu_0 = \frac{\beta\omega_0}{2}. \tag{12.68}$$

Finally, the expression for the magnetization is

$$M_z = \frac{|e|}{2mc}\left[\sum_{n=1}^{\infty}\frac{4n\pi\theta\nu_c}{(\nu_c^2 + \theta^2 - n^2\pi^2)^2 + 4n^2\pi^2\nu_c^2} + \frac{\nu_c}{\theta^2 + \nu_c^2}\right.$$
$$\left. - \frac{1}{2}\left[\frac{\sinh(2\nu_c)}{\sinh^2\nu_c + \sin^2\theta}\right]\right]. \tag{12.69}$$

Equation (12.69) has the correct dissipation-free Landau result:

$$m_z^0 = \frac{|e|}{2mc}\left[\frac{1}{\nu_c} - \coth(\nu_c)\right]. \tag{12.70}$$

It is instructive to rewrite the above result in a suggestive form by splitting off a Landau-like (lifetime-broadened) result as

$$M_z = \frac{|e|}{2mc}\left\{\sum_{n=1}^{\infty}\frac{4n\pi\theta\upsilon_c}{(\upsilon_c^2 + \theta^2 - n^2\pi^2)^2 + 4n^2\pi^2\upsilon_c^2} + \mathrm{Re}\left[\frac{1}{\nu_c - i\theta} - \coth(\nu_c - i\theta)\right]\right\}. \tag{12.71}$$

The last two terms involve the Landau answer with the levels modified by the so-called Matsubara frequency (Sakurai 1994). The first term (under the summation over n) provides a nontrivial correction to the Landau expression. Employing the summation formula:

$$\coth z = \sum_{n=-\infty}^{\infty}\frac{1}{z + in\pi} = \frac{1}{z} + 2\sum_{n=1}^{\infty}\frac{1}{z^2 + n^2\pi^2}, \tag{12.72}$$

we find a more compact formula as

$$M_z = -2K_B T_B \left(\frac{e}{mc}\right)^2 \sum_{n=1}^{\infty} \frac{1}{(\upsilon_n + \gamma)^2 + \omega_c^2}, \tag{12.73}$$

where

$$\upsilon_n = 2KTn\pi. \tag{12.74}$$

This form immediately entails a generalization to the non-Ohmic case (see next chapter) wherein the friction acquires a frequency dependence. We find (Li et al. 1990, 1996; Bandyopadhyay and Dattagupta 2006):

$$M_z = -2K_B TB \left(\frac{e}{mc}\right)^2 \sum_{n=1}^{\infty} \frac{1}{[\upsilon_n + \gamma(i\upsilon_n)]^2 + \omega_c^2}, \tag{12.75}$$

We shall return to the discussion of the above results, their dependence on damping, and relevance to coherence-to-decoherence transition in Chapter 16.

References

Bohr, N. 1972. Ph.D. thesis (1911). In *Collected Works*, Vol. 1. Amsterdam: North-Holland.

Caldeira, A. O. and A. J. Leggett. 1983. Quantum tunneling in a dissipative system. *Physica A* 121, 587.

Chang, L. D. and S. Chakravarty. 1985. Dissipative dynamics of a two-state system coupled to a heat bath. *Phys. Rev.* B31, 154.

Darwin, G. C. 1930. The diamagnetism of the free electron. *Proc. Cambridge Philos. Soc.* 27, 86.

Dattagupta, S. and J. Singh. 1997. Landau diamagnetism in a dissipative and confined system. *Phys. Rev. Lett.* 79, 961.

Dattagupta, S. and J. Singh 1996. *Pramana J. Phys.* 47, 211.

Dirac, P.A.M. 1958. *The Principles of Quantum Mechanics*, 4th ed. London: Oxford University Press.

Ford, G. W., M. Kac, and P. Mazur. 1965. Statistical mechanics of assemblies of coupled oscillators. *J. Math. Phys.* 6, 504.

Landau, L. D. 1930. Diamagnetism of metals. *Z. Phys.* 64, 629.

Laughlin, R. B. 1981. Theory of polarization of crystalline solids. *Phys. Rev.* 23, 5632.

Li, X. L., G. W Ford, and R. F. O'Connell. 1990. Charged oscillator in a heat bath in the presence of a magnetic field. *Phys. Rev. A* 42, 4519.

Li, X. L., G. W. Ford, and R. F. O'Connell. 1990. Magnetic-field effects on the motion of a charged particle in a heat bath. *Phys. Rev. A* 41, 5287.
Peierls, R.E. 1979. *Surprises in Theoretical Physics.* Princeton: Princeton University Press.
Peierls, R. E. 1933. Quantum theory of solids. *Z. Phys.* 81, 186.
Sakurai J.J. 1994. *Modern Quantum Mechanics* (Rev. Ed.). Reading, MA: Addison Wesley.
van Leeuwen, H. J. 1921. Problèms de la théorie électronique du magnétisme. *J. Phys. Radium* 2, 361.
van Vleck, J.H. 1932. *Theory of Electric and Magnetic Susceptibilities.* London: Oxford University Press.
Zwanzig, R. 1973. Nonlinear generalized Langevin equations. *J. Stat. Phys.* 9, 215.

13

Path Integral Treatment of Quantum Diffusion

13.1 Introduction

The reader may gather from Chapters 4 to 6 that the sequence in which we developed the classical formalism of diffusion was to first introduce the equation of motion for dynamical variables. This was achieved through Langevin equations for position and momentum. Diffusive effects were incorporated through heat bath-induced thermal noise and systematic damping parameters. This approach follows the Heisenberg picture of quantum mechanics in which the operators evolve in time whereas the wave function stays constant. Contrast this with the Schrödinger picture in which the wave functions are taken to satisfy the time-dependent Schrödinger equation while the operators remain frozen in time.

Chapter 5, following Chapter 4 on Langevin equations, was devoted to the Smoluchowski, Fokker–Planck, and Kramers equations for the probability distribution function, which is the phase space analogue of the Schrödinger wave function. In developing quantum diffusion, we follow the same pathway, i.e., first discuss quantum Langevin equation (QLE) as we do in Chapter 12, and then explain in this chapter the quantum version of something close to a Fokker–Planck equation.

The entity that plays the role of the probability distribution function in quantum mechanics is the von Neumann density operator $\rho(t)$. The operator can be constructed from the time-dependent Schrödinger wave function $|\Psi(t)>$ as $|\Psi(t)> \, <\Psi(t)>|$. While $|\Psi(t)>$ obeys the Schrödinger equation, $\rho(t)$ satisfies the Liouville equation:

$$\frac{\partial \rho}{\partial t} = -\frac{i}{\hbar}[\mathcal{H}, \rho(t)], \tag{13.1}$$

that can be constructed easily from the underlying Schrödinger equation:

$$\frac{\partial}{\partial t}|\Psi(t)> = -\frac{i}{\hbar}[\mathcal{H}, \Psi(t)>. \tag{13.2}$$

Equation (13.1) is the exact analogue of the phase space density function in classical mechanics (Goldstein 1964). However, it is a far cry from describing diffusive processes as it exhibits unitary evolution associated with a Hamiltonian $\mathcal{H}$ that is hermitian. How then do we go about converting Equation (12.1) into something that can incorporate quantum diffusion?

Our strategy here is very similar to what we did in Section 12.2 in deriving from first principles a quantum Langevin equation from a system-plus-bath Hamiltonian $\mathcal{H}$ of Equation (12.1). We start from the solution of Equation (13.1),

$$\rho(t) = e^{-i\mathcal{H}t/\hbar}\rho(o)\; e^{-i\mathcal{H}t/\hbar}, \tag{13.3}$$

and integrate out (as shown below) the bath degrees of freedom exactly as in Section 12.1. Our initial step is to write the matrix elements of $\rho(t)$. In the coordinate representation:

$$\begin{aligned} < qQ\,|\,\rho(t)\; q'Q' > \; &= \int dq''\, dq'''\, dQ''\, dQ'''\;\; K(q,Q,t;\; q'',Q'',o) \\ K^{*}(q',Q',t;q''',Q''',o) &< q''Q''\,|\,\rho(o)\,|\,q'''Q''' >, \end{aligned} \tag{13.4}$$

where the qs denote coordinates of the system of interest and Qs represent coordinates of the bath. The Ks are propagators that have functional integral forms as

$$K(q,Q,t;q'',Q'',O) = < qQ\,|\,e^{-i\mathcal{H}t/\hbar}\,|\,q''Q'' > = \iint Dq\; DQ\;\; \exp\left(\frac{i}{\hbar}S\,[q,Q]\right), \tag{13.5}$$

and we have a similar expression for K^{*}. All the functional integrals are evaluated over paths $q(t)$, $Q(t')$, with end points $q(t) = q$, $q(o) = q''$, $Q(t) = Q$, $Q(o) = Q''$ and the action S is given by

$$S = \int_{0}^{t} \mathcal{L}\; dt', \tag{13.6}$$

where $\mathcal{L}$ is the Lagrangian associated with the Hamiltonian in Equation (12.1). As in Section 12.2, we assume that initially the system is decoupled from the bath and that the bath continues to remain in thermal equilibrium (and hence is only weakly disturbed by the system). Thus we begin from a factorized initial condition:

$$\rho(o) = \rho_s(o) \bullet \rho_B(o), \tag{13.7}$$

where $\rho_B(o)$ is given by Equation (12.13). In accordance with our stated objective, we want to trace out or integrate out all the bath coordinates to derive a reduced density operator ρ_S that contains information about the influence of the bath on the system alone:

$$\rho_S(q,q',t) = \int dq'' dq''' J(q,q',t;\ q'',q''',o)\ \rho_s(q'',q''',o), \tag{13.8}$$

where

$$J(q,q',t;\ q'',q''',o) = \int\int Dq\ Dq' \exp\frac{i}{\hbar}\ S_s[q]\ \exp\ -\ \frac{i}{\hbar} S_s(q').\mathfrak{F}[q,q']. \tag{13.9}$$

The so-called influence functional (Feynman and Vernon 1963) is given by

$$\mathfrak{F}\ [q,q'] = \int dQ'' dQ''' \rho_B(Q'',Q''';o)$$

$$\cdot \int\int DQ\ DQ' \exp\frac{i}{\hbar}\ (\underset{I}{S}[q,Q] - \underset{I}{S}[q',Q'] + \underset{B}{S}[Q] - \underset{B}{S}\ [Q'])\ . \tag{13.10}$$

Here $S_{I(B)}$ is the appropriate action for the interaction (bath) term in the Lagrangian $\mathfrak{L}$. Similarly, S is the action for the system part of the Lagrangian.

13.2 Basic Model

To describe quantum diffusion, we stay with the simple model (Zwanzig 1973) discussed in Chapter 4 in the context of classical Langevin equation and adaptation of the simple model (Ford et al. 1965, Ullersma 1966) to quantum Langevin equations in Chapter 12. Thus, referring to Equation (12.1), the different terms in the Lagrangian can be read out as

$$\mathfrak{L}_s = \frac{1}{2} m(\overset{0}{q})^2 - U(q) + q^2 \sum_j \frac{C_j^2}{m_j \omega_j^2}, \tag{13.11}$$

$$\mathfrak{L}_I = q \sum_j C_j\ q_j, \tag{13.12}$$

$$\mathfrak{L}_B = \sum_j \frac{1}{2}\ m_j[(\overset{0}{q}_j)^2 - \omega_j^2 q_j^2]. \tag{13.13}$$

We may note that the manifest translational invariance of the Hamiltonian in Equation (12.1), or in particular its equivalent form in Equation (12.4), led to a counter term, i.e., the last term in Equation (13.11) that can be absorbed in the system potential $U(q)$. Specifically, if $U(q)$ were to represent a harmonic oscillator, the counter term (also quadratic in q) would yield a renormalized oscillator frequency. The validity of this model toward a transparent discussion of quantum dissipation has been elaborated earlier (Caldeira and Leggett 1983). We first dispense with the heat bath density matrix in equilibrium that, because of the harmonic nature of the bath Lagrangian $\mathfrak{L}_B$, has the usual expression:

$$\rho_B(Q_1, Q_2; o) = \prod_j \rho_B^{(j)}(Q_{1j}, Q_{2j}; o), \tag{13.14}$$

where

$$\rho_B^{(j)}(Q_{ij}, Q_{2j}; o) = \frac{m_j \omega_j}{2\pi\hbar \; \sinh(\hbar\beta\omega_j)} .$$

$$\exp\left\{ \frac{-m_j\omega_j}{2\hbar \; \sinh(\hbar\beta\omega_j)} [(Q_{1j}^2 + Q_{2j}^2)\cosh(\hbar\beta\omega_j) - 2Q_{1j}Q_{2j}] \right\}. \tag{13.15}$$

This plus the Lagrangian forms of $\mathfrak{L}_I$ and $\mathfrak{L}_B$ in Equations (13.12) and (13.13) allow us to exactly evaluate the influence functional in Equation (13.10) as (Feynman and Vernon 1963, Feynman and Hibbs 1965):

$$\mathfrak{F}[q, q'] = \exp - \frac{1}{\hbar} \int_0^t \int_0^\tau \{[q(\tau) - q'(\tau)] \; i\alpha_I(\tau - \tau') \; [q(\tau') + q'(\tau'))$$

$$+ [q(\tau) - q'(\tau)] \; \underset{R}{\alpha}(\tau - \tau') \; [q(\tau) - q'(\tau')]\} d\tau \; d\tau'. \tag{13.16}$$

where

$$\alpha_R(\tau - \tau') = \sum_j \frac{C_j^2}{2m_j\omega_j} \coth(\hbar\beta\omega_j)\cos[\omega_j(\tau - \tau')], \tag{13.17}$$

$$\alpha_I(\tau - \tau') = \sum_j \frac{C_j^2}{2m_j\omega_j} \sin\omega_j(\tau - \tau'). \tag{13.18}$$

A comparison with Equations (12.27) and (12.28) leads us to anticipate that α_R and α_I are connected with the symmetrized correlation function and the thermally averaged commutator of the noise operator $\hat{\eta}(t)$ (see below for explicit connections). We are now set to work with the reduced density operator in Equation (13.8) wherein the influence functional in Equation (13.16) is substituted in the propagator for the density operator given by Equation (13.9).

13.3 Ohmic Dissipation and Classical Limit

Following the lead in Section 12.2 we replace the summation over j by an integral, weighted by the spectral density $j(\omega)$ of bath oscillators that has the Ohmic dissipation form of Equation (12.19) (Caldeira and Leggett 1983). This allows us to write Equations (13.17) and (13.18) in the forms of Equations (12.27) and (12.28), as noted earlier:

$$\alpha_R(\tau-\tau') = \frac{2m\gamma}{\pi}\int_0^D d\omega\ \omega\ \coth\left(\frac{1}{2}\hbar\beta\omega\right)\cos[\omega(\tau-\tau')], \tag{13.19}$$

$$\alpha_I(\tau-\tau') = \frac{2m\gamma}{\pi}\int_0^D d\omega\,\omega\,\sin\,[\omega\,(\tau-\tau')]. \tag{13.20}$$

Note that we have retained the upper limit as the cut-off frequency D of bath oscillators to facilitate the computation of the propagator in Equation (13.8); see below. Eventually, we will let D go to infinity. [We do not discuss here the interesting issues of cut-off dependence of various physical attributes (Ingold 1998).] In Equation (13.20), the integral over ω can be easily evaluated as

$$\int_0^D d\omega\,\omega\sin\,[\omega\ (\tau-\tau')] = -\frac{d}{d(\tau-\tau')}\left(\frac{\sin D(\tau-\tau')}{(\tau-\tau')}\right). \tag{13.21}$$

Hence,

$$\int_0^\tau d\tau'\quad \alpha_I(\tau-\tau')\ [q(\tau')+q'(\tau')]$$

$$= \frac{2m\gamma}{\pi}\int_0^\tau d\tau'(-)\frac{d}{d(\tau-\tau')}\cdot\left[\frac{\sin\ D(\tau-\tau')}{(\tau-\tau')}\right](q(\tau')+q'(\tau')),$$

which, upon integration by parts, yields

$$\frac{2m\gamma}{\pi}\left\{\int_0^{\tau} d\tau' \frac{\sin\ D(\tau-\tau')}{(\tau-\tau')}(\dot{q}(\tau')+\mathring{q}'(\tau')-\frac{\sin(D\tau)}{\tau}(q(o)+q'(o))\right\}. \tag{13.22}$$

At this stage, we let D go to infinity and write the following representation for the delta function as

$$\underset{D\to\infty}{Lim}\frac{\sin D(\tau-\tau')}{(\tau-\tau')}=\pi\,\delta(\tau-\tau'). \tag{13.23}$$

Equation (13.22) then leads to

$$\int_0^{\tau} d\tau'\,\alpha_I(\tau-\tau')\,[(q(\tau')+q'(\tau')]=2m\gamma\,[(\mathring{q}(\tau)+\mathring{q}'(\tau))-\delta(\tau)\,(q(o)+q'(o)]. \tag{13.24}$$

The influence functional in Equation (13.16) can then be written as

$$\mathfrak{F}[q,q')=\exp-\frac{i}{\hbar}\left\{2m\gamma\int_0^{t}(q\mathring{q}+q\mathring{q}'-q'\mathring{q}-q'\mathring{q}')-(q^2(o)-q'^2(o)\right\}$$

$$\times\exp-\frac{i}{\hbar}\left\{\frac{2m\gamma}{\pi}\int_0^{t}\int_0^{\tau}d\tau\,d\tau'\int_0^{D}d\omega\,\omega\coth\left(\frac{1}{2}\beta\hbar\omega\right).\cos\,[\omega(\tau-\tau')]\,[q(\tau')-q'(\tau)][q(\tau')-q'(\tau')]\right\}. \tag{13.25}$$

Finally, the propagator in Equation (13.9) reads

$$J(q,q',t;\ q'',q''',o)=\int\int Dq\,Dq'\exp\ \frac{i}{\hbar}\left\{\underset{R}{S}[q]-\underset{R}{S}[q']-2m\gamma\int_0^{t}(q\mathring{q}-q'\mathring{q}'+q\mathring{q}'-q'\mathring{q})\right\}$$

$$\times\exp-\frac{2m\gamma}{\hbar\pi}\int_0^{D}d\omega\,\omega\coth\left(\frac{1}{2}\beta\hbar\omega\right)\int_0^{t}d\tau\int_0^{\tau}d\tau'[q(\tau)-q'(\tau)].\cos[\omega(\tau-\tau')]\,[q(\tau')-q'(\tau')], \tag{13.26}$$

where the renormalized action is given by

$$S_R=\int d\tau\left[\frac{1}{2}m(\dot{q})^2-U(q)+\ \frac{1}{2}m(\Delta\omega)^2q^2\right]. \tag{13.27}$$

The last expression within the square parentheses owes its existence to the counter term described earlier.

13.4 Master Equation in High-Temperature Limit

Our aim in this section is to analyze the semi-classical limit of Equation (13.26) to make contact with classical diffusion according to the Fokker–Planck equation covered in Chapter 5 (Caldeira and Leggett 1983). To this end, we consider high temperatures ($\beta \to o$) in which case *coth* (x) can be replaced by $(x)^{-1}$ when $x \to$ o. The integral over ω can then be easily carried out. Using as before the delta function representation of Equation (13.23), we find

$$J(q,q',t;q'',q''',o) = \int\int Dq\, Dq' \exp\frac{i}{\hbar}\left\{S_R[q] - S_R[q'] - 2m\gamma\int_0^t (q\dot{q} - q'\dot{q}' + q\dot{q}' - q'\dot{q})\right\} \tag{13.28}$$

Following the methodology of deriving the Schrödinger equation from the functional integration formalism (Feynman and Hibbs 1965) we now attempt to deduce a master equation for $\tilde{\rho}$. This requires that we convert Equation (13.8) into an integro-differential equation. Considering then an infinitesimal time element Δt, we find from Equation (13.8),

$$\tilde{\rho}_s(q,q',t+\Delta t) = \int\int dq''dq''' J(q,q',t+\Delta t;q'',q''',t)\; \tilde{\rho}_s(q'',q''',t). \tag{13.29}$$

Referring to Equation (13.28), the integral from o to t can be replaced by an integral from t to $t + \Delta t$ as far as the evaluation of the first term within the integrand (13.29) is concerned. Now,

$$\dot{q} \approx \frac{q(t+\Delta t) - q(t)}{\Delta t} = \frac{x}{\Delta t},\quad \dot{q}' = \frac{x'}{\Delta t}, \tag{13.30}$$

and

$$\int_t^{t+\Delta t} d\iota\, f(q(\iota)) \approx \Delta t\frac{q(t+\Delta t) + q(t)}{2} = \Delta t\left(q + \frac{x}{2}\right). \tag{13.31}$$

Our task now is to write J up to first order in Δt, extend the limits of the integrals over x and x' to $-\infty \to \infty$, and evaluate the Fresnel-type integrals by the saddle-point method (for details, see Caldeira and Leggett 1983). Equating terms of order Δt from both sides of Equation (13.31) in much the same manner as in the derivation of the Chapman–Kolmogorov equation in Chapter 2, we find

$$\frac{\partial}{\partial t}\tilde{\rho} = -\frac{\hbar}{2im}\left(\frac{\partial^2}{\partial q_1^2} - \frac{\partial^2}{\partial q_2^2}\right)\tilde{\rho} - \gamma(q_1 - q_2)$$

$$\cdot\left(\frac{\partial}{\partial q_1} - \frac{\partial}{dq_2}\right)\tilde{\rho} + \frac{1}{i\hbar}[U_R(q_1) - U_R(q_2)]\,\tilde{\rho} - \frac{2m\gamma K_B T}{\hbar^2}(q_1 - q_2)^2\tilde{\rho}. \quad (13.32)$$

In Equation (13.32), $\underset{R}{U}(q)$ is the renormalized potential energy given by

$$U_R(q) = U(q) - \frac{1}{2}m(\Delta\omega)^2 q^2. \quad (13.33)$$

We rewrite Equation (13.32) in the operator from

$$\frac{\partial}{\partial t}\tilde{\rho} = \frac{1}{i\hbar}[\mathcal{H}_{R,}\ \tilde{\rho}] + \frac{\gamma}{i\hbar}\{[q,\tilde{\rho}p] + [pq,\tilde{\rho}] + [\tilde{\rho}q,p]\} + \frac{2m\gamma K_B T}{\hbar^2}[[q,\tilde{\rho}],q], \quad (13.34)$$

where $\mathcal{H}_R$ is the renormalized Hamiltonian of the system alone. To directly relate Equation (13.34) to a Fokker–Planck-like classical equation, we consider the Wigner distribution (or Wigner transform of $\tilde{\rho}$) defined as

$$\mathcal{W}(q,p,t) = \frac{1}{2\pi\hbar}\int_{\infty}^{\infty} dq' \exp\left(\frac{ipq'}{\hbar}\right) < q - \frac{q'}{2}\,|\,\tilde{\rho}\,|\,q + \frac{q'}{2} >. \quad (13.35)$$

We find (Dekker 1977)

$$\frac{\partial \mathcal{W}}{\partial t} = -\frac{\partial}{\partial q}(p\mathcal{W}) + \frac{\partial}{\partial p}\left(\frac{\partial U}{\partial q}\mathcal{W}\right) + 2\gamma\frac{\partial}{\gamma p}(p\mathcal{W}) + 2m\gamma K_B T\frac{\partial^2}{\gamma p^2}\mathcal{W}. \quad (13.36)$$

In the limit $\hbar \to 0$, $\mathcal{W}$ reduces to the classical phase space distribution that obeys the Smoluchowski–Fokker–Planck–Kramers equation [cf. Equation (5.18)].

13.5 Application to Dissipative Diamagnetism

The problem of quantum dissipation and its effect on diamagnetism was already explained in Chapter 12 on the basis of quantum Brownian motion. The diamagnetic moment was evaluated from the asymptotic (t going to infinity) limit of a time-dependent expectation value of the orbital magnetic moment operator. The rationale of this approach was that because the fluctuation–dissipation relation is built into the system-plus-bath method, the asymptotic results must match with what we expect in equilibrium statistical mechanics.

No wonder then that this scheme has been dubbed by Kadanoff (2000) as the Einstein approach to statistical mechanics. Since the Einstein approach is embedded in the essential concepts of Brownian motion, it is intuitive and closely related to experiments as measurements are usually made in the time domain. By assuming ergodicity, time-averaged results are believed to give way to what one would calculate from the usual equilibrium statistical mechanics based on a Gibbs measure.

It is then of interest to learn directly what the Gibbs approach would predict for dissipative diamagnetism and compare the expressions with those derived from the Einstein approach (Bandyopadhyay and Dattagupta 2006). The functional integral method provides a convenient setting for posing this problem, inter alia combining thermodynamics with dynamics when dynamics is described by quantum Langevin equation. In a way, working with the path integral approach is tantamount to employing the Schrödinger picture of quantum mechanics as opposed to the Heisenberg picture inherent in the Langevin equations. Recall that the partition function in Gibbsian statistical mechanics can be written as

$$Z = \int D[x]\exp\left[-\frac{A_e[x]}{\hbar}\right], \tag{13.37}$$

where $A_e[x]$ is the effective Euclidean action and the functional integral is over all paths with periods $\hbar\beta$. The thermodynamic Helmholtz free energy is then given by (Huang 1987):

$$F = -\frac{1}{\beta}\ln Z. \tag{13.38}$$

The diamagnetization is calculated by taking the derivative of F with respect to the magnetic field B, applied along the laboratory-fixed z axis as

$$M_z = -\frac{\partial F}{\partial B}. \tag{13.39}$$

It may be recalled that in the functional integral approach, the canonical density operator $exp(-\beta\mathcal{H})$ is made to look like a time-development operator $exp(-i\mathcal{H}t/\hbar)$ by an analytic continuation or Wick rotation of time t to $t = -i\hbar\beta$. That procedure allows us to rewrite the Euclidean action as

$$A_e = \int_0^{\hbar\beta} d\tau\ [L_s(\tau) + L_B(\tau) + L_I(\tau)]\ , \tag{13.40}$$

where the subscripts S, B, and I represent the system, bath, and interaction, respectively. The corresponding Lagrangians are enumerated as [cf. Equations (13.11) through (13.13)]:

$$L_s(\tau) = \frac{m}{2}[\dot{x}^2(\tau) + \omega_0^2 \quad x^2(\tau) - i\omega_c(x(\tau) \times \dot{x}(\tau))_z], \tag{13.41}$$

where $\omega_c = \dfrac{eB}{mc}$ is the cyclotron frequency,

$$L_B(\tau) = \sum_{j=1}^{N} \frac{1}{2} m_j \left[\dot{x}_j^2(\tau) + \omega_j^2 x_j^2(\tau)\right],$$

$$L_I(\tau) = \sum_{j=1}^{N} \frac{1}{2} m_j \omega_j^2 \left[\dot{x}^2(\tau) - 2x_j(\tau).\, x(\tau)\right]. \tag{13.42}$$

Exploiting the fact that the paths have imaginary time periodicity $x(\hbar\beta) = x(0)$, we perform Fourier series expansions of system and bath variables as follows:

$$x(\tau) = \sum_n \tilde{x}(\upsilon_n) e^{-i\upsilon_n \tau}, \tag{13.43}$$

and

$$\vec{x}_j(\tau) = \sum_n \tilde{x}_j(\upsilon_n) e^{-i\upsilon_n \tau}. \tag{13.44}$$

In the above, the bosonic Matsubara frequencies υ_n are given by

$$\upsilon_n = \frac{2\pi n}{\hbar\beta}, \qquad n = 0, \ \pm 1, \ \pm 2, \ldots. \tag{13.45}$$

Following the detailed analysis of Weiss (1999) and employing Equation (13.42) the system part of the action can be expressed as

$$A_e^s = \frac{m}{2}\hbar\beta \sum_n \left[\left(\upsilon_n^2 + \omega_0^2\right)\left(\tilde{\mathfrak{X}}(\upsilon_n).\, \tilde{\mathfrak{X}}^*(\upsilon_n)\right) + \omega_c \ \upsilon_n \left(\tilde{\mathfrak{X}}(\upsilon_n) \times \tilde{\mathfrak{X}}^*(\upsilon_n)\right)_z\right]. \tag{13.46}$$

Next, the contribution of the interaction term to the action is

$$A_e^{B-I} = \frac{m}{2}\hbar\beta \sum_n \xi(\upsilon_n) \ \ (\tilde{\mathfrak{X}}(\upsilon_n). \ \ \tilde{\mathfrak{X}}^*(\upsilon_n)), \tag{13.47}$$

where

$$\xi(\upsilon_n) = \frac{1}{m} \sum_{j=1}^{N} m_j \ \omega_j^2 \frac{\upsilon_n^2}{\left(\upsilon_n^2 + \omega_j^2\right)}. \tag{13.48}$$

Defining the bath spectral function as

$$j(\omega) = \frac{\pi}{2} \sum_{j=1}^{N} m_j \, \omega_j^2 \, \delta(\omega - \omega_j), \tag{13.49}$$

we may re-express Equation (13.48) as

$$\xi(\upsilon_n) = \frac{2}{m\pi} \int_0^\infty d\omega \, \frac{j(\omega)}{\omega} \frac{\upsilon_n^2}{\left(\upsilon_n^2 + \omega^2\right)}. \tag{13.50}$$

Then Equation (13.41) leads us to

$$A_e = \frac{m}{2} \hbar\beta \sum_n \left[\left(\upsilon_n^2 + \omega_0^2 + \upsilon_n \, \tilde{\gamma}(\upsilon_n)\right) \left(\tilde{\mathfrak{X}}(\upsilon_n) . \tilde{\mathfrak{X}}^*(\upsilon_n)\right) + \omega_c \, \upsilon_n \left(\tilde{\mathfrak{X}}(\upsilon_n) \times \tilde{\mathfrak{X}}^*(\upsilon_n)\right)_z \right], \tag{13.51}$$

where the memory function can be defined as

$$\tilde{\gamma}(\upsilon_n) = \frac{2}{m\pi} \int_0^\infty d\omega \, \frac{j(\omega)}{\omega} \frac{\upsilon_n^2}{\left(\upsilon_n^2 + \omega^2\right)}. \tag{13.52}$$

Finally, we obtain the effective action in a separable form as

$$\begin{aligned} A_e = \frac{m}{2} \hbar\beta \sum_n \Big[& \left(\upsilon_n^2 + \omega_0^2 + \upsilon_n \, \tilde{\gamma}(\upsilon_n) + i\omega_c \, \upsilon_n\right) \left(\tilde{z}_+(\upsilon_n) \tilde{z}_+^*(\upsilon_n)\right) \\ & + \left(\upsilon_n^2 + \omega_0^2 + \upsilon_n \, \tilde{\gamma}(\upsilon_n) - i\omega_c \upsilon_n\right) \left(\tilde{z}_-(\upsilon_n) \tilde{z}_-^*(\upsilon_n)\right) \Big], \end{aligned} \tag{13.53}$$

where the relevant normal modes are

$$\begin{aligned} \tilde{z}^+(\upsilon_n) &= \frac{1}{\sqrt{2}} (x(\upsilon_n + i\tilde{y}(\upsilon_n)), \\ \tilde{z}^-(\upsilon_n) &= \frac{1}{\sqrt{2}} (x(\upsilon_n) - i\tilde{y}(\upsilon_n)). \end{aligned} \tag{13.54}$$

We are now set to derive the partition function as

$$Z = \frac{2\pi}{m\beta} \Pi_n \, \left[(\upsilon_n^2 + \omega_0^2 + \upsilon_n \, \ddot{\gamma}(\upsilon_n))^2 + \omega_c^2 \upsilon_n^2 \right]^{-1}, \tag{13.55}$$

from which the free energy can be deduced:

$$F = \frac{1}{\beta} \ln\left(\frac{m\beta\omega_0^4}{2\pi}\right) + \frac{2}{\beta} \sum_{n=1}^{\infty} \ln \, \left[\left(\upsilon_n^2 + \omega_0^2 + \upsilon_n \, \tilde{\gamma}(\upsilon_n)\right)^2 + \omega_c^2 \upsilon_n^2 \right]. \tag{13.56}$$

The first term is independent of the magnetic field because of the confining parabolic potential and will have no effect on diamagnetism, which is given by

$$M_z = -\sum_{n=1}^{\infty} \frac{\frac{4}{\beta B}\omega_c^2 \upsilon_n^2}{\left[\left(\upsilon_n^2 + \omega_0^2 + \upsilon_n \ \tilde{\gamma}(\upsilon_n)\right)^2 + \omega_c^2 \upsilon_n^2\right]}. \tag{13.57}$$

The result is manifestly negative as one would expect for diamagnetism. This equation exactly matches the result derived by Li et al. (1966). It can also be shown to be identical to the one derived from the quantum Langevin equation (Chapter 12) if we invoke Ohmic dissipation, for which

$$M_z = -\frac{B}{k_B T}\left(\frac{e\hbar}{mc}\right)^2 \sum_{n=1}^{\infty} \frac{1}{\left[n\pi + \frac{\upsilon_0^2}{n\pi} + \zeta\right]^2 + \upsilon_c^2}, \tag{13.58}$$

where υ_0 and υ_c have been defined in Equation (12.68), and

$$\zeta = \frac{1}{2}\hbar\beta\tilde{\gamma}(\upsilon_n). \tag{13.59}$$

References

Bandyopadhyay, M. and S. Dattagupta. 2006a. Dissipative diamagnetism—A case study for equilibrium and nonequilibrium statistical mechanics. *J. Stat Phys.* 123, 1273.

Bandyopadhyay, M. and S. Dattagupta. 2006b. Landau–Drude diamagnetism: Fluctuation, dissipation and decoherence. *J. Phys. Cond. Matt.* 18, 10029.

Caldeira, A. O. and A. J. Leggett. 1983. Quantum tunneling in a dissipative system. *Physica A.* 121, 587.

Feynman, R. P. and F. L. Vernon. 1963. Quantum statistical mechanics. *Ann. Phys.* (N.Y.) 24, 118.

Ford, G. W., M. Kac, and P. Mazur. 1965. Statistical mechanics of assemblies of coupled oscillators. *J. Math. Phys.* 6, 504.

Goldstein, H. 1964. *Classical Mechanics.* Reading, MA: Addison Wesley.

Ingold, G. L. 1998. In *Quantum Transport and Dissipation.* Weinheim: John Wiley & Sons.

Kadanoff, L. P. 2000. *Statistical Physics: Statics, Dynamics, and Renormalization.* Singapore: World Scientific.

Li, X. L., G. W. Ford, and R. F O'Connell. 1990. Charged oscillator in a heat bath in the presence of a magnetic field. *Phys. Rev. A* 42, 4519.

Li, X. L., G. W. Ford, and R. F. O'Connell. 1990. Magnetic-field effects on the motion of a charged particle in a heat bath. *Phys. Rev. A* 41, 5287.

Ullersma, P. 1966. An exactly solvable model for Brownian motion. I. Derivation of the Langevin equation. *Physica* 32, 27.

Zwanzig, R. 1973. Nonlinear generalized Langevin equations. *J. Stat. Phys.* 9, 215.

14

Quantum Continuous Time Random Walk Model

14.1 Introduction

The continuous time random walk (CTRW) framework (Montroll and Weiss 1965, Montroll and Scher 1973) was described in our earlier Chapters 6 through 9. The CTRW scheme is conceptually appealing and at the same time mathematically flexible to facilitate incorporation of various generalizations. Thus, although CTRW is usually set in the context of a Poisson process-driven pulse sequence leading to stationary Markov systems, incorporation of non-Poissonian pulsing with concomitant waiting-time distributions was seen in Chapter 7 to yield anomalous diffusion.

This technique has widespread applications to diffusion in disordered or glassy systems, ionic conduction, and other processes. Conversely, CTRW also provides a convenient way to include systematic evolution between pulses; that evaluation is described by a hermitic Hamiltonian.

That idea was discussed in detail in the context of molecular rotation spectroscopy of the infrared or Raman type in which a molecule rotates freely due to its inertia (characterized by its moment of inertia) until a pulse (or collision in a gas or liquid phase) randomly disrupts the systematic rotation (Dattagupta 1987). This competition between systematics (free rotational dynamics for molecules) and diffusion (occasioned by collisions, say) is at the heart of random processes.

In quantum systems, free evolution is governed again by a hermitian Hamiltonian leading to unitary dynamics. Unitary status is at the core of the quantum phase coherence responsible for phenomena like interference and information processing. Is there a CTRW-like scheme that can include coherent evolution interrupted by decohering heat bath-induced fluctuations?

In the CTRW premise, the streaming operator of coherent evolution is replaced by a propagator treated as a functional integral in Chapter 13. We are therefore interested in exploring the development of a formalism that allows a quantum system to be subjected to an ongoing renewal process in which Hamiltonian dynamics (represented in the Heisenberg or Schrödinger picture) is subjected to random pulsing by the environment.

As one would imagine, each such pulse would be expected to cause quantum transitions between the energy levels of the free Hamiltonian. Hence the pulses would be accompanied by interaction terms in the Hamiltonian that would contain operators for the system of interest. The bath part of the interaction will have to be incorporated in terms of effective classical fields. Unlike the quantum Langevin (Chapter 12) or functional integral situation (Chapter 13), the treatment is not fully quantum mechanical. Only the part pertaining to the system is handled quantum mechanically. The remainder, influenced by the bath, is dealt with by random rates and strengths of pulses.

14.2 Formulation

With this background, the quantum CTRW problem is posed as follows. We write the full Hamiltonian in an explicitly time-dependent form as (Clauser and Blume 1971):

$$\mathcal{H}(t) = \mathcal{H}_0 + \sum_{i=1}^{n} V_i \delta(t - t_i), \tag{14.1}$$

where $\mathcal{H}_0$ is the free Hamiltonian, t_i represents the instants at which the pulses occur, and V_i (model forms of which will be specified later) is the coupling operator that mimics the interaction with the environment. The t_is are assumed to be randomly distributed—most often as Poissonian—and V_is will contain certain random fields specified below.

A concept inherent in Equation (14.1), similar to the kinetic theory of gases, is that systems usually remain free under the Hamiltonian dynamics of $\mathcal{H}_0$, but occasionally are disturbed by collision-like interactions with the environment that throw a system instantaneously [because of the delta functions in Equation (14.1)] into a new quantum state from which it evolves again, and so on and on. Our use of pulses of infinitesimally short duration may appear to over-simplify the state of a system but it follows the spirit of the so-called impact approximation of the classical kinetic theory of gases (Huang 1987).

If we work in the Schrödinger mode, the free evolution will be dictated by the propagator that can be written as

$$\mathcal{U}_0(t) = \exp(i\mathcal{L}_0 t), \tag{14.2}$$

where $\mathcal{L}_0$ is the Liouville operator associated with the Hamiltonian $\mathcal{H}_0$. For any arbitrary operator A, $\mathcal{L}_0$ is defined as (Blume 1968):

$$\mathcal{L}_0 A = [\mathcal{H}_0, A] = \mathcal{H}_0 A - A\mathcal{H}_0, \tag{14.3}$$

whereas for a non-hermitian $\mathcal{H}_0$,

$$\mathcal{L}_0 A = (\mathcal{H}_0)^+ A - A\mathcal{H}_0. \tag{14.4}$$

The Liouville space description is convenient for averaging over only one time-ordered series (unlike two in the usual Schrödinger propagation) for heat bath-induced fluctuations. A similar Liouvillian exists for V_i which is designated $\mathcal{L}_i$.

The meaning of the Hamiltonian in Equation (14.1) is clear. Starting from an arbitrary initial state characterized by the density operator $\rho(t = 0)$, the system evolves until time t_1 in a given realization of the pulse sequence. At t_1, a pulse "hits" the system, causing it to "jump" to another arbitrary state. This process is mathematically incorporated by a transition matrix $\mathcal{T}_i$, which is given by $\exp(i\mathcal{L}_i)$. This transition matrix is reminiscent of the T matrix in the Lippman–Schwinger formulation of perturbation theory (Fetter and Walecka 2003). The probability that a pulse occurs between times t_i and $t_i + dt_i$ is, for a Poissonian process, $\exp(-\lambda\ \ t_i)\ \ dt_i$. The system then proceeds until time t_2 when another pulse changes the quantum state and the process is renewed. Thus, as before, we can write down the zero-collision, one-collision, two-collision, and n-collision terms as given below (Dattagupta 1984).

14.2.1 Zero-Collision Term

If there is no collision in time t, the probability of which is $\exp(-\lambda\, t)$, the averaged density operator (average carried out over an underlying Poisson distribution of pulses) is given by

$$(\rho(t))_{av} = \exp(-\lambda t)\exp(-i\mathcal{L}_0 t).\ \ \rho(0). \tag{14.5}$$

14.2.2 One-Collision Term

In this sequence of events, only one collision is assumed to occur, at time t_1 during the interval 0 to t. As argued in earlier chapters, the averaged density operator will then be given by

$$(\rho(t))_{av} = \int [\exp(i\mathcal{L}_0(t - t_1).\exp\,(-\lambda(t - t_1)(\mathcal{T}_1)_{av}\exp(-\mathcal{L}_0 t_1)\exp(-\lambda t_1)\,dt_1.\,\rho(0). \tag{14.6}$$

14.2.3 Two-Collision Term

Similarly, the two-collision term, when one collision occurs at t_1 followed by another at tine t_2, can be written as ($t_2 > t_1$):

$$(\rho(t))_{av} = \int\int [\exp(i\mathcal{L}_0(t-t_2)\cdot\exp(-\lambda(t-t_2))(\mathfrak{T}_2)_{av}\exp(i\mathcal{L}_0(t-t_1)\cdot\exp(-\lambda(t-t_1))$$
$$(\mathfrak{T}_1)_{av}\exp(-i\mathcal{L}_0 t_1)\exp(-\lambda t_1)\,dt_1.\,\rho(0), \tag{14.7}$$

and so on. In the above, we tacitly assumed that pulses are uncorrelated and hence the averaging over the stochastic fields embedded in V_is is also independent. This means that every pulse causes only an averaged $\mathfrak{T}$ matrix and hence the subscript i on $\mathfrak{T}$ can be dropped.

Summing all the contributions and employing the usual use of the convolution theorem, the Laplace transform of the averaged density operator can be expressed as

$$(\tilde{\rho}(s))_{av} = [sI - i\mathcal{L}_0 - \lambda(\mathfrak{T}_{av} - I)\}^{-1}\cdot\rho(0). \tag{14.8}$$

Equation (14.8) implies that when we revert back to the time space, the density matrix obeys the master equation:

$$(\dot{\rho}(t))_{av} = -i\,[\mathcal{H}_0, \rho(t)] + R\,\rho(t), \tag{14.9}$$

where the relaxation matrix R is given by

$$R = \lambda\,[(\mathfrak{T})_{av} - I]. \tag{14.10}$$

Furthermore, if we work up to the second order in V_i, tantamount to the familiar Born approximation of quantum optics (Agarwal 1971, 1974), we have

$$(\mathfrak{T})_{av} = I - i(\mathcal{L}_i)_{av} - \frac{1}{2}(\mathcal{L}_i^2)_{av} + \ldots \tag{14.11}$$

$(V_i)_{av}$, and hence $(\mathcal{L}_i)_{av}$ can be taken to be zero without loss of generality. Even if the average were not equal to zero, the latter could have been lumped with $\mathcal{L}_0$, and hence the first order correction (average of deviation from the mean) would still vanish. The master equation then takes the double commutator Lindblad form (Gorini et al. 1976):

$$\dot{\rho}(t) = -i\,[\mathcal{H}_0, \rho(t)] - \lambda/2\,([V_i,[V_i,\rho(t)]])_{av}. \tag{14.12}$$

14.3 Applications

14.3.1 Quantum Harmonic Oscillator

In this case,

$$\mathcal{H}_0 = \hbar\omega\,(a^+a + 1/2). \tag{14.13}$$

The coupling term can be chosen as

$$V_i = f_i(a + a^+). \tag{14.14}$$

The rationale behind choosing this particular form of V_i is as follows. Recall that for Brownian dynamics as per the Langevin equation (Chapter 4), the momentum that is envisaged to be the primary observable feels the brunt of heat bath-driven fluctuations. Thus, the heat bath effects characterized by the random noise and systematic friction terms appear directly only in the momentum equation.

In the kinetic theory language, one assumes that the tagged Brownian particles undergo random momentum-changing collisions, the cumulative effect of which is captured by the damping term. The collisions are envisaged to be weak in some sense so that the frictional term is taken to be linearly proportional to the instantaneous momentum. In quantum mechanics, the momentum is an operator and any momentum-changing events must be accompanied by transitions caused by operators that are diagonal in the momentum representation. The obvious candidate for such transition operators is the position operator which, in terms of creation and annihilation operators, is just the bracketed term in Equation (14.14)! In the CTRW picture, f_i measures the random impulse due to the ith collision.

Because we are attempting to stay close to the Langevin picture, it makes sense in the quantum version of CTRW to treat f_i to second order to simulate the weakness of the collisions. Therefore, the transition matrix can be written as

$$(\mathfrak{J})_{av} = I - 1/2[(\mathcal{L}_I)^2]_{av} + \dots. \tag{14.15}$$

Using then the properties of Liouville operators, the master equation can be expressed as

$$\dot{\rho}(t) = -i\,[\mathcal{H}_0, \rho(t)] - \lambda/2\ (f^2)_{av}\{[(a^\times + (a+)^\times]^2\}\,\rho(t), \tag{14.16}$$

where we dropped the i suffix on f since the averaging process obliterates the dependence on i. By working the commutators out, Equation (14.16) can be rewritten:

$$\rho(t) = -i[\mathcal{H}_0, \rho(t)] - \gamma\{2[a^+, [a,\rho(t)]] + [a, [a,\rho(t)]] + [a^+, [a^+, \rho(t)]]\}, \quad (14.17)$$

where

$$\gamma = \lambda/2(f^2)_{av}. \quad (14.18)$$

It is instructive to consider the relaxation to equilibrium of the average occupation number. It is easy to find the appropriate equation that reads

$$\frac{d}{dt} < n(t) = -2\gamma < n(t) > . \quad (14.19)$$

Equation (14.19) immediately triggers one important issue, namely that $<n(t)>$ relaxes to zero as time t goes to infinity. This is, however, not expected at finite temperatures when a detailed balance implies that the probability of transition, say from state $|n-1>$ to $n>$, is related to the probability of transition from state $|n>$ to $|n-1>$ by the respective Boltzmann factors. The reason for this problem is not hard to find. The operator V has been taken to be hermitian and hence the transitions effected by $(\mathfrak{I})_{av}$ are temperature-independent. Therefore, the model is only as good as relaxations at infinite temperatures are concerned. We will return to the finite temperature considerations below.

At this stage, it is pertinent to separate the relaxation matrix R into two parts: one that causes relaxation of the off-diagonal components of $\rho(t): R_{od}$ and the other responsible for the relaxation of the diagonal components R_d:

$$R = R_{od} + R_d, \quad (14.20)$$

where

$$R_{od} = \gamma\{ a^\times a^\times + (a^+)^\times (a^+)^\times \}, \quad (14.21)$$

and

$$R_d = \gamma\{ a^\times (a^+)^\times + (a^+)^\times a^\times \}. \quad (14.22)$$

This splitting will be useful in our attempt later to consider the finite-temperature generalization of the scheme.

14.3.2 Spin Relaxation

From the example of the harmonic oscillator whose quantum states span an infinite dimensional Hilbert space, we turn our attention now to a spin-half particle in an external magnetic field subjected to heat bath-induced fluctuations. We are thus led to study a problem in a finite dimensional Hilbert space, in this case, two or four if we decide to work in the corresponding Liouville space. The terms in the Hamiltonian are now given by

$$\mathcal{H}_0 = -\gamma H\, s_z, \tag{14.23}$$

and

$$V = f\{(s)^+ + (s)^-\}. \tag{14.24}$$

Again the choice of the coupling term is dictated by the consideration that we want operators that can cause transitions among the eigenstates of the free Hamiltonian, i.e., the eigenstates of s_z, and that is why we utilize the ladder operators $(s)^+$ and $(s)^-$. Following the procedure covered earlier, we find for the master equation:

$$\frac{d}{dt}\rho(t) = -\mathrm{i}\, H[s_z, \rho(t)] - \gamma\, \{2[s^+, [s^-, \rho\,(t)]]\ ^+\ [s^-, s^-, \rho(t)]] + [s^+, [s^+, \rho'\,(t)]]\}. \tag{14.25}$$

Again the relaxation matrix separates into

$$R_{od} = \gamma\, \{(s^-)^\times . (s^-)^\times + (s^+)^\times . (s^+)^\times\}, \tag{14.26}$$

and

$$R_d = \gamma\, \{ (s^-)^\times . (s^+)^\times + (s^+)^\times . (s^-)^\times\}. \tag{14.27}$$

The diagonal part R_d is responsible for the Bloch relaxation of the averaged z component of the spin, given by

$$\frac{d}{dt}\langle s_z\rangle = -4\,\gamma\, \langle s_z\rangle, \tag{14.28}$$

which suggests that $\langle s_z\rangle$ relaxes to zero asymptotically where in point of fact it should go to $\tanh(H/KT)$, the equilibrium value. We will fix this problem below.

14.4 Finite Temperature Effects

As we mentioned earlier the hermiticity of the transition operator V is responsible for the breakdown of the detailed balance condition. This implies that in order to treat irreversibility at finite temperatures we should introduce nonhermitian interactions in the present scheme. We may contrast the situation with the treatments in Chapters 12 and 13 wherein irreversibility was achieved by projecting out the degrees of freedom of the external system from the equations of motion governed by a full Hamiltonian (or Lagrangian) that retains hermiticity and hence unitary evolution.

It is important at this stage to dispense with a red herring of probability conservation. One might worry that the presence of nonhermitian terms will lead to nonconservation of probability, but we want to stress that what we require is conservation of probability only in the averaged sense. If the quantum state of a system immediately prior to a Poissonian ith pulse is $|\psi\rangle$, the pulse causes the system to jump to the state $\exp(-iV_i)\,|\psi\rangle$. Hence, conservation of average probability implies

$$\langle\psi|\ [\exp(iV_i^+).\ \exp(-iV_i)]_{av}\ |\psi\rangle = \langle\psi|\psi\rangle, \tag{14.29}$$

where $(V_i)^+$ differs from V_i, in general. Clearly, for models chosen earlier in Section 14.3, V_i^+ equals V_i and hence Equation (14.29) is trivially satisfied. The point is, even when V_i is nonhermitian, Equation (14.29) must be valid for the sake of consistency. For dealing with the finite temperature effects, we first consider the harmonic oscillator case.

14.4.1 Relaxation of the Harmonic Oscillator at Finite Temperatures

We would like to treat the two terms in the relaxation matrix separately, because only the diagonal part causes energy relaxation while the off-diagonal part leads to dephasing. Recall that

$$R_{od} = \gamma\ \{ (a)^\times \cdot (a)^\times + (a^+)^\times \cdot [(a^+)]^\times \}, \tag{14.30}$$

and

$$R_d = \gamma\ \{ (a)^\times . (a^+)^\times + (a^+)^\times . (a)^\times \},$$

which we want to rewrite as

$$R_d = \gamma\ \{ [(a^+)^+]^\times \cdot (a^+)^\times + (a^+)^\times \cdot (a)^\times \}, \tag{14.31}$$

where we used the trivial fact that $a = (a^+)^+$. We now replace a and a^+ by their temperature interaction representations:

$$a(\beta) = \sqrt{2Z}.\exp(\beta\,\mathcal{H}_0/4)\,a\,\exp(-\beta\mathcal{H}_0/4),$$

and likewise

$$a^+(\beta) = \sqrt{2Z}.\exp(\beta\,\mathcal{H}_0/4)\,a^+\,\exp(-\beta\mathcal{H}_0/4),$$

and the reformulated R_{od} and R_d become

$$R_{od} = \gamma\{[a(\beta)^\times . a(\beta)^\times + [a^+(\beta)]^\times . [a^+(\beta)^\times\}, \tag{14.32}$$

$$R_d = \gamma\{[(a^+(\beta))^+]^\times . [a^+(\beta)]^\times + [(a(\beta))^+]^\times[a(\beta)]^\times\} \tag{14.33}$$

Note that $[a(\beta)]^+$ is not the same as $a^+(\beta)$! Similarly, $[a^+(\beta)]^+$ does not equal $a(\beta)$. Rather,

$$[a(\beta)]^+ = a^+(-\beta), \quad \text{and} \quad [a^+(\beta)]^+ = a(-\beta). \tag{14.34}$$

Using the definition of Liouville operators for nonhermitian operators [cf. Equation (14.4)], we can now derive the desired master equation for the averaged density operator $\bar{\rho}(t)$ as (Dattagupta 1984):

$$\frac{d}{dt}\bar{\rho}(t) = -i\omega_0\,[a^+a,\bar{\rho}] - \gamma\,\{a^+a\bar{\rho} - 2a\bar{\rho}a^+ + \bar{\rho}\,a^+a + a^2\bar{\rho} - a\bar{\rho}a - a^+\bar{\rho}a^+ + \bar{\rho}\,(a^+)^2\}$$
$$-\gamma\langle n(\omega)\rangle_0\;\{2[a^+,[a,\bar{\rho}]] + [a^+,[a^+,\bar{\rho}]] + [a,[a,\bar{\rho}]]\}. \tag{14.35}$$

This equation is identical to the master equation derived by Agarwal (1971) using a system-plus-bath approach and within the so-called Born–Markov approximation familiar in quantum optics. It can be easily verified from Equation (14.35) that the correct approach to equilibrium is maintained in that

$$\frac{d}{dt}\langle n(t)\rangle = -2\,\gamma\,(\langle n(t)\rangle - \langle n(\omega)\rangle_0). \tag{14.36}$$

The Markovian nature is inherently built-in here because of the assumption that the pulses are driven by a Poisson process and also have zero duration of occurrence. The Born-like approximation is also implied in our assumption about the smallness of the impulses f.

14.4.2 Phase Space Dynamics and Free Particle Limit

The quantum Brownian motion of the oscillator becomes more transparent when we write the master equation in the phase space of the position and momentum coordinates. We find

$$\frac{d}{dt}\bar{\rho}(t) = \frac{-i}{\hbar}\left[\left(\frac{p^2}{2m} + \frac{1}{2} m\,\omega_0^2\,q^2\right), \bar{\rho}\right] - i\gamma([q,p\ \bar{\rho}] + [q,\bar{\rho}\ p])$$
$$- m\gamma\hbar\,\omega_0 \coth\left(\frac{1}{2}\beta\hbar\omega_0\right)[q,[q,\bar{\rho}]]. \tag{14.37}$$

In this form, it is easy to recover the master equation for a free quantum particle in Brownian motion. Thus taking the $\omega = 0$ limit and rewriting Equation (14.37) in the momentum representation, we obtain

$$\frac{\partial}{\partial t}\bar{\rho}\,(p,p';t) = \left[-\frac{i}{2m\hbar}\left(p^2 - p'^2\right) + \gamma\left(\frac{\partial}{\partial p} + \frac{\partial}{\partial p'}\right)(p + p') + D_0\left(\frac{\partial}{\partial p} + \frac{\partial}{\partial p'}\right)^2\right]\bar{\rho}(p,p';t), \tag{14.38}$$

where

$$\bar{\rho}(p,p';t) = \langle p|\ \bar{\rho}(t)\ |p'\rangle, \tag{14.39}$$

and

$$D_0 = 2m\gamma KT. \tag{14.40}$$

Equation (14.38) was derived for a free particle by Kumar (1984) and employs the essential ideas of CTRW.

14.4.3 Spin Relaxation at Finite Temperatures: Bloch–Redfield Equations

As in the harmonic oscillator case, we construct the operation of the relaxation matrix on $\bar{\rho}(t)$ as

$$R\bar{\rho}(t) = -\gamma\,\{[s^+(\beta)]^\times\,[s^+(\beta)]^\times + [s^-(\beta)]^\times\,[s^-(\beta)]^\times\}\,\bar{\rho}(t) \ + \ \gamma\{[s^+(-\beta)]^\times\,[s^-(\beta)]^\times$$
$$+ [s^-(-\beta)]^\times\,[s^+(\beta)]^\times\}\,\bar{\rho}(t)$$
$$= -\gamma\,\{[s^+(\beta),[\{s^+(\beta), \bar{\rho}(t)]] + [s^-(\beta), [s^-(\beta), \bar{\rho}(t)]] + [s^+\ (-\beta), [s^-(\beta), \bar{\rho}(t)]]$$
$$+ [s^-\ (-\beta), [s^+(\beta), \bar{\rho}(t)]]\}\ . \tag{14.41}$$

In the above,

$$s^{+}(\beta) = \sqrt{2Z}\ .\ \exp\ [\beta H/4]\ s^{+}\ \exp\ [-\beta H/4]\,,$$

and similarly,

$$s^{-}(\beta) = \sqrt{2Z}\cdot\exp\ [\ \beta\ \ H/4]\cdot s^{-}\ \exp\,[-\beta\ \ H/4]\,, \tag{14.42}$$

where the partition function in the present case is given by

$$Z = 2\cosh\ [\beta\ H/2]. \tag{14.43}$$

The full master equation can then be written as (Bloch 1957, Redfield 1957, 1965):

$$\frac{d}{dt}\bar{\rho}(t) = -iH\ [s_z,\ \bar{\rho}(t)] + R\ \bar{\rho}(t), \tag{14.44}$$

where R is given by Equation (14.41). As before, the relaxation of the expectation of s_z is given by the last two terms of the relaxation matrix R as in Equation (14.41). After some algebra, we find

$$\frac{d}{dt}<s_z(t)> = -2\gamma<s_z(t)> - \gamma\ \tanh\ (\beta H/2), \tag{14.45}$$

which evidently is consistent with the equilibrium result for $<s_z>_{eq}$, that reads:

$$<s_z>_{eq} = 1/2.\ \tanh\ (\beta\ \ H/2). \tag{14.46}$$

References

Abrikosov, A.A., L.P. Gorkov, and E. Dzyloshinsky. 1963. *Methods of the Quantum Field Theory in Statistical Physics*. Englewood Cliffs: Prentice Hall.

Agarwal, G. S. 1971. Brownian motion of a quantum oscillator. *Phys. Rev.* A4, 739.

Bloch, F. 1957. Generalized theory of relaxation. *Phys. Rev.* 105, 1206.

Clauser, M. J. and M. Blume. 1971. Nondiabetic effects in the stochastic theory of line-shape. *Phys. Rev.* B3, 583.

Dattagupta, S. 1984. Brownian motion of a quantum system. *Phys. Rev.* A30, 1525.

Dattagupta, S. 1987. *Relaxation Phenomena in Condensed Matter Physics*. Orlando: Academic Press (new edition by Levant Books, Kolkata, 2011).

Fetter, A.L. and J.D. Walecka. 2003. *Quantum Theory of Many-Particle Systems*. New York: Dover.

Gorini, V., A. Kossakowski, and E. C. G. Sudarshan. 1976. Completely positive dynamical semigroups of N-level systems. *J. Math. Phys*. 17, 821.

Huang, K. 1987. *Statistical Mechanics*, 2nd ed. New York: John Wiley & Sons.

Kumar, D. 1984. Brownian motion of a quantum particle. *Phys. Rev.* A29, 1571.

Montroll, E. and G. H. Weiss. 1965. Random walks on lattices. II. *J. Math. Phys*. 6, 167.

Montroll. E. W. and H. Scher. 1973. Random walks on lattices. IV. Continuous-time walks and influence of absorbing boundaries. *J. Stat. Phys*. 9(2), 101.

Redfield, A. G. 1957. On the theory of relaxation processes. *IBM J. Res. Dev.* 1, 19.

Redfield, A. G. 1965. Spin lattice relaxation of coupled spin 1–spin 1/2 system in anisotropic phases. *Adv. Magn. Reson*. 1, 1.

15

Quantum Jump Models

15.1 Introduction

In our scheme of developing classical diffusion, we naturally began from the view of continuum diffusion in which the position and momentum of a tagged, diffusing particle are continuous variables—which is what the spirit of Brownian motion entails. That indeed was our plan in Chapters 1 through 7. However, we noted in Chapter 7 that in order to describe diffusive jump motion in solids we need discrete jump models in which momentum plays no role. These discrete models were constituted on the basis of stationary Markov processes that were also at the core of continuum diffusion and dealt with via Langevin equations or Fokker–Planck equations.

One question automatically arises. What is the analogous method of treatment when the diffusing entity is a subatomic particle, e.g., an electron, muon, proton, or an abstract construct such as a quantum spin? A subatomic particle, because of its tiny mass, has an extended de Broglie wavelength and therefore its diffusion in a medium dominated by other elementary quantum excitations, e.g., phonons, magnons, etc., must be described by quantum diffusion.

However, in contrast to the quantum Langevin model, we will restrict the system Hamiltonian within a discrete Hilbert space. The other point to emphasize, that also appeared earlier, is that quantum dissipation must be described by what may be called quantum friction, the genesis of which can be traced to the degrees of freedom of the surrounding quantum bath.

As used for the quantum Langevin equation, we will utilize a first-principles approach in which the total system is split into three distinct parts: (1) a system Hamiltonian $\mathcal{H}_S$, (2) an interaction Hamiltonian $\mathcal{H}_I$, and (3) a bath Hamiltonian $\mathcal{H}_B$. This is known as a system-plus-bath approach (SBA). The idea is to trace (or average) out the bath degrees of freedom such that the system dynamics in a projected Hilbert space exhibits quantum diffusion. Recall that the same methodology was also employed in our functional integral treatment of quantum diffusion in Chapter 13, wherein we derived an equation for the reduced density operator by tracing the full density operator selectively over the bath degrees of freedom.

We will adopt an identical philosophy in this chapter and inter alia consider suitable limits in which our quantum models will start resembling the classical jump models, mostly described in terms of the telegraph process (Chapter 6). With this background, we write the total Hamiltonian as

$$\mathcal{H} = \mathcal{H}_S + \mathcal{H}_I + \mathcal{H}_B, \tag{15.1}$$

where the various terms have been explained above. In all cases discussed below, the bath will be taken as a collection of quantum harmonic oscillators or their second quantized version of bosonic creation and annihilation operators. The basis of the bath Hamiltonian is of course already familiar to us from the work of Zwanzig (1973) for classical diffusion and that of Ford et al. (1965), introduced in earlier chapters. We now discuss various applications of Equation (15.1).

15.1.1 Spin Lattice Relaxation in Solids

This issue is of great importance in various resonance techniques such as nuclear magnetic resonance, electron paramagnetic resonance, muon spectroscopy, and others in which the subsystem involves the spin of a nucleus, electron, or muon while the lattice relaxations are effected by thermal phonons of acoustic or optic type. The generic structures are

$$\mathcal{H}_s = H\sigma_Z, \quad \mathcal{H}_I = \sigma_X \sum_k g_k (b_k + b_k^+), \tag{15.2}$$

where b_k (and b_k^+) are bosonic annihilation (creation) operators for the qth phonon mode. The σ_X is off-diagonal in the representation in which σ_Z is diagonal and thus causes quantum transitions between the eigenstates of σ_Z. These transitions are, however, modulated by the phonon bath. An important feature of this and all other models discussed below—and the rationale behind the splitting effected in Equation (15.1)—is that $\mathcal{H}_I$ and $\mathcal{H}_B$ commute with each other and hence can be simultaneously diagonalized.

15.1.2 Quantum Tunneling in Symmetric Double Well

There are numerous physical situations in which the trapped motion of a quantum particle can be described in terms of a simplified picture of a double well. A case in point is a hydrogen (a proton, in most cases, as the hydrogen is stripped off its electron in a solid) or a muon tunneling from a defect site to another or an electron undergoing a chemical transfer reaction as in hemoglobin, a superconducting squid junction, or a quantum two-state system. In this case,

$$\mathcal{H}_s = \Delta\,\sigma_X, \quad \mathcal{H}_I = \sigma_Z \sum_k (b_k + b_k^+), \quad \mathcal{H}_B = \sum_k \hbar\omega_k b_k^+ b_k, \tag{15.3}$$

where the two eigenstates |+> and |-> of σ_Z represent the two minima of the double well. The system Hamiltonian then describes coherent tunneling between the two minima, σ_X being fully off-diagonal in the σ_Z representation, with Δ the overlap matrix element or, when divided by $\hbar$, is the tunneling frequency. The coupling with the bath represented by H_I, leads to decoherence of the tunneling process.

Coherent tunneling works like a quantum clock. Thus if a wave packet is prepared so that it is localized in one of the minima initially, tunneling will ensure that it is delocalized eventually. However, an intriguing phenomenon for tunneling in a symmetric well can be studied within Equation (15.3). Beyond a certain critical strength of the heat bath interaction, the wave packet is permanently localized at its initial position at temperature $T = 0$. This critical strength of the coupling constant beyond which the symmetry breaking transition occurs coincides with the so-called Tolouse limit of the equivalent Kondo Hamiltonian (Weiss 1999).

15.1.3 Tunneling in Asymmetric Double Well

While Section 1.5.1.2 refers to a situation in which the two wells are symmetric, a bias can be easily introduced into the model by adding a term $\epsilon\sigma_Z$ to $\mathcal{H}_S$. The bias ϵ is a realistic mimicking of asymmetry caused, for instance, in the problem of $\mathcal{H}$ tunneling in superconducting (or normal) niobium metal because of the presence of other (immobile) defects such as oxygen or carbon (Dattagupta et al. 1989). Thus $\mathcal{H}_S$ now reads

$$\mathcal{H}_S = \epsilon\sigma_Z + \Delta\sigma_X. \tag{15.4}$$

15.1.4 Dephasing of Qubit

A qubit is a set of two quantum dots whose states are ordinarily entangled. It has become a standard paradigm for studying quantum information processes. Environment-induced decoherence is an impediment for storage of quantum information. A reasonable model for assessing the decoherence processes is to expand the Hamiltonian $\mathcal{H}_S$ in Equation (15.4) as

$$\mathcal{H}_S = \epsilon\sigma_Z + \Delta\sigma_X + \tau_x(\xi_\epsilon\sigma_Z + \zeta_\Delta\sigma_Z), \tag{15.5}$$

whereas the interaction Hamiltonian is given by

$$\mathcal{H}_I = \tau_Z\sum_k g_k(b_k + b_k^+). \tag{15.6}$$

The physics behind the choice of the above forms can be understood easily. The interaction with the bath causes decoherence as far as the system

Hamiltonian is concerned. Indeed if one introduces a unitary transformation—a rotation in the τ spin space about the y axis by an angle of $\pi/2$ such that $\tau_X \to -\tau_Z$ and $\tau_Z \to \tau_X$, and works in the representation in which τ_Z is diagonal, one can easily see that the interaction with the bath occasions spontaneous spin flips of τ_Z, much like in the Glauber kinetics of the Ising model (Glauber 1963). Therefore, in this form, it is most convenient to compare the predictions of the quantum model with a corresponding jump model governed by a telegraph process in which the bath causes random fluctuations in the site energy and the bond energy. The latter terminology becomes self-evident if one views the system Hamiltonian in the tight-binding language:

$$\mathcal{H}_S = (\epsilon + \tau_x\zeta_\epsilon)(|L><L| - |R><R|) + (\Delta + \tau_x\zeta_\Delta)(|L><R| + |R><L|), \quad (15.7)$$

where $|L>$ and $|R>$ represent the left and the right dots of the qubit, respectively.

15.2 Formalism

With the introductory remarks behind us, we are ready to set up the mathematical formulation of the problem. First we will discuss the treatment in the time space employing a cumulant expansion scheme and then employ a similar theory in the frequency domain using a resolvent method.

15.2.1 Cumulant Expansion

Our starting point is the solution of the von Neumann–Liouville equation for the density operator [cf. Equation (13.1)]. We want to rewrite it in the interaction picture:

$$\rho_I(t) = \exp\left[\frac{i}{\hbar}(\mathcal{H}_S + \mathcal{H}_B)t\right]\rho(t)\exp\left[-\frac{i}{\hbar}(\mathcal{H}_S + \mathcal{H}_B)t\right]. \quad (15.8)$$

The density operator equation can then be written as

$$i\hbar\frac{\partial}{\partial t}\rho_I(t) = \left[\mathcal{H}_I(t), \rho_I(t)\right], \quad (15.9)$$

where

$$\mathcal{H}_I(t) = \exp\left[\frac{i}{\hbar}(\mathcal{H}_S + \mathcal{H}_B)t\right]\rho(t)\exp\left[-\frac{i}{\hbar}(\mathcal{H}_S + \mathcal{H}_B)t\right]. \quad (15.10)$$

At this stage we employ the Liouville operator associated with the interaction Hamiltonian defined in Chapter 14 [cf. Equation (14.3)]. The above equation then reads

$$\frac{\partial}{\partial t}\rho_I(t) = -i\mathcal{L}_I(t)\,\rho_I(t), \tag{15.11}$$

the solution of which can be expressed as (upon suppressing $\hbar$ within the definition of the Liouville operator):

$$\rho_I(t) = \exp_+\left(-i\int_0^t \mathcal{L}_I(t')\,dt'\right)\rho(0). \tag{15.12}$$

In the above, $\exp_+(....)$ denotes time-ordering wherein the operators are placed from left to right as their time arguments decrease. Combining with Equation (15.8), we have for the total density operator

$$\rho(t) = \exp\ [-i(\mathcal{L}_S + \mathcal{L}_B)t]\ \exp_+\left(-i\int_0^t \mathcal{L}_I(t')\,dt'\right)\rho(0). \tag{15.13}$$

We now construct the reduced density operator (see Chapter 13) by tracing ρ over the bath degrees of freedom thus obtaining

$$\rho_S(t) = \exp(-i\mathcal{L}_S t)Tr_B\left[\exp_+\left(-i\int_0^t \mathcal{L}_I(t')\,dt'\right)\right]. \tag{15.14}$$

Finally, introducing the operator

$$\tilde{\rho}_S(t) = \exp\left(\frac{i}{\hbar}\mathcal{L}_S t\right)\rho_S(t)\ \exp\left(-\frac{i}{\hbar}\mathcal{L}_S t\right), \tag{15.15}$$

we have

$$\tilde{\rho}_S(t) = Tr_B\left[\exp_+\left(-i\int_0^t \mathcal{L}_I(t')\,dt'\right)\rho(0)\right]. \tag{15.16}$$

Various simplifications ensue when we assume the factorized initial condition for the density operator [see Chapter 13, Equation (13.7)] and also invoke the property responsible for calling the environmental system a heat bath—its ability to remain in thermal equilibrium at a fixed temperature T. The perturbation in the form of the coupling H_I is then assumed to be switched on at time $t = 0$. Thus,

$$\rho(0) = \rho_S(0) \otimes \frac{1}{Z_B} \exp\,(-\beta \mathcal{H}_B), \tag{15.17}$$

where Z_B is the partition function of the heat bath. We may now write

$$\tilde{\rho}_S(t) = \left[\exp_+\left(-i\int_0^t \mathcal{L}_I(t')dt'\right)\right]_{av} \rho_S(0)]. \tag{15.18}$$

In terms of matrix elements,

$$\left[\exp_+\left(-i\int_0^t \mathcal{L}_I(t')dt'\right)\right]_{av} = \sum_{n_b, n_b'}\left(n_b n_b \,|\exp_+\left(-i\int_0^t \mathcal{L}_I(t')dt'\right)|\, n_b' n_b'\right) < n_b' \,|\, \rho_B \,|\, n_b' > \tag{15.19}$$

where $| n_b >$ and $| n_b' >$ designate the eigenstates of $\mathcal{H}_B$. Employing the cumulant expansion theorem and retaining terms up to second order in the interaction, we can develop a perturbation series (Dattagupta and Puri 2004):

$$\tilde{\rho}_S(t) = \left\{\exp_+\left(-i\int_0^t dt' < \mathcal{L}_I(t') >_{av} - \int_0^t dt' \int_0^t dt'' [< \mathcal{L}_I(t')\mathcal{L}_I(t'') >_{av} - < \mathcal{L}_I(t') >_{av} < \mathcal{L}_I(t'') > \right\} \rho_S(0). \tag{15.20}$$

We should underscore the difference between the above expansion and the usual perturbation method in that the above expansion is carried out in the exponent so that terms up to arbitrary orders in the perturbation can be handled, albeit only a class of diagrams is summed, in the language of diagrammatic many-body theory (Kubo 1962). In Equation (15.20), the angular brackets represent average over the heat bath. Additionally, the plus sign (+) subscript refers to time ordering discussed earlier. Thus,

$$< \mathcal{L}_I(t')\mathcal{L}_I(t'') >_{av} = \sum_{n_b, n_b'} (n_b n_b \,|\, \mathcal{L}_I(t')\mathcal{L}_I(t'') \,|\, n_b' n_b') < n_b' \,|\, \rho_B \,|\, n_b' >. \tag{15.21}$$

We assume $(\mathcal{L}_I)_{av}$ to be zero. This is done without loss of generality. Even if the former did not vanish, the average quantity could always be added to the system Hamiltonian by redefining it. The two distinct forms in which the results are discussed are covered below.

Form 1 — We first assume the system to be stationary, i.e., invariant under time translation so that correlation functions depend only on time differences that allow one of the time integrations to be dispensed with, and then extend the upper limit of the integral to infinity. The latter assumption is tantamount to the fact that bath-induced fluctuations die out well within timescales of importance for the system. In that case,

$$\tilde{\rho}_S(t) = \exp\left(-t\int_0^\infty d\tau < \mathcal{L}_I(\tau)\,\mathcal{L}_I(0) >_{av}\right)\rho_S(0) = \exp(-\hat{R}t)\,\rho_S(0). \tag{15.22}$$

where $\hat{R}$ is the so called relaxation matrix, defined by

$$\hat{R} = \int_0^\infty d\tau < \mathcal{L}_I(\tau)\,\mathcal{L}_I(0) >_{av}. \tag{15.23}$$

From the above, we automatically arrive at the master equation:

$$\frac{\partial}{\partial t}\rho_S(t) = -\frac{i}{\hbar}[(\mathcal{L}_S, \rho_S(t)] - e^{-i\mathcal{H}_S t/\hbar}\;\hat{\mathrm{R}}\,e^{-i\mathcal{H}_S t/\hbar}\rho_S(t), \tag{15.24}$$

which subsumes through the first term on the right the systematic dynamics generated by the subsystem Hamiltonian. Employing the definition of angular brackets and the properties of Liouville operators (Chapter 13), we can write

$$\begin{aligned}\hat{R}\,\tilde{\rho}_S(t) &= \frac{1}{\hbar^2}\int_0^\infty dt\; Tr_B([V_I(\tau), [V_I(0), \rho_B\tilde{\rho}_S(t)]]) \\ &= \frac{1}{\hbar^2}\int_0^\infty dt\; Tr_B([V_I(\tau)\,V_I(0)\,\rho_B\tilde{\rho}_S(t) + \rho_B\tilde{\rho}_S(t)\,V_I(0)V_I(\tau) \\ &\quad - V_I(\tau)\,\rho_B\tilde{\rho}_S(t)\,V_I(0) - V_I(0)\rho_B\tilde{\rho}_S(t)\,V_I(\tau)].\end{aligned} \tag{15.25}$$

Form 2 — The strategy now is to work from the outset with the time derivative of the reduced density matrix and develop approximation schemes therein. Thus we have from Equation (15.20):

$$\frac{\partial}{\partial t}\hat{\rho}_S(t) = -\int_0^t d\tau(<\mathcal{L}_I(t)\ \ \mathcal{L}_I(\tau)>_{av} - <\mathcal{L}_I(t)>_{av}<\mathcal{L}_I(\tau)>_{av})\tilde{\rho}_S(t). \quad (15.26)$$

As we showed clearly in our introductory section, the interaction term has a built-in structure that is a direct product of operators that work separately in the Hilbert spaces of the system and the bath. Thus the generic form can be represented as

$$\mathcal{H}_I = \hbar\sum_{j=1}^{n} S_j B_j, \quad (15.27)$$

where S_j is a system operator while B_j is a bath operator. Note that

$$\mathcal{H}_I(t) = \ \hbar\sum_{j=1}^{n} S_j(t) B_j(t), \quad (15.28)$$

where

$$S_j(t) = \exp\left(\frac{i}{\hbar}\mathcal{H}_S t\right) S_j \exp\left(-\frac{i}{\hbar}\mathcal{H}_S t\right), \quad (15.29)$$

and

$$B_j(t) = \exp\left(\frac{i}{\hbar}\mathcal{H}_B t\right) B_j \exp\left(-\frac{i}{\hbar}\mathcal{H}_B t\right). \quad (15.30)$$

To calculate the integrand of Equation (15.26), we introduce the eigenvalue equation for $\mathcal{H}_s$ and also use the closure property, as

$$\mathcal{H}_S\,|n> = E_S^n\,|n>, \quad 1 = \sum_n |n>\ <n|. \quad (15.31)$$

Thus,

$$<n|<\mathcal{L}(t)\mathcal{L}_I(\tau)>_{av}\ \tilde{\rho}_S(t)\,|m> = \sum_{n_b, n_b'}\sum_{n'm'}(nn_b, mn_b\,|\,\mathcal{L}_I(t)\mathcal{L}_I(t)]\ n'n_b', m', n_b')$$
$$<n_b'\,|\,\rho_B\,|\,n_b'>\ <n'\,|\,\tilde{\rho}_S(t)\,|\,m'>. \quad (15.32)$$

Further,

$$LHS = \sum_{jk}\sum_{n'm'}\{<B_j(t-\tau)B_k(o)>_{av}\ [<n|S_j(t)|m'><m'|S_k(\tau)|n'><n'|\tilde{\rho}_S(t)|m>$$

$$-<m'|S_j(t)|m><n|S_k(\tau)|n'><n'|\tilde{\rho}_S(t)|m'>+<B_k(0)B_j(t-\tau)>_{av}$$

$$[<n'|S_j(t)|m><m'|S_k(\tau)|n'><n|\tilde{\rho}_S(t)|m'>-<n|S_j(t)|n'>$$

$$<m'|S_k(\tau)|m><n'|\tilde{\rho}_S(t)|n'>]\} \tag{15.33}$$

In operator form,

$$<\mathcal{L}_I(t)\ \mathcal{L}_I(\tau)>_{av}\ \tilde{\rho}_S(t) = \sum_{jk}\{<B_j(t-\tau)\ B_k(0)>_{av}\ [S_j(t)\ S_k(\tau)\tilde{\rho}_S(t)$$

$$-S_k(\tau)\tilde{\rho}_S(t)\ S_j(t) - <B_k(0)B_j(t-\tau)>_{av}\ [S_j(t), \tilde{\rho}_S(t)\ S_k(\tau)S_j(t)]\}. \tag{15.34}$$

Now comes the important assumption that bears on the Markovian nature of the underlying processes. Because the bath fluctuations are short lived in comparison to the system timescales, the argument t of $\rho_S(t)$ under the integral can be replaced by τ. The import of this assumption is that for $t > \tau$, the integrand is zero anyway because the bath fluctuations would have withered away! Hence we write:

$$\int_0^t d\tau <\mathcal{L}_I(t)\ \mathcal{L}_I(\tau)>_{av}\ \tilde{\rho}_S(t) = \int_0^t d\tau\ e^{i\mathcal{H}_S t/\hbar}\sum_{jk}\{<B_j(t-\tau)B_k(0)>_{av}$$

$$[S_j, e^{-i\mathcal{H}_S(t-\tau)/\hbar}S_k\rho_S(\tau)\,e^{i\mathcal{H}_S(t-\tau)/\hbar}] - <B_k(0)B_j(t-\tau)>$$

$$[S_j, e^{-i\mathcal{H}_S(t-\tau)/\hbar}\ \rho_S(\tau)\ S_k\ e^{-i\mathcal{H}_S(t-\tau)/\hbar}]\}\ e^{-i\mathcal{H}_S t/\hbar}, \tag{15.35}$$

where we used Equations (15.15) and (15.29). Substituting Equation (15.35) into Equation (15.26) and noting Equation (15.15), we finally arrive at the promised master equation:

$$\frac{\partial}{\partial t}\rho_S(t) = -\frac{i}{\hbar}[\mathcal{H}_S, \rho_S(t)] - \int_0^t d\tau \sum_{j,k=1}^{n}\{<B_j(t-\tau)B_k(0)>_{av}\ [Sj, e^{-i\mathcal{H}_S(t-\tau)}S_k\rho_S(\tau)e^{+i\mathcal{H}_S(t-\tau)/\hbar}]$$

$$-<B_k(0)B_j(t-\tau)>_{av}\ [S_j, e^{-i\mathcal{H}_S(t-\tau)/\hbar}\rho_S(\tau)\ S_k\ e^{-i\mathrm{H}_S(t-\tau)/\hbar}]\}. \tag{15.36}$$

15.2.2 Resolvent Expansion

Our stratagem now is to transit to the frequency domain at the beginning by taking the Laplace transform of the von Neumann–Liouville equation and write it as

$$\tilde{\rho}(s) = (s + i\mathcal{L})^{-1}\ \rho(t = 0). \tag{15.37}$$

Concomitantly, the Laplace transform of the reduced density operator can be expressed as

$$\tilde{\rho}_S(s) = Tr_B[(s + i\mathcal{L})^{-1}\rho_B]\,\rho_S(t = 0), \tag{15.38}$$

using the factorization property of the initial density operator [cf. Equation (15.17)]. Upon making straightforward manipulations we can write:

$$\tilde{\rho}_S(s) = \frac{1}{s + i\mathcal{L}_S}\left[1 + (\tilde{G}(s))_{av}\frac{1}{s + i\mathcal{L}_S}\right]\rho_S(t = 0), \tag{15.39}$$

where

$$\tilde{G}(s) = \mathcal{L}_I\sum_{n=0}^{\alpha}(-i)^{n+1}\left[\frac{1}{s + i\mathcal{L}_S + \mathcal{L}_B}\right]\mathcal{L}_I, \tag{15.40}$$

and

$$(\tilde{G}(s))_{av} = \sum_{n_n, n_b}(n_b n_b\,|\,\tilde{G}(s)\,|\,n_b' n_b') < n_b'\,|\,\rho\,|\,n_b' > . \tag{15.41}$$

The above equation can be further simplified as

$$\tilde{\rho}(s) = \left[\frac{1}{s + i\mathcal{L}_S + (\tilde{G}^c(s))_{av}}\right]\rho_S(t = 0), \tag{15.42}$$

where

$$(\tilde{G}^c(s))_{av} = (\tilde{G}(s))_{av}\sum_{n=0}^{\infty}(-1)^{n+1}\left[\frac{1}{s + i\mathcal{L}_S}(\tilde{G}(s))_{av}\right]^n. \tag{15.43}$$

with the superscript c alluding to connected terms in diagrammatics (Kubo 1962). It is now possible to return to the time domain to facilitate comparison. It is easy to find

$$\frac{\partial}{\partial t}\rho_S(t) = i\mathcal{L}_S\rho(t) - \int_0^t d\tau\,[G^c(t-\tau)]_{av}\,\rho_S(\tau), \tag{15.44}$$

which is the desired master equation. If we treat the interaction Hamiltonian to second order as we did in the cumulant method, we would arrive at

$$\tilde{\rho}_S(s) \simeq \left[\frac{1}{s + i\mathcal{L}_S + \tilde{\Sigma}(s)}\right]\rho_S(t=0), \tag{15.45}$$

where $\tilde{\Sigma}(s)$ is the so-called self energy given by

$$\tilde{\Sigma}(s) = \left[\mathcal{L}_I \frac{1}{s + i(\mathcal{L}_S + \mathcal{L}_B)}\mathcal{L}_I\right]_{av}. \tag{15.46}$$

Note that the s dependence of the self energy lets us to treat non-Markovian status from which we can go to the Markovian limit by setting s equal to zero. Thus,

$$\tilde{\Sigma}(s) = \tilde{\Sigma}(o) = \int_0^\infty dt\{\mathcal{L}_I \exp[-i(\mathcal{L}_S + \mathcal{L}_B)t\mathcal{L}_I]\}_{av}. \tag{15.47}$$

15.3 Polaronic Transformation

In all the cases introduced in Section 15.1, the linear form of the coupling term enables itself to be eliminated by a unitary transformation familiar in polaron physics. The operator for that transformation (Holstein 1959, Lang and Firsov 1962, Firsov 1975, and Mahan 1990) is given by

$$\mathfrak{S} = \exp\left[-\sum_q \frac{g_q}{2\omega_q}(b_q - b_q^{\dagger})\sigma_Z\right]. \tag{15.48}$$

The transformed Hamiltonian in the case of Equation (15.3), for instance, is then

$$\tilde{\mathcal{H}} = \mathfrak{S}\mathcal{H}\mathfrak{S}^{-1} = \epsilon\,\sigma_Z + \frac{1}{2}\Delta(\sigma^+ B_- + \sigma^- B_+) + \sum_k \hbar\omega_k b_k^+ b_k, \tag{15.49}$$

where

$$B_{\pm} = \exp\left[\pm\sum_q \frac{g_q}{2\omega_q}(b_q - b_q^+)\right]. \tag{15.50}$$

Depending on the problem at hand, the system part of the operator in the exponent in $\mathfrak{S}$ has a different interpretation. For instance, in the spin boson case and its generalizations, the operator is σ_Z itself as in Equation (15.48). For the qubit case where we are dealing with an expanded Hilbert space of the system, the relevant operator is τ_Z.

We mentioned earlier that both the cumulant and the resolvent expansions incorporate the interaction Hamiltonian either in the exponent or in the resolvent, thus allowing perturbation treatments of all orders. The additional and significant point of the polaronic transformation is that the original coupling constant g_q is elevated to the exponent, again facilitating calculations for handling strong coupling cases. This becomes apparent, as shown below, when the cumulant scheme with incorporated polaronically transformed Hamiltonian automatically leads to the so-called dilute bounce gas approximation of the functional treatment of the spin boson model (Leggett et al. 1987).

In the language of the system-plus-bath approach, we find the following decomposition upon using the polaronic transformation:

$$\mathcal{H} = \mathcal{H}_S + \mathcal{H}_I + \mathcal{H}_B,$$

where

$$\mathcal{H}_s = 0, \quad \mathcal{H}_I = \frac{1}{2}\Delta\ (\sigma^+ B_- + \sigma^- B_+), \quad \mathcal{H}_B = \sum_k \hbar\omega_k\, b_k^+\, b_k. \tag{15.51}$$

Further, we may identify, with respect to Equation (15.27):

$$S_1 = \sigma^+, \quad S_2 = \sigma^-, B_1 = \frac{1}{2}\Delta\, B_-, \quad B_2 = \frac{1}{2}\Delta\, B_+. \tag{15.52}$$

Using the result of the cumulant expansion theorem [cf. Equation (15.36)], we find for the equation of motion of $<\sigma_z(t)>$:

$$\frac{d}{dt} < \sigma_z(t) > \equiv Tr\left[\frac{\partial \rho_S(t)}{\partial t}\sigma_Z\right] = -\Delta^2 \int_0^t d\tau\ [\Phi(\tau) + \Phi(-\tau)] < \sigma_z(t-\tau) >, \tag{15.53}$$

where $\Phi(\tau)$ is the correlation function of bath operators:

$$\Phi(\tau) = < B_- (0) B_+(\tau) >_{av} = < B_+ (0) B_-(\tau) >_{av} \tag{15.54}$$

In deriving Equation (15.53), we employed the following result:

$$< B_\pm(t) \ B_\pm(0) >_{av} = 0 \ . \tag{15.55}$$

The solution of Equation (15.53) turns out to be consistent (Aslangul et al. 1986) with the so-called dilute bounce gas approximation (DBGA) of a path integral treatment of the spin boson model (Leggett et al. 1987).

15.3.1 Asymmetric Spin Boson Model

For many practical applications such as defect tunneling between two sites, a symmetric double well is an oversimplification of the problem. In such cases, one of the two defect sites may be deeper in energy than the other, and require the introduction of an asymmetry parameter in the spin boson model. This, in spin language, amounts to adding a Zeeman term to the *Z* component of the spin, yielding

$$\mathcal{H} = \epsilon \sigma_z + \Delta \sigma_x + \sigma_z \sum_k g_k (b_k + b_k^+) + \sum_k \omega_k b_k^+ b_k . \tag{15.56}$$

Apart from the problem of tunneling in an asymmetric double well, Equation (15.56) has another application to the problem of spin lattice relaxation, wherein the spin is supposed to be under the influence of cross magnetic fields, one along the *z* direction, the other along *x*. Upon applying an identical polaronic transformation as in the original spin boson model, we find for the transformed Hamiltonian:

$$\tilde{\mathcal{H}} = \epsilon \sigma_z + \frac{1}{2} \Delta (\sigma^+ B_- + \sigma^- B_+) + \sum_k \hbar \omega_k b_k^+ b_k . \tag{15.57}$$

Evidently, the asymmetry term remains unaffected by the unitary (polaronic) transformation because σ_z commutes with $\mathcal{S}$. The system-plus-bath decomposition now has a nonzero term $\tilde{\mathcal{H}}_S$, in contrast to Equation (15.51):

$$\tilde{\mathcal{H}}_S = \epsilon \sigma_Z . \tag{15.58}$$

Apart from the expectation of σ_z (calculated above for the spin boson problem), one other quantity of physical interest related to the structure factor in a scattering measurement is the symmetrized correlation function:

$$C(t) = \frac{1}{2}\ [<\sigma_z(0)\sigma_z(t)>_{eq} + <\sigma_z(t)\sigma_z(0)>_{eq}]. \tag{15.59}$$

Because the structure factor is usually measured in the frequency domain, it is convenient to work with the Laplace transform of $C(t)$, defined by

$$\tilde{C}(s) = \int_0^\infty dte^{-st}\ C(t), \quad \text{Re }(s) > 0. \tag{15.60}$$

For the same reason, it is prudent to employ the resolvent expansion scheme of Section 15.2.2. Writing the trace as a weighted average of the equilibrium Gibbs measure:

$$\rho_{eq} = \frac{1}{Z} \exp(-\beta\mathcal{H}) \tag{15.61}$$

,

wherein the unitary transformation $\mathfrak{S}$ commutes with σ_z. We ignore the interaction term in comparison to the Boltzmann energy KT, and find the factorized form of the equilibrium density matrix:

$$\tilde{\rho}_{eq} \simeq \frac{1}{2}\ [1 - \tanh\ (\beta \in)\ \sigma_z]\rho_B, \tag{15.62}$$

where

$$\rho_B = \frac{1}{Z_B}\ \exp\ \left(-\beta\sum_k \hbar\omega_k\, b_k^+ b_k\right). \tag{15.63}$$

Collecting all these terms, we find

$$\tilde{C}(s) = \frac{1}{2}\sum_{\mu\mu'}(\mu + <\sigma_z>)\mu'(\mu\mu\,|\,[\tilde{U}(s)]_{av}\,|\,\mu'\mu'), \tag{15.64}$$

where $\tilde{U}(s)$ is the Laplace transform of the time development operator $U(t)$ defined by

$$U(t) = \exp(i\tilde{\mathfrak{L}}t), \tag{15.65}$$

where $\tilde{\mathfrak{L}}$ is the Liouvillian associated with the full Hamiltonian $\tilde{\mathcal{H}}$. We may now use the resolvent expansion of Section 15.2.2 to express $[\tilde{U}(s)]_{av}$ as

$$[\tilde{U}(s)]_{av} = \frac{1}{s - i\mathfrak{L}_s + \tilde{\Sigma}(s)}, \tag{15.66}$$

where the self-energy $\tilde{\Sigma}(s)$ is now a 4×4 matrix (the basic Hilbert space is 2×2 but the corresponding Hilbert space of the Liouvillian is $(2)^2 \times (2)^2$). Using the properties of Liouville operators and after some algebra (Dattagupta et al 1989, Dattagupta and Puri 2004), we find:

$$\tilde{\Sigma}(s) \simeq \Delta^2 \begin{pmatrix} a_{11} & a_{12} & 0 & 0 \\ a_{21} & a_{22} & 0 & 0 \\ 0 & 0 & a_{33} & a_{34} \\ 0 & 0 & a_{43} & a_{44} \end{pmatrix}, \tag{15.67}$$

where the rows and columns are labeled $++$, $--$, $+-$, and $-+$, respectively, and

$$\begin{aligned} a_{11} &= \tilde{\Phi}_{-+}(s_+) + \tilde{\Phi}'_{-+}(s_-), \\ a_{12} &= -\tilde{\Phi}_{-+}(s_+) - \tilde{\Phi}'_{-+}(s_-), \\ a_{21} &= -\tilde{\Phi}_{+-}(s_-) - \tilde{\Phi}'_{+-}(s_+), \\ a_{22} &= \tilde{\Phi}_{+-}(s_-) + \tilde{\Phi}'_{+-}(s_+), \\ a_{33} &= \tilde{\Phi}_{-+}(s) + \tilde{\Phi}'_{+-}(s), \\ a_{34} &= -\tilde{\Phi}_{--}(s) - \tilde{\Phi}'_{--}(s), \\ a_{43} &= -\tilde{\Phi}_{++}(s) - \tilde{\Phi}'_{++}(s), \\ a_{44} &= \tilde{\Phi}_{+-}(s) + \tilde{\Phi}'_{-+}(s), \\ s_\pm &= s \pm 2i \in, \end{aligned} \tag{15.68}$$

with

$$\begin{aligned} \Phi_{+-}(t) &= < B_+(0) > B_-(t) >_{av}, \\ \tilde{\Phi}_{-+}(t) &= < B_-(0) > B_+(t) >_{av}, \\ \tilde{\Phi}_{++}(t) &= < B_+(0) > B_+(t) >_{av}, \\ \tilde{\Phi}_{--}(t) &= < B_-(0) > B_-(t) >_{av}. \end{aligned} \tag{15.69}$$

Additionally, all primed quantities are obtained by replacing the argument t by $-t$. We remind the reader that the angular brackets represent thermal averages with respect to the bath states alone. We now calculate the correlation functions in the Ohmic dissipation model in which the spectral density discussed earlier takes the form:

$$J(\omega) = 2\sum_q g_q^2 \delta\ (\omega - \omega_q) = K\omega^s e^{-\omega/D}, \tag{15.70}$$

It is then straightforward to find that $\Phi_{++}(t) = \Phi_{--}(t) = 0$, and $\Phi_{+-}(t) = \Phi_{-+}(t) = \Phi(t)$, given by

$$\Phi(t) = \exp\left\{-\sum_q \frac{4g_q^2}{\omega_q^2}\left[\coth\left(\frac{\hbar\beta\omega_q}{2}\right)(1-\cos(\omega_q t)) + i\sin(\omega_q t)\right]\right\}. \tag{15.71}$$

For $\beta D >> 1$, the above expression can be reduced to the closed form,

$$\Phi(t) = \exp[i\pi K\,\text{sgn}(t)]\left[\frac{\pi}{\hbar\beta D\sinh(\pi|t|/\beta}\right]^{2K}. \tag{15.72}$$

From this the Laplace transform, $\tilde{\Phi}(s)$ can be computed as

$$\tilde{\Phi}(s) = \frac{e^{i\pi K}}{D}\left(\frac{2\pi}{\hbar\beta D}\right)_{2K-1}\frac{\Gamma(1-2K)\Gamma(K+s\beta/2\pi)}{\Gamma\ (1-K+s\beta/2\pi)}, \tag{15.73}$$

where $\Gamma(x)$ is the gamma function of argument x. Finally, the Laplace transform of the correlation function reads

$$\tilde{C}(s) = \frac{s - i<\sigma_z>\sin\ (\pi K)\ [\tilde{J}(s+2i\in) - \tilde{J}(s-2i\in)]}{s\{s+\cos\ (\pi K)\ [\tilde{J}(s+2i\in)+\tilde{J}(s-2i\in)]\}}, \tag{15.74}$$

where

$$\tilde{J}(s) = \frac{2\Delta^2}{D}\left(\frac{2\pi}{\beta D}\right)^{2K-1}\frac{\Gamma(1-2K)\Gamma(K+s\beta/2\pi)}{\Gamma\ (1-K+s\beta/2\pi)}. \tag{15.75}$$

For further discussion, it is useful to define an effective tunneling frequency as

$$\tilde{\Delta} = [\cos(\pi K)\Gamma(1-2K)]^{1/(2-2K)}\,\Delta\left(\frac{2\Delta}{D}\right)^{K/(1-K)}, \tag{15.76}$$

in terms of which

$$\tilde{J}(s) = \tilde{\Delta}\left(\frac{\beta\tilde{\Delta}}{\pi}\right)^{1-2K} \frac{\Gamma(K + s\beta/2\pi)}{\cos(\pi K)\ \Gamma\ (1 - K + s\beta/2\pi)}. \tag{15.77}$$

Evidently, the result above also subsumes the symmetric case ($\in = 0$) for which

$$\tilde{C}(s) = [s + 2\cos\ (\pi K)\ \tilde{J}(s)].^{-1} \tag{15.78}$$

We will examine the results below in the context of muon diffusion in metals and also make a comparison with the results for the classical telegraph process.

15.4 Qubit

The qubit Hamiltonian was introduced in Equation (15.5). As already stated, the customary polaronic transformation [cf. Equation (15.48)] now reads:

$$\mathfrak{S} = \exp\left[-\sum_q \frac{g_q}{\omega_q}(b_q - b_q^+)\ \tau_Z\right]. \tag{15.79}$$

The transformed Hamiltonian can then be written as

$$\tilde{\mathcal{H}} = (\in \sigma_Z + \Delta\sigma_X) + \frac{1}{2}(\tau^- B_+ + \tau^+ B_-)\ (\zeta_\in \sigma_Z + \zeta_\Delta \sigma_x) + \sum_k \hbar\omega_k b_k^+ b_k. \tag{15.80}$$

In the system-plus-bath decomposition, we can identify various Hamiltonian terms as

$$\begin{aligned} \tilde{\mathcal{H}} &= \in \sigma_Z + \Delta\sigma_X \\ \mathcal{H}_I &= \frac{1}{2}(\tau^- B_+ + \tau^+ B_-)(\zeta_\in \sigma_Z + \zeta_\Delta \sigma_x) \\ \mathcal{H}_B &= \sum_k \hbar\omega_k b_k^+ b_k. \end{aligned} \tag{15.81}$$

While the full Hamiltonian in Equation (15.80) has been studied recently (Kamil and Dattagupta 2012), we shall consider a simpler case in which the asymmetry term (and fluctuation therein) is ignored. The simplified Hamiltonian terms then read:

$$\mathcal{H}_S = \Delta\sigma_X, \quad \mathcal{H}_I = \frac{1}{2}\zeta_\Delta\sigma_X(\tau^- B_+ + \tau^+ B_-), \quad \mathcal{H}_B = \sum_k \hbar\omega_k b_k^+ b_k. \tag{15.82}$$

Note that the subsystem Hamiltonian $\mathcal{H}_S$ now commutes with the interaction term $\mathcal{H}_I$, implying no energy transfer between the former and the bath. If we were to replace the effect of the bath by a classical noise, as would be represented by a telegraph process, the alternate form of the stochastic Hamiltonian would have read:

$$\mathcal{H}(t) = \Delta\sigma_X + f(t)\zeta_\Delta\sigma_X \tag{15.83}$$

Such cases in the line shape literature are designated *adiabatic* (Blume 1968). In the context of quantum decoherence to be covered in Chapter 16, Equation (15.83) falls into the category of adiabatic decoherence (Mozyrsky and Privman 1998) and dissipation-less decoherence (Gangopadhyay et al. 2001).

We can make a rotation in the σ space about the y axis by an angle $\pi/2$ by means of which $\sigma_X \to -\sigma_Z$. That does not change the physics and we can continue to work in the usual representation in which $\mathcal{H}_s$ is diagonal. The reduced density operator for the qubit in the Laplace-transformed space is again given by Equation (15.45):

$$\tilde{\rho}_s(s) - \left[\frac{1}{s + i\mathcal{L}_q + \tilde{\Sigma}(s)}\right]\rho_q(t=0), \tag{15.84}$$

where $\mathcal{L}_q$ is the qubit Liouville operator associated with $\mathcal{H}_S$ in Equation (15.85) where we may replace σ^x with $^-\sigma_z$ for the reasons stated above.

Unlike the spin boson case (Equation 15.56) in which the subsystem Hamiltonian does not commute with the interaction term, the matrix of the self energy $\tilde{\Sigma}(s)$ (labeled + +, – –, + –, and – +) is now:

$$\tilde{\Sigma}(s) = \begin{pmatrix} 0 & 0 & 0 & 0 \\ 0 & 0 & 0 & 0 \\ 0 & 0 & b_1(s) & 0 \\ 0 & 0 & 0 & b_2(s) \end{pmatrix}, \tag{15.85}$$

where

$$b_1(s) = 4\int_{\Delta}^{2} [\Phi'(s_+) + \Phi(s_+)]$$

$$b_2(s) = 4\int_{\Delta}^{2} [\Phi'(s_-) + \Phi(s_-)], \tag{15.86}$$

$$s_{\pm} = s \pm 2i\Delta,$$

and $\tilde{\Phi}(s)$ is the Laplace transform of $\Phi(t)$, defined in Equation (15.71) with $\tilde{\Phi}'(s)$ obtained by replacing t in $\Phi(t)$ by $-t$. It is then fairly straightforward (because only diagonal matrices are involved) to evaluate all the elements of the density operator for the qubit. We find:

$$(\tilde{\rho}_q)_{++}(s) = (\tilde{\rho}_q)_{--}(s) = \frac{1}{2s}, \tag{15.87}$$

and

$$(\tilde{\rho}_q)_{+-}(s) = \frac{1}{2}\left(\frac{1}{s + 2i\Delta + b_1(s)}\right),$$

$$(\tilde{\rho}_q)_{-+}(s) = \frac{1}{2}\left(\frac{1}{s - 2i\Delta + b_2(s)}\right). \tag{15.88}$$

The diagonal elements of the density operator stay constant (one half each). The off-diagonal elements oscillate with (the dominant) frequency 2Δ and asymptotically as $t \to \infty$ and decay to zero at a rate $b_1(s = 0)$. There is a caveat. The bases $1\pm>$ we employed here are the eigenstates of a pseudo spin operator σ_Z and are not the physical states of the quantum dot. The latter are given by (Kamil and Dattagupta 2012):

$$|L> = \frac{1}{\sqrt{2}}(1+> +1->),$$

$$|R> = \frac{1}{\sqrt{2}}(1+> -1->), \tag{15.89}$$

where $|R>$ and $|L>$ refer to the right and left dots of the qubit. The states $1\pm>$ are called the bonding (and anti-bonding) states in the theory of chemical bonding. Transformed to the dot basis, the diagonal elements of

ρ_q remain the same as in Equation (15.87). However, the off-diagonal elements do not decay to zero but reduce to finite values at long times. This phenomenon, called partial decoherence, will be discussed in detail in the next section.

At this stage it is our limited aim to compare the expressions for the off-diagonal elements of the density operator in Equation (15.88) with the corresponding terms for the telegraphic process [when the qubit Hamiltonian is interpreted as in Equation (15.83)]. We note that $b_{1,2}(s)$ is governed by the Laplace transform of the correlation function which, in the Ohmic dissipation model, is given by Equation (15.73). As stated earlier, the relaxation behavior is dictated by the Markovian limit in which $b_{1,2}(s)$ can be replaced by their frequency independent expressions $b_{1,2}(s = 0)$. In that situation (cf. Section 15.3),

$$\tilde{\Phi}(s=0) = \frac{e^{i\pi K}}{D}\left(\frac{2\pi}{\hbar\beta D}\right)^{2K-1} \frac{\Gamma(1-2K)\Gamma(K)}{\Gamma(1-K)}. \tag{15.90}$$

Because the telegraph process would have led to identical expressions for the qubit density operator [as in Equations (15.87) and (15.88)] it is tempting to compare Equation (15.90) with the jump rate λ for the telegraph process (Chapter 8). While this is reasonable, it is relevant to note that for classical cases, the rates are usually associated with Arrhenius expressions for jump rates across energy barriers.

The temperature dependence indicated in Equation (15.90) is far from being exponential—indeed, it is a power law. The power law temperature dependence of jump rates in the context of H or muon diffusion in metals can be attributed to electron hole excitations near the Fermi surface (Sections 15.5 and 15.6 below). The closest one can approach exponential temperature dependence of the jump rate is when jumps are mediated by phonon-assisted processes at high temperatures. Recall that the general expression for the correlation function (before invoking Ohmic dissipation) reads from Equation (15.71).

$$\Phi(t) = \exp\left\{-2\int_0^\infty d\omega \frac{j(\omega)}{\omega^2}\left[\coth\left(\frac{\hbar\beta\omega}{2}\right)(1-\cos\omega t) + i\sin\omega t\right]\right\}. \tag{15.91}$$

Joining the two trigonometric functions,

$$\Phi(t) = \exp\left\{-2\int_0^\infty d\omega \frac{j(\omega)}{\omega^2}\left[\coth\left(\frac{\hbar\beta\omega}{2}\right) - \frac{\cos(\omega(t - i\hbar\beta/2))}{\sinh(\hbar\beta\omega/2)}\right]\right\}. \tag{15.92}$$

By making a change of variable form $t - i\hbar\beta/2$ to t in Equation (15.92), the Laplace transform $\tilde{\Phi}(s)$ can be written as

$$\tilde{\Phi}(s) = \exp\left(-\frac{i\hbar\beta s}{2}\right)\int_{-i\hbar\beta/2}^{\infty} dt\, e^{-st} \exp\left\{-2\int_0^{\infty}\frac{d\omega\, j(\omega)}{\omega^2}\left[\coth\left(\frac{\hbar\beta\omega}{2}\right) - \frac{\cos(\omega t)}{\sinh(\hbar\beta\omega/2)}\right]\right\}. \tag{15.93}$$

In the high-temperature limit (Flynn and Stoneham 1970, Dattagupta and Schober 1998), the integrand in the second term in Equation (15.93) can be replaced by its short-time value as

$$\Phi(s) = \exp\left(-\frac{i\hbar s\beta}{2}\right)\int_{-i\hbar\beta/2}^{\infty} dt\, e^{-st} \exp\left\{-2\int_0^{\infty}\frac{d\omega\, j(\omega)}{\omega^2}\times\left[\tanh\left(\frac{\beta\hbar\omega}{4}\right) + \frac{\omega^2 t^2}{2\sinh(\hbar\beta\omega/2)}\right]\right\}, \tag{15.94}$$

Further, in this regime, i.e., low β, $\tanh(X)$ and $\sinh(X)$ can be replaced by the argument X and hence the integrals in Equation (15.94) can be carried out easily, yielding:

$$\tilde{\Phi}(s) = \exp\left[-\beta\left(a + \frac{is\hbar}{2}\right)\right]\int_{-i\hbar\beta/2}^{\infty} dt \;\; e^{-st} e^{-4at^2/\beta\hbar}. \tag{15.95}$$

Thus if we ignore the s dependence, the dominant temperature dependence is indeed exponential but the barrier height a has a different interpretation:

$$a = \frac{1}{2}\int_0^{\infty}\frac{d\omega\hbar\, j(\omega)}{\omega}. \tag{15.96}$$

The quantity a is called the coincidence energy in the context of quantum diffusion of point defects (Flynn and Stoneham 1970). The result (15.96), however, is generally true for any phononic spectral density, whether the phonons are acoustic or optic.

15.5 Neutron Structure Factor for H Diffusion in Metals

As an application of quantum tunneling in an asymmetric double well, we consider neutron scattering from a hydrogen atom tunneling from one defect site to another in the normal metallic state of niobium (Nb). The two trapped sites are caused by other (more massive and therefore immobile oxygen or nitrogen interstitial) defects that yield an asymmetric double well.

Hydrogen, as a light particle, can undergo tunneling between the two minima of the double well. This tunneling process is quantum-coherent but is impeded by quantum dissipation occasioned by the environment that causes a charge coupling between the electrons of Nb and the proton (in the stripped state of the tunneling hydrogen).

The environment consists of conduction electrons of Nb. As noted earlier, it can be viewed as a bosonic bath (as envisaged in the spin boson model) in view of the Schwinger bosonization of electron hole pairs. The most convenient spectroscopic tool to probe hydrogen is neutron scattering, discussed in Chapter 8. Recall that the parameter measured is the incoherent structure factor $S(k,\omega)$ [cf. Equation (8.2)].

At ultralow temperatures only the two lowest tunneling states are accessible. Thus the minima locations of the double well can be mapped into the two states of a pseudo spin operator σ_Z, thus,

$$r = \hat{i}x_0\ \sigma_Z, \tag{15.97}$$

where $\hat{i}$ is the unit vector along the x axis and $2x_0$ is the separation between the two minima of the double well. Because of the Pauli spin properties of σ_Z, the structure factor can be re-expressed (apart from some form factors) in terms of the Laplace transform of the symmetrized correlation function (Dattagupta et al. 1989, Dattagupta and Puri 2004):

$$C(s) = \frac{1}{2}\int_0^{\infty} dt\ e^{-st} < \sigma_z(0)\ \sigma_z(t) + \sigma_z(t)\,\sigma_z(0) >_{eq}. \tag{15.98}$$

We can directly transplant the results for the spin correlation function of asymmetric tunneling from Section 15.3 above. However, as in the textbook discussion of the heat capacities of metals (Ashcroft and Mermin 1976), where the low-temperature contributions are split into fermionic and a bosonic contributions (due to electrons and phonons, respectively), we proceed likewise by incorporating into the spectral density of the bath oscillators the phononic (non-Ohmic) term along with the Ohmic term that mimics the conduction electron effect. Thus,

$$J(\omega) = [K\omega + G(\omega)^{2m+1}]\ \exp(-\omega/D), \tag{15.99}$$

where G is the damping constant appropriate for phonons (wherein the index m takes different values depending on whether the phonons are acoustic or optic).

At very low temperatures when the heat bath effects have not set in, tunneling remains coherent. It manifests as two inelastic peaks in the structure factor, the heights of which are in the ratio of (detailed balance) Bose factors. However, our focus of attention is incoherent tunneling that makes the two inelastic lines merge into a single quasi-elastic line centered around $\omega = 0$. In the incoherent tunneling regime, the diffusive phenomenon, although fully quantum mechanical, can nevertheless be described in terms of a Markovian process. This implies that the appropriate memory kernel can be assumed to be frequency-independent. Thus,

$$C(s) = [s + \tilde{\Phi}(s = 0)]^{-1}. \quad (15.100)$$

The quantum jump rate of the hydrogen can then be expressed as

$$\lambda = \tilde{\Phi}(s = 0)]. \quad (15.101)$$

It is relevant to note that had we chosen to describe the jump diffusion of hydrogen in classical terms, it would have been associated with temperature-assisted jumps across an Arrhenius barrier (a two-state jump process described in Chapter 6). However, it may be stressed again that although the expressions for the structure factor now look (deceptively) similar, the underlying physics principles are quite distinct. For example, classical methods can never treat coherent tunneling (and concomitant occurrence of inelastic peaks)—not included here—but even in the incoherent regime, the jump rate is much richer in structure. In order to assess this, we write λ as

$$\lambda = \frac{\Delta^2}{4} \int_{-\infty}^{\infty} dt\ e^{-\Delta(t)}, \quad (15.102)$$

where

$$\Delta(t) = \Delta_{el}(t) + \Delta_{ph}(t). \quad (15.103)$$

The decay function $\Delta(t)$ can be split into electronic and phononic parts as

$$\Delta(t) = \Delta_{el}(t) + \Delta_{ph}(t), \quad (15.104)$$

yielding after a few simplifying steps (Grabert 1992):

$$\Delta_{el} = \frac{\Gamma(K)}{\Gamma(1-K)} \frac{\tilde{\Delta}}{2} \left(\frac{2\pi}{\beta \tilde{\Delta}} \right)^{2K-1}, \quad (15.105)$$

and

$$\Delta_{ph} = \left(\frac{\pi\beta}{16E_c} \right)^{1/2} \tilde{\Delta}^{\,2}\, e^{-\beta E_c}$$

where $\tilde{\Delta}$ is the renormalized tunneling frequency:

$$\tilde{\Delta} = [\cos(\pi K)\Gamma(1-2K)]_{1/(2-2K)}\, \Delta \left(\frac{2\Delta}{\omega_c} \right)^{K/(1-K)}. \quad (15.106)$$

Evidently the electronic contribution has no classical counterpart but the phononic part resembles a classical Arrhenius form, although E_c, the so-called coincidence energy, is far from the usual meaning of a barrier height but is instead given by

$$Ec = \frac{G}{2} \int_0^{\infty} d\omega\, \omega^{2m} \exp(-\omega/D). \quad (15.107)$$

15.6 Spin Relaxation of Muon Diffusion in Metals

To further examine the applicability of quantum jump models and their relations to classical telegraph processes, we now consider the case of a positive muon undergoing jump diffusion in a metal. A muon is about 300 times heavier than an electron and occupies the middle position in the hierarchy of "quantumness." It is flanked on both sides by an electron and a proton (or hydrogen). Like hydrogen, the muon also occupies an interstitial position. In the incoherent tunneling regime, when successive jumps are uncorrelated, the jump process can be approximately viewed in terms of tunneling in a double well.

The issue at hand is similar to hydrogen in niobium, but unlike that situation, the well can be taken as symmetric (Section 15.3). Another important distinction concerns the measurement techniques. While the hydrogen is probed through its nuclear interaction with an incoming neutron from a thermal reactor, the muon, because of its half-integer spin, is made to interact with an applied magnetic field. Thus the method is very much like spin relaxation studies carried out by nuclear magnetic resonance (NMR; Abraham 1961).

While the external field triggers coherent spin rotation of the muon, the coherence is disturbed by internal magnetic fields occasioned by phonons, other electrons, and other particles. We make a very simplifying assumption (that nevertheless captures the essential ideas) that the internal fields (denoted H) are identical in magnitude but oppositely oriented to the two minima of the double well.

The underlying Hamiltonian that describes the process outlined above can be written (Qureshi and Dattagupta 1993) as

$$\mathcal{H} = HI_Z\sigma_Z + \Delta\sigma_x + \sigma_Z\sum_k g_k(b_k + b_k^+) + \sum_k \hbar\omega_k\, b_k^+ b_k \tag{15.108}$$

where I_z (for $I = 1/2$) is the spin angular momentum of the muon in the direction in which the internal magnetic fields are taken to act. As customary, σ_Z represents a pseudo spin operator, the eigenstates (+, −) of which are the two minima of the well. The Hamiltonian above is a slight generalization of the spin boson Hamiltonian for asymmetric tunneling [cf. Equation (15.4)] in which the asymmetry term is replaced by the dynamic field term HI_z. We make the usual polaronic transformation that reduces the Hamiltonian to

$$\mathcal{H} = HI_z\sigma_Z + (B_+\sigma^- + B_-\sigma^+) + \sum_k \hbar\omega_k\ b_k^+ b_k\ . \tag{15.109}$$

In this form, it is convenient to assess the Hamiltonian in terms of an effective stochastic model. Because $\sigma^{\pm}$ causes transitions between the eigenstates of σ_Z, we may use an interaction picture in which the Hamiltonian above may be viewed as explicitly stochastic:

$$\mathcal{H} = H(t)I_z, \tag{15.110}$$

where $H(t)$ is a two-state telegraph process in which $H(t)$ jumps at random between the values $\pm H$. The randomness of $H(t)$ has its genesis in the quantum fields $B_\pm$, that are further excited by the bath Hamiltonian.

As in the neutron scattering of hydrogen, we compute a spin correlation function, but of the transverse component I_x. Following our earlier treatment of the polaronically transformed Hamiltonian through the resolvent method (that gives identical results as the dilute bounce gas approximation of an underlying path integral formulation), we find:

$$\tilde{\mathrm{C}}_{\mathrm{xx}}\ (\mathrm{s}) = \left[\mathrm{s} + \frac{\mathrm{H}^2}{\mathrm{s} + 2\cos(\pi\mathrm{K})\tilde{\mathrm{J}}(\mathrm{s})}\right], \tag{15.111}$$

where $\tilde{\mathrm{J}}(\mathrm{s})$ is again given by Equation (15.77). We now have a jump rate that is frequency-dependent and given by $\lambda(s) = 2\cos\pi K\ \tilde{J}(s)$. The frequency dependence of $\lambda(s)$ would yield non-Lorentzian line shapes. To make a comparison with the prediction of the stochastic models, we ignore the frequency dependence and write the high-temperature form of λ as

$$\lambda(K,\beta) \simeq \frac{4\Delta^2}{\omega_c}\left(\frac{2\pi}{\beta\omega_c}\right)_{2K-1} \frac{\Gamma(1-2K)\Gamma(K)\cos(\pi K)}{\Gamma(1-K)}. \tag{15.112}$$

Note that we have taken only the electronic contribution to the jump rate (and omitted the phononic contribution included in the neutron scattering case) and restricted ourselves to Ohmic dissipation. Consequentially the temperature dependence of the rate can be read as

$$\lambda(K,T) \sim (T)^{2K-1}, \tag{15.113}$$

which is a power law distinct from the Arrhenius form of exponential dependence. This analysis again underscores the point that quantum jump models are richer in structure than their classical cousins for predicting different line shapes and also providing intricate temperature dependence of the relaxation rates related to jump rates.

References

Ashcroft, N.W. and N.D. Mermim. 1976. *Solid State Physics.* New York: Holt, Rinehart & Winston.

Blume, M. 1968. Stochastic theory of lineshape: Generalization of the Kubo–Anderson model. *Phys. Rev.* 174, 351.

Dattagupta, S. and H. R. Schober. 1998. Neutron scattering and diffusion. *Phys. Rev.* B57, 7606.

Dattagupta, S. and S. Puri. 2004. *Dissipative Phenomena in Condensed Matter: Some Applications.* Heidelberg: Springer.

Dattagupta, S., H. Grabert, and R. Jung. 1989. *J. Phys. Condens. Matter* 1, 1405.

Firsov, Y.A. 1975. *Polarons.* Moscow: Nauka.

Flynn, C. P. and A. M. Stoneham 1970. Electrical phenomena in oxide films. *Phys. Rev.* B1, 3966, 19.

Ford, G.W., M. Kac, and P. Mazur. 1965. *J. Math. Phys.* 6, 504.

Gangopadhyay, G., M. S. Kumar, and S. Dattagupta. 2001. On dissipationless decoherence. *J. Phys.* A34, 5485.

Glauber, R. J. 1963. Time dependent statistics of the Ising model. *J. Math. Phys.* 4, 294.

Holstein, T. 1959a. Studies of the polaron motion. Part I: The molecular-crystal model. *Ann. Phys.* 8, 325.

Holstein, T. 1959b. Studies of the polaron motion. Part II: The "small" polaron. *Ann. Phys.* 8, 343.

Lang, I. J. and Y. A. Firsov. 1962. Polarons. *Sov. Phys. JETP* 16, 1301.

Kamil, E. and S. Dattagupta. 2012. *Pramana J. Phys.* 79, 357.

Kubo R. 1962. In ter Haar, D., Ed., *Fluctuation Relaxation and Resonance in Magnetic Systems.* Edinburgh: Oliver & Boyd, p. 23.

Lang, I.J. and Y.A. Firsov. 1962. Polarons. *Sov. Phys. JETP* 16, 1301.

Leggett, A. J., S. Chakravarty, A. T. Dorsey, M. P. A. Fisher, A. Garg, and W. Zwerger. 1987. Dynamics of the dissipative two-level system. *Rev. Mod. Phys.* 59, 1.

Mahan, G.D. 1980. *Many-Particle Physics.* New York: Plenum Press.

Mozyrsky, D. and V. Privman, 1998. Adiabatic decoherence. *J. Stat. Phys.* 91. 787.

Qureshi, T. and S. Dattagupta, 1993. Quantum diffusion of muons in metals. *Phys. Rev.* B47, 1092.

Weiss, U. 1999. *Quantum Dissipative Systems.* Singapore: World Scientific.

Zwanzig, R. 1973. Nonlinear generalized Langevin equations. *J. Stat. Phys.* 9, 215.

16

Quantum Diffusion: Decoherence and Localization

16.1 Introductory Comments

Quantum mechanics rests on the axiom of wave particle duality. A point particle in the classical sense is a wave packet with a de Broglie wavelength. The latter increases as the particle mass and the temperature decrease. Thus quantum mechanics predicts an extended wave packet for tiny subatomic particles, especially when the temperature is low.

The bigger the wavelength, the more significant the quantum effects. We saw evidence of this in Chapter 15 in the context of tunneling behaviors of muons and hydrogen atoms. "Tunneling" is a euphemism for the extended nature of a quantum wave—the tail of the wave function extends to regions otherwise forbidden classically and that constitutes tunneling. Accompanying the wave nature of a quantum particle is the phase coherence of the Schrödinger wave function. The essential ideas of coherence are already familiar to us in the context of interference of electromagnetic waves.

The purpose of this concluding chapter is to remark on how quantum diffusion, the subject of the second half of this book, leads to decoherence of an otherwise coherent quantum system. These considerations are of great significance in the modern context of information processing and quantum computing.

One other important area of condensed matter physics in which quantum diffusion plays a central role is localization. The Schrödinger waves of electrons in a perfect crystal are extended as governed by Bloch waves. Again the underlying reason is constructive interference of electron waves scattered by almost rigid nuclei. However, if a crystal is spatially disordered because of ubiquitous imperfections, multiple scattering, tantamount to momentum-changing collisions, occur (in the simplified Langevin picture) and thus yield quantum diffusion.

The concomitant dissipation that we designated as quantum friction is of course at the heart of the material property of resistance. However, if the disorder is strong (or the dimension is lower) the Schrödinger waves are not endowed with Gaussian tails; they are localized "lumps" that led Anderson to term it the absence of quantum diffusion. Quantum localization is another manifestation of destructive interference of electron waves in a strongly incoherent regime. We will make some brief remarks on this important aspect of quantum diffusion as well.

The basic idea of quantum interference and phase coherence can be introduced by the rudimentary example of an electron wave characterized by a wave vector k scattered by two fixed scatterers at positions x_1 and x_2. The intensity of the scattered intensity is the square of the mod of the scattered amplitude, that is, the linear superposition of the two scattered components as

$$\mathrm{I}\,(\mathbf{k}) = |\mathrm{A}(\mathbf{k})\,[\exp\,(\mathrm{i}\mathbf{k}.\mathbf{x}_1) + \exp\,(\mathrm{i}\mathbf{k}.\mathbf{x}_2)]|^2 \tag{16.1}$$

The intensity can be split into two parts:

$$\mathrm{I}\,(\mathbf{k}) = \mathrm{I}_1(\mathbf{k}) + \mathrm{I}_2(\mathbf{k}), \quad \mathrm{I}_1(\mathbf{k}) = 2|\mathrm{A}(\mathbf{k})|^2, \quad \mathrm{I}_2(\mathbf{k}) = 2|\mathrm{A}(\mathbf{k})|^2 \cos(\mathbf{k}.\,\mathbf{r}), \tag{16.2}$$

where $\boldsymbol{r}$ is the vector distance between the two rigid scatterers. Note that $I_1(\boldsymbol{k})$ would have been simply obtained by adding the two probabilities (the classical contribution). The second term, however, leads to periodically oscillating (in space) of the intensity with minima and maxima governed by the magnitude of the argument of the cosine function. Such trigonometric functions are the hallmarks of quantum phase coherence, whether spatial or temporal; see below. However, if we have a frozen and random distribution of scattering pairs, as may obtain in a glassy system, we must perform an average over the values of $|\mathbf{r}|$. If the latter is governed by a normal distribution, it is well known that the cosine function will go into a Gaussian function $\exp[-1/2\,(k^2 < r^2 >)]$, leading to localization.

A different example of localization is what we encountered in Chapter 15 in the context of tunneling in a symmetric double well and its decoherence as described by the spin boson model. At ultra-low temperatures when the relevant Hilbert space of the double well shrinks to the two lowest states of the well, the mean position of a particle positioned at one of the two minima is given by $< x(t) > = a \cos\,(\Delta t)$ where Δ is the so-called tunneling frequency that may be given by the Wentzel–Kramers–Brillouin (WKB) approximation (Messiah 1961):

$$\Delta \sim [(8\omega_0 V_0)/\pi]^{1/2} \exp(-2V_0/\omega_0),\ \hbar = 1, \tag{16.3}$$

where V_0 is the barrier height and ω_0 the small oscillation frequency in each well. Again, a trigonometric function characterizes coherence which is what quantum tunneling is all about. On the other hand, when quantum diffusion sets in, it may be accentuated by a bosonic bath, as discussed earlier in the context of hydrogen or muon diffusion in metals.

When the coupling with the environment exceeds a certain value (as we saw in the quasi-elastic scattering regime for the hydrogen problem), the mean position is exponentially damped, $< x(t) >= a\exp[-(\lambda t)$, where the damping constant depends on the tunneling frequency Δ, the temperature T, and the K coupling with the environment. This localization transition in a symmetric double well as the coupling constant K in the spin boson model is tuned, is known in other contexts as the quantum Zeno paradox (Misra and Sudarshan 1977) and the watched-pot effect (Simonius 1978).

Localization can occur from spatial or temporal (i.e., thermal) disorder, the latter characterizing dissipative tunneling. Quantum localization, in the Anderson sense, combines both these aspects as we saw in the CTRW model of quantum Brownian motion of a free particle (Chapter 14). In the diffusive regime, the probability (or diagonal component of the underlying density matrix) is $\exp[-(x - x_0)^2/2Dt]$ leading to exponential localization.

With this background we shall quantify coherence-to-decoherence transition, occasioned by quantum diffusion in certain specific situations as we already introduced in this text.

16.2 Landau to Bohr–van Leeuwen Transition of Diamagnetism

As we know, when a system of so-called nondegenerate electrons is subjected to an external magnetic field, their collective response is given by the diamagnetic moment calculated by Landau. The expression reads (Landau 1930, van Vleck 1932):

$$M_z^0 = \frac{|e|}{2mc}\left[\frac{1}{\upsilon_c} - \coth(\upsilon_c)\right]. \tag{16.4}$$

Recalling that υ_c equals $\hbar\beta\Omega_c/2$, Equation (16.4) vanishes in the classical limit $\hbar \to 0$, $T \to \infty$, which is the Bohr–van Leeuwen result. Since the Landau result emerges as a consequence of coherent cyclotron motion of an electron's velocity vector due to Lorentz force and the Bohr–van Leeuwen theorem is entirely classical, we ask whether we can transit to the classical arena from the Landau regime when we tune the coupling constant K or the corresponding quantum friction. Note that the exact result derived earlier from quantum Langevin equations (Chapter 12) and the path integral formulation (Chapter 13) for dissipative diamagnetism can be written as

$$M_z = -\frac{2K_BT}{B}\;\Omega_c^2\sum_{n=1}^{\infty}\frac{1}{(\upsilon_n+\gamma)^2+\Omega_c^2}, \tag{16.5}$$

where

$$\upsilon_n = 2KTn\pi. \tag{16.6}$$

The remarkable aspect of the exact relation of Equation (16.5) is that the friction coefficient appears explicitly in an equilibrium formula derivable from thermodynamics. This is not normally the case in weak coupling treatments of system-plus-bath approaches to quantum dissipative systems. Indeed, diamagnetism is a thermodynamic property (it can be derived from the Gibbsian partition function, Chapter 13) and also has attributes of a transport property (implicit in the Langevin treatment of the average orbital moment, Chapter 12).

In that sense, diamagnetism is reminiscent of the Drude formula for electrical conductivity (Ashcroft and Mermin 1976). Although diamagnetism is an asymptotic property, it depends on dissipative parameters such as the mean free times of collisions. Motivated by this comparison we rewrite Equation (16.5) as

$$-\frac{mc}{|e|}M_z = \frac{1}{\upsilon_c}\sum_{n=1}^{\infty}\frac{1}{1+(\bar{\mu}_n + r)^2}, \tag{16.7}$$

where we introduce the dimensionless parameter:

$$\bar{\mu}_n = \frac{n\pi}{\upsilon_c}. \tag{16.8}$$

The quantity r is a scaled resistance—the ratio of the Drude resistivity R_D to the Hall resistivity R_H:

$$r = R_D/R_H, \quad R_D = M\gamma/n_e e^2, \quad R_H = B/n_e ec, \tag{16.9}$$

where n_e is the number of conduction electrons per unit volume. Note that $r = \gamma/\upsilon_c$, where γ was introduced earlier as dimensionless dissipation. We are now set to examine the central question posed at the beginning of this section by plotting the left side of Equation (16.7) as a function of r for two values of the dimensionless cyclotron frequency.

As one would expect, a larger value of r implies stronger decoherence and therefore a more rapid transition from the Landau-to-Bohr–van Leeuwen regions. The rapidity of transition, of course, depends on the value of the cyclotron frequency. A larger value would make coherence persist longer, as exemplified in Figure 16.1.

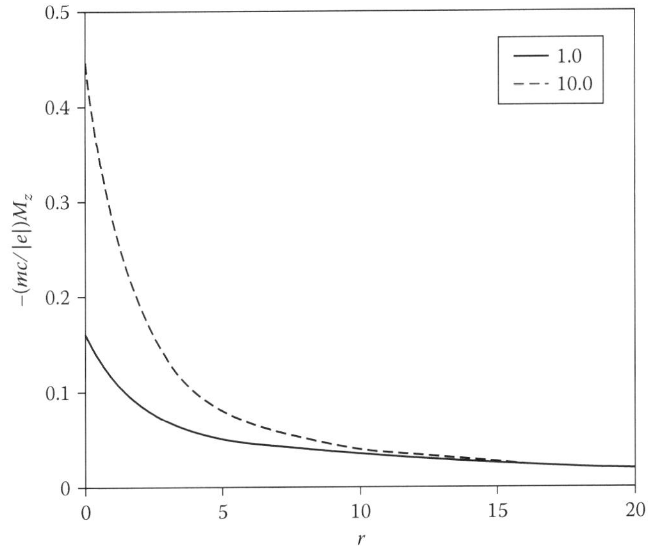

FIGURE 16.1
Dimensionless magnetization $-(mc/|e|)M_z$ versus scaled resistance r for two distinct values of reduced cyclotron frequency $\nu_c = 1.0$ and 10.0. The case $r = 0$ corresponds to the Landau limit and $r = \infty$ corresponds to the Bohr–van Leeuwen limit.

16.3 Harmonic Oscillator in Quantum Diffusive Regime

In this and the next subsection, we provide criteria to quantify the crossover from coherence to decoherence. We note that the criteria are not universal in that they depend on the specific physical quantity focused on. We utilize zero temperature ($T = 0$) wherein quantum effects are the most prominent. Insofar as our generic model of quantum dissipation is concerned, decoherence at $T = 0$ involves zero point fluctuations of the quantum harmonic oscillators that constitute the environment. Averaging then over the ground quantum state at $T = 0$ denoted by angular brackets, the equation of motion for the damped oscillator reads

$$\frac{d^2}{dq^2} < q(t) > + \gamma \frac{d}{dt} < q(t) > + \frac{k}{m} q < t > = \frac{\xi(t)}{m}. \tag{16.10}$$

The quantity of interest we examine first is the Fourier transform of the positional correlation function (Dattagupta and Puri 2004, Egger 1997):

$$S(\omega)\,(1/|\omega|)\,C(\omega), \tag{16.11}$$

where

$$C(\omega) = \mathrm{Re} < q(t)q(0) > . \tag{16.12}$$

The result is

$$S(\omega) = \gamma\omega_0\, [(\omega_0^2 - \omega^2)^2 + \gamma^2\omega^2]^{-1}. \tag{16.13}$$

The signature of coherence embedded in $S(\omega)$ is the occurrence of two inelastic peaks (observed in neutron scattering for H in Nb) at $\omega_m = \pm\,\omega_0(1-k^2/2)^{1/2}$ for weak damping where k is a dimensionless parameter defined by

$$\kappa = \gamma/\omega_0. \tag{16.14}$$

Note that κ can be viewed as a crossover parameter for coherence-to-decoherence transition. When $\kappa \to \kappa_c$ (quantified below), $S(\omega)$ is expected to yield a quasi-elastic line. Since the latter is centered around $\omega = 0$, we expand $S(\omega)$:

$$S(\omega) = \left(\frac{\kappa}{\omega_0^2}\right)\left[1 + (2-\kappa^2)\frac{\omega^2}{\omega_0^2} + O(\omega^4)\right]. \tag{16.15}$$

Hence, the curvature of $S(\omega)$ is positive, implying coherence for $\kappa < \sqrt{2}$, but changes sign as κ crosses the critical value $\kappa_c = \sqrt{2}$. Evidently that is also the value of κ for which the location of the inelastic peaks approaches zero frequency. Interestingly and in accordance with these facts, we find that the criterion for decoherence is different if we look at a quantity other than the correlation function, i.e., the decay of the mean position <q(t)> from a non-equilibrium initial value of q_0. The corresponding physical attribute, again measurable in an experiment, is

$$P(t) = < q(t) > / q_0 = \frac{[\cos(\Omega t - \xi)]}{\cos\xi} \exp[-\gamma t/2], \tag{16.16}$$

where

$$\Omega = \left[\omega_0^2 - \frac{\gamma^2}{4}\right]^{1/2}, \quad \xi = \tan^{-1}\left(\frac{\gamma}{2\Omega}\right). \tag{16.17}$$

The signature of coherence is now-damped oscillatory behavior for weak damping, ($\gamma^2 < 4\omega_0^2$ *or* $\kappa < 2$), that crosses over to relaxation behavior at a critical value $\kappa_c{}^* = 2$, such that

$$\frac{\kappa_c}{\kappa_c{}^*} = \frac{1}{\sqrt{2}}, \tag{16.18}$$

connecting the crossover criteria for the two quantities $S(\omega)$ and $P(t)$.

16.4 Decoherence in Spin Boson Model

Our next example is the symmetric $\in = 0$ spin boson model at $T = 0$. Recall that the physical quantities relevant for neutron scattering and muon spin rotation are, respectively, the equilibrium correlation function:

$$C(t) = < \sigma_z(t)\sigma_z(0) >_{eq}, \tag{16.19}$$

and the decay function:

$$P(t) = \frac{1}{2}[1 + < \sigma_z(t) >], \quad P(t = 0) = 1. \tag{16.20}$$

As before, we define $S(\omega) = C(\omega)|\omega|^{-1}$ which now has the low-frequency expansion:

$$S(\omega) = K\pi\chi_0^2[1 + g(K)\chi_0^2\omega^2 + O(\omega^4)], \tag{16.21}$$

where K is the damping coefficient, χ_0 is the static susceptibility, and $g(K)$ is a dimensionless parameter. The function $g(K)$ was calculated earlier by renormalization group methods (Costi and Kieffer 1996) and form factor methods (Lesage et al. 1996), and found to switch sign at $K_c = 2/3$.

Finally, the transition of $P(t)$ from coherent to incoherent relaxation occurs within a dilute bounce gas approximation at $K_c{}^* = 1$, according to Weiss (1993) and is an exact result—the so-called Toulouse limit (Leggett et al. 1987). Again, as in the oscillator case, we examine the critical ratio $K_c / K_c{}^*$:

$$\frac{K_c}{K_c{}^*} = \frac{2}{3}, \tag{16.22}$$

which, not surprisingly, is different from the corresponding result for the harmonic oscillator case. This underscores the fact that there is no universal limit for coherence-to-decoherence transition. Each limit is specific to the physical attribute under consideration. Memory effects and preparation of initial states are crucial for quantum phenomena.

16.5 Dissipationless Decoherence

Certain models exhibit no energy exchange between the system of interest and the heat bath. That implies no dissipation although decoherence may occur. The issue can be illustrated in the context of classical diffusion processes with the aid of a simple example from line shape studies (Dattagupta 1987). Consider a nuclear spin I subject to a stochastic field $H(t)$ in addition to a laboratory field H_0, both assumed to be along the space-fixed z axis, such that the Hamiltonian can be written in the explicit time-dependent form as

$$\mathcal{H}(t) = H_0 I_z + H(t) I_z. \tag{16.23}$$

Note that $\mathcal{H}(t)$ commutes with itself at all times and hence may be diagonalized once and for all in a representation in which I_z is diagonal. However if we consider the precessional motion of a transverse component of the spin, say I_x, as in a nuclear magnetic resonance experiment, the frequency of Larmor precession around the z axis may be described by a jump-diffusive process of telegraph, Kubo–Anderson, kangaroo, or other type (Chapter 6). The result is a dephasing of the transverse component of the nuclear spin in the absence of any energy exchange between the nuclear spin and the surrounding bath.

In proper quantum mechanical terms and in pursuance of our generic model of quantum diffusion in which the system of interest is embedded in a bosonic heat bath, the underlying physics may be described by a Hamiltonian (Gangopadhyay et al. 2001):

$$\mathcal{H} = \mathcal{H}_s + \mathcal{H}_s \sum_k g_k \,(b_k + b_k^+) + \sum_k \hbar\omega_k \, b_k^+ b_k. \tag{16.24}$$

We employ the now familiar unitary transformation [cf. Equation (15.48)] that now reads:

$$\mathcal{S} = \exp\left[\mathcal{H}_s \sum_k \frac{g_k}{\omega_k} \,(b_k^+ - b_k)\right]. \tag{16.25}$$

The transformed Hamiltonian can be written

$$\tilde{\mathcal{H}} = \mathcal{S}\mathcal{H}\mathcal{S}^{-1} = \mathcal{H}_s + \sum_k \hbar\omega_k b_k^+ b_k, \tag{16.26}$$

The von Neumann density operator (in the Schrödinger picture) can be written

$$\rho(t) = \mathcal{S}^{-1} e^{-i\tilde{\mathcal{H}}t} \,\mathcal{S}\rho_B \rho_s(0)\, \mathcal{S}^{-1} e^{i\tilde{\mathcal{H}}t} \mathcal{S}. \tag{16.27}$$

where we performed the usual factorization approximation in that the system is initially viewed as decoupled from the bath such that

$$\rho(0) = \rho_B \otimes \rho_s(0), \quad \rho_B = \frac{1}{Z_B}\exp(-\beta \mathcal{H}_B). \tag{16.28}$$

As stated earlier we can work in a representation in which $\mathcal{H}_s$ is diagonal:

$$\mathcal{H}_s \,|\, n> = E_n \,|\, n>, \tag{16.29}$$

and hence the off-diagonal elements (that are the only relevant elements now as shown at the outset in terms of the transverse component of the nuclear spin in the Heisenberg representation),

$$\rho_{nm}(t) = \mathcal{S}_n^{-1} e^{-i\tilde{\mathcal{H}}_n t} \mathcal{S}_n \rho_B \rho_{s,nm}(0)\ \mathcal{S}_m^{-1}\ e^{i\tilde{\mathcal{H}}_m t} \mathcal{S}_m \tag{16.30}$$

where

$$\mathcal{S}_n = \exp\left[E_n \sum_k \frac{g_k}{\omega_k}(b_k^+ - b_k) \right], \tag{16.31}$$

and

$$\tilde{\mathcal{H}}_n = E_n + \sum_k \hbar\omega_k b_k^+ b_k. \tag{16.32}$$

We are now ready to write the matrix elements of the reduced density matrix (after tracing over the bath states) as

$$\rho_{s,nm}(t) = Tr_B\ [\rho_{nm}(t)]. \tag{16.33}$$

Upon employing the cyclic property of the trace over the bath states, the time evolution of the bosonic operators b_k^+ and b_k, the treatment of the commutator as a c number, and some algebra, we find

$$\rho_{nm}(t) = \exp[-i(E_n - E_m)t]\ \exp[i(E_n^2 - E_m^2)\eta(t)] \times \exp[-(E_n - E_m)^2\ \gamma(t)]\,\rho_{s,nm}(0). \tag{16.34}$$

where

$$\eta(t) = -\sum_k \left(\frac{g_k}{\omega_k}\right)^2 \sin(\omega_k t), \tag{16.35}$$

and

$$\gamma(t) = \sum_k \left(\frac{g_k}{\omega_k}\right)^2 \sin^2\left(\frac{\omega_k t}{2}\right) \coth\left(\frac{\beta\hbar\omega_k}{2}\right). \tag{16.36}$$

Equation (16.34) yields a master equation:

$$\frac{d}{dt}\rho_{nm}(t) = \left[-i(E_n - E_m) + i\dot{\eta}(t)\ \left(E_n^2 - E_m^2\right) - \dot{\gamma}(t)\ (E_n - E_m)^2\right]\rho_{nm}(t). \tag{16.37}$$

Apart from the crucial presence of the $\dot{\eta}$ term, the master equation is of the Born–Markov–Lindblad form (Gorini et al. 1976). The explicit expressions for $\eta(t)$ and $\gamma(t)$ can be derived by going to a continuum of bath states:

$$\sum_k g_k^2\ f(\omega_k) \to \int_0^\infty d\omega\ j(\omega)\ f(\omega). \tag{16.38}$$

Using an Ohmic form of the spectral density, as before:

$$j(\omega) = K\omega\ e^{-\omega/\omega_c}, \tag{16.39}$$

we find

$$\eta(t) = K\tan^{-1}(\omega_c t), \quad \dot{\eta}(t) = \frac{K\omega_c}{1+(\omega_c t)^2}. \tag{16.40}$$

On the other hand, an analytical formula is not available for $\gamma(t)$ and it must be evaluated numerically. Our results are illustrated for the case of a harmonic oscillator for which $\mathcal{H}_s = \hbar\omega_0\ [(a + a^+) + 1/2]$. Decoherence is characterized by the survival probability:

$$P(t) = Tr\ [\rho_s(0)\ \rho_s(t)]. \tag{16.41}$$

where, for instance, the initial state may be assumed to be coherent. Choosing the coherent state as the Fock state |n = 1>, $P(t)$ is plotted as a function of log $(\omega_c t)$ for two different values of the ratio $\omega_0/D = 0.1$ and 100, respectively. The other chosen parameters are $(\hbar\beta\omega_0)^{-1} = 0.1$ and $\pi K\omega_0^2 = 0.1$. The oscillations in $P(t)$ are signatures of coherence. At higher temperatures, however, the oscillations die out when the system enters an incoherent regime.

16.6 Retrieving Quantum Information Despite Decoherence

One of the very active areas of research today is information that can be processed in quantum computers. Such information is usually stored in qubits represented as linear superpositions of two basis states:

$$|\psi_0> = \cos\alpha\,|L> + e^{i\gamma}\sin\alpha\,|R>, \tag{16.42}$$

where α and γ are real, and L and R represent the left and right dots, respectively. A single electron, put in one of the two dots initially, can exist in a superposed state given by Equation (16.42). The Hamiltonian of the qubit in a tight-binding language (Ashcroft and Mermin 1976) can be written as

$$\mathcal{H}_q = \epsilon_L\, a_L^+ a_L + \epsilon_R\, a_R^+ a_R - (J_{LR} a_L^+ a_R + H.C.), \tag{16.43}$$

where ϵ_L and ϵ_R are dot energies while J_{LR} is inter-dot tunneling and both may be tuned. The essential ideas of our discourse can be captured by the simple case in which ϵ_L is considered the same as ϵ_R, the so-called symmetric case. Apart from a constant energy term, the Hamiltonian reduces to

$$\mathcal{H}_q = -(J_{LR} a_L^+ a_R + J_{LR}{}^*\, a_R^+ a_L). \tag{16.44}$$

Quantum computation has the prerequisite of quantum coherence such that the state stored in each qubit can stay stable. However, a qubit is a miniature system inevitably in contact with its environment that acts as a bath in the sense of diffusion and dissipation—the central theme of this book. The result of this coupling is usually total decoherence and loss of information encapsulated by the asymptotic form of the reduced density operator that reduces to a diagonal form with constant coefficients. Recall that the reduced density operator is obtained by tracing the full density operator over the bath quantum states, thus $\rho_{s,nm}(t = \infty) = \delta_{nm}/2$ in the fully mixed state, independent of the basis comprising the underlying Hilbert space.

We will discuss in this subsection certain models of decoherence similar to the one described in Section 16.5 in which the system Hamiltonian commutes with the total Hamiltonian and no dissipation ensues. We will show that by exploiting certain inherent symmetry we can maintain partial decoherence. One crucial difference between the present and the previous analysis is that now we will treat the bath in terms of jump diffusion models of the telegraph process type. Our model Hamiltonian may then be constructed as

$$\mathcal{H}_q\ (t) = \mathcal{H}_q\ + V\ f(t)\ , \tag{16.45}$$

where

$$V = -(\zeta_J a_L^{+} a_R + \zeta_J{}^{*}\ a_R^{+} a_L). \tag{16.46}$$

Here ζ_J is a parameter and *f(t)* is a telegraph process. The condition under which $[H_q, V] = 0$ can be easily ascertained as

$$J_{LR}\ \zeta_J{}^{*} - J_{LR}{}^{*}\ \zeta_J = 0, \quad \text{or} \quad \zeta_J / J_{LR} = \zeta_J{}^{*} / J_{LR}{}^{*} = x, \tag{16.47}$$

where x is a real number. The last condition implies that if J_{LR} is written as $|J|\ \exp(i\theta)$, the same θ must appear as the phase of ζ_J, i.e., $\zeta_J = |\ \zeta|\ \exp(i\theta)$. The parameters J_{LR} and ζ_J can be made complex by inserting an Aharonov–Bohm flux on the two dots (Aharony et al. 2010). Further, the stipulation on the ratio of J_{LR} and ζ_J can be achieved by tuning gate voltages. The condition (16.47) entails that $\mathcal{H}_q$ and V can be simultaneously diagonalized among the basis states designated the bonding and anti-bonding states in the chemical physics literature, given by

$$|\pm\rangle = (\,|L\rangle\ \mp\ \exp(-i\ \theta)\ |R\rangle)/\sqrt{2}. \tag{16.48}$$

The corresponding eigenvalues are

$$E_{q,\pm} = \pm\ J, \quad (V)_{\pm} = \pm\ \zeta_J. \tag{16.49}$$

Because of telegraphic noise, the diagonal elements of the density operator (in the $|\pm\rangle$ basis) asymptotically become independent of time and the off-diagonal elements decay to zero, implying decoherence. However, the crucial point is that when transformed to the dot basis according to Equation (16.48), the off-diagonal elements also approach constant values, thus ensuring partial coherence.

We first calculate the density operator in the diagonal basis employing now familiar formulation of the telegraph process (Chapter 8). Recall that the density operator obeys the master equation (Blume 1968):

$$\frac{d}{dt}\ \rho\ (t) = -i\ [(\mathcal{H}_q\ + V\ F), \rho(t)] + W\ \rho\ (t), \tag{16.50}$$

where F is a fully diagonal 2×2 matrix in the linear vector space defined by the dichotomic telegraph process, and given by

$$F = \begin{pmatrix} 1 & 0 \\ 0 & -1 \end{pmatrix}, \tag{16.51}$$

and W is the so-called relaxation matrix that upon incorporating detailed balance can be constructed (Chapter 8) as

$$W = \lambda\,(\mathfrak{J} - I), \qquad \mathfrak{J} = \begin{pmatrix} p_1 & p_1 \\ p_2 & p_2 \end{pmatrix}, \tag{16.52}$$

where p_1 and p_2 are the a priori probabilities of the occurrence of two stochastic states $|1)$ and $|2)$ of the telegraph process. Taking the matrix elements of Equation (16.50) among the + and – states, we find

$$\frac{d}{dt} < \mu\,|\,(\rho)\,|\,\upsilon > = -i\,\{\,[(\mathcal{H}_q)_\mu - (\mathcal{H}_q)_\upsilon] + [(V)_\mu - (V)_\upsilon]\,F + W\} < \mu|\,\rho(t)]\,\upsilon>. \tag{16.53}$$

Upon Laplace transformation and taking the stochastic average of ρ (that yields the reduced density operator ρ_S), the above equation takes the familiar form

$$<\mu\,|\,\tilde{\rho}_S\,(s)\,|\,\upsilon> = \sum_{ab} p_a\,(b\,|\,\{\,(s+\lambda\,)\cdot I + i\,[(\mathcal{H}_q)_\mu - (\mathcal{H}_q)_\upsilon]\cdot I + i[(V)_\mu - (V)_\upsilon]\cdot F$$

$$+ W\,\}^{-1}|\ a)< \mu\,|\,\rho_S\ (0)\,|\,\upsilon> \tag{16.54}$$

where $\rho_S(0)$ is the initial density operator for the qubit alone and p_a is the a priori occupational probability of the stochastic state $|a)$ (a,b = 1 or 2). We define

$$[\tilde{U}(s)]_{av} = \sum_{ab} p_a\ (b\,|\,[\tilde{U}(s)\,|\,a),$$

$$[\tilde{U}_0\ (s)]_{av} = \sum_{a} p_a\ \{(s+\lambda\,)\cdot I + i\,[(\mathcal{H}_q)_\mu - (\mathcal{H}_q)_\upsilon]\cdot I + i[\,(V)_\mu - (V)_\upsilon]\cdot F_a\}^{-1}. \tag{16.55}$$

Using the usual manipulation (Chapter 8),

$$[\tilde{U}(s)]_{av} = [\tilde{U}_0(s)]_{av}]\ \{1 - \lambda\ [\tilde{U}_0(s)]_{av}\}^{-1}. \tag{16.56}$$

After straightforward algebra and inverse Laplace transformation, we deduce (Aharony et al. 2010, Itakura and Tokura 2003):

$$\rho_S^{++}(t) = \rho_S^{++}(0),\ \rho_S^{--}(t) = \rho_S^{--}(0), \tag{16.57}$$

for the diagonal elements, and

$$\rho_S^{+-}(t) = (A^+ e^{\alpha^+ t} + A^- e^{\alpha^- t})\,\rho^{+-}(0), \tag{16.58}$$

for the off-diagonal elements, where

$$A^{\pm} = \frac{\pm(\lambda - 4i\zeta_J \Delta p) + \sqrt{\lambda^2 - 16\zeta_J^2 - 8i\lambda\zeta_J\Delta p}}{2\sqrt{\lambda^2 - 16\zeta_J^2 - 8i\lambda\zeta_J\Delta p}}, \tag{16.59}$$

and

$$2\alpha^{\pm} = -\lambda \pm \sqrt{\lambda^2 - 16\zeta_J^2 - 8i\lambda\zeta_J\Delta p}\,,\quad \Delta p = p_1 - p_2. \tag{16.60}$$

Therefore $\rho_S^{+-}(t)$ oscillates and decays to zero but as emphasized earlier the important components are the physical dot states to which we must transform the respective elements of the density matrix. The real parts of $\alpha^{\pm}$ yield two decay rates, the longer one similar to $(T_2)^{-1}$ of nuclear magnetic resonance (NMR; Slichter 1963). However the other familiar timescale T_1 is actually infinity in the present case because of the absence of energy transfer with the bath. Transforming to the dot basis, we find an interesting result (Aharony et al. 2010):

$$\rho_S\ (t = \infty) = \begin{pmatrix} 1/2 & e^{i\theta}\,\mathrm{Re}[e^{-i\theta}\rho_{LR}(0)] \\ e^{-i\theta}\,\mathrm{Re}[e^{-i\theta}\rho_{LR}(0)] & 1/2 \end{pmatrix}, \tag{16.61}$$

where

$$\rho_{LR}(0) = \langle L|\psi_0\rangle\langle\psi_0|R\rangle = e^{-i\gamma}\sin(2\alpha)/2 \quad \text{[Equation (16.42)]}. \tag{16.62}$$

As claimed earlier, the off-diagonal elements of the density operator asymptotically go to nonzero values, implying partial decoherence. It is evident that partial decoherence is basis-dependent. To find a more satisfactory definition of partial decoherence in a basis-independent manner, we consider the quantity $[1 - Tr\ \rho^2] \equiv 2\ (\rho_{LL}\rho_{RR} - |\rho_{LR}|^2)$ (Aharony et al. 2010). The latter goes to one half for a fully mixed state but approaches a value exceeding one half for partial decoherence. This phenomenon is generic for models of dissipationless decoherence (also known as adiabatic decoherence; Mozyrsky and Privman 1998) in which the system part of the Hamiltonian commutes with the total Hamiltonian. It is also present when the telegraphic process $f(t)$ is treated more generally in terms of bosonic fields as in spin boson situations (Section 15.3).

16.7 Localization of Electronic States in Disordered Systems

The idea of electron localization in a random potential was first discussed by Anderson (1958) in an article titled "Absence of Diffusion in Certain Random Lattices." The connection of localization and quantum diffusion was further elucidated in Nagaoka and Fukuyama (1982). Today the localization concept plays a critical role in a plethora of important topics of the condensed phase such as metal–insulator transition, magnetism, and high-temperature superconductivity (Ramakrishnan 1983).

Broadly speaking, the interest in localization arises from the nature of motion that can be assessed from the eigenstates that represent solutions of the Schrödinger equation of a quantum particle in a static random potential, as cited early in this chapter. If the allowed regions of motion do not overlap, a classical particle would be expected to be trapped and that quantum tunneling would ensure that such islands of classically forbidden regions would be bridged by quantum tunneling. However, if these islands are randomly distributed because of the frozen randomness of the underlying potential, the interferences of the various tunneling paths (measured by tunneling amplitudes) will greatly reduce the probability of transitions from one classically allowed region to another, because the probability is the mod-square of the sum of complex amplitudes. The probability vanishes for sufficiently far regions if a system is highly disordered. Thus quantum interference terms inhibit quantum diffusion and eventually lead to localization of electronic states.

The model treated by Anderson is of the tight-binding type. The site energies are taken to be randomly distributed within a range –w to + w. Thus the hopping of an electron from site to site will be influenced by this randomness. By considering the on-site Green's function and the probability distribution of its randomness, Anderson showed that the stay-put probability of an electron at a tight-binding site is nonvanishing when disorder exceeds a certain critical value, i.e., for $(w/ZJ) > n_c$, where J is the hopping term, Z is the coordination number of the lattice, and n_c depends on electron energy and spatial dimensionality. The best numerical estimate for n_c is ~2.3 for electron energy at the center of the energy band ($E = 0$) and in three dimensions. Thus for stronger disorder when the ratio w/ZJ exceeds 2.3, the electron does not diffuse and is spatially localized. Although this number is

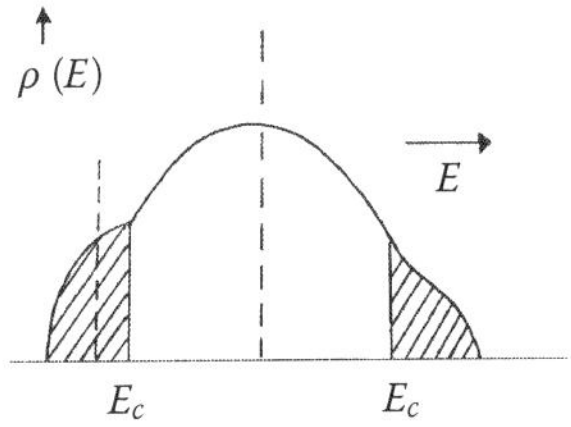

FIGURE 16.2
The density of states ρ(E) is plotted as a function of energy E. Shaded areas indicate localized regions. Dashed lines exhibit two cases: $E_F < E_c$ (insulator) and $E_F > E_c$ (metal).

specific to the Anderson model, the effect is general and occurs in all situations of random potentials.

Mott and Davis (1979) were the first researchers to worry about the connection of localization and physical properties. The underlying physics can be assessed from Figure 16.2. The shaded regions indicate localized states at the edges of the energy E_c that defines the onset of the conduction bands. The density of electronic states in a random static potential versus electron energy E is plotted. Because an electron can conduct only when its energy E exceeds its Fermi energy E_F, two situations can be distinguished: (1) E_F is inside the localized region (left shaded area) $E_F < -E_C$ or (2) E_F is in the middle of the band $E_F > -E_C$. The former case corresponds to an insulator and the latter yields a metal.

The demarcating energy is called the mobility edge because at temperature $T = 0$, all states are filled within the Fermi sphere $E_F < -E_C$. The low energy excitations are all localized and the system is an insulator. However, when E_F exceeds $-E_C$, we have a disordered metal. Mott named the consequent transition from an insulator to a metal the Anderson transition during which the (quantum) diffusion becomes more sluggish and eventually absent.

The Mott concept can be stated in physical terms. Recall that the solution of the Schrödinger equation of an electron in a periodic lattice involves scattering of electron waves from lattice sites. When a crystal is aperiodic because of disorder, scattering events will be random, characterized by random amplitudes and random phases. In physical terms, the electron can be seen to undergo random momentum-reversing collisions that constitute the essence of Brownian motion. Diffusive behavior sets in, characterized by a diffusion constant D written as

$$D = l v_F / d, \tag{16.63}$$

where l is the mean free path of electron collisions, v_F is the Fermi velocity, and d is the spatial dimensionality. The other important physical attribute is the Drude dc (frequency-independent) conductivity given by

$$\sigma = ne^2 l / (v_F \cdot m) = ne^2 l / (hk_F), \tag{16.64}$$

where k_F is the Fermi wave vector and n is the electron density. Thus both D and the sigma scale as l, the mean free path. As the collisions become more frequent (as disorder strengthens) l decreases but not below the a interatomic spacing. Thus the minimum diffusivity can be expressed as

$$D_{\min} = v_F a/d = h/\pi d, \tag{16.65}$$

where in the last term on the right side we estimated a from the minimum uncertainty relation:

$$(\hbar k_F/m)a = \hbar. \tag{16.66}$$

On the other hand, σ contains n which is dimension-dependent. Thus using Equation (16.66) and the relation between the number density and the Fermi momentum (Huang 1987), we find:

$$\text{in three dimensions: } \sigma_{\min} = (e^2 k_F)/(\hbar\pi^2), \tag{16.67}$$

$$\text{and in two dimensions: } \sigma_{\min} = e^2/h. \tag{16.68}$$

The interpretation of Equation (16.68) is that in the metallic state a nonzero minimum diffusion constant suddenly drops to zero at the Anderson transition, as in a first order phase transition. Because D_{min} depends on $\hbar$, its occurrence is of purely quantum origin, i.e., quantum diffusion. The numerical value of D_{min} is estimated to be nearly 1 cm^2/sec. The related attribute of minimum conductivity is also of quantum origin, but in three dimensions depends on material properties and in two dimensions is universal and has the dimension of conductance. Typically, for a metal with $k_F \sim (10)^8\,(\mathring{A})^{-1}$, $\rho_{\max} = (\sigma_{\min})^{-1}$ is about 1000 μOhm/cm, while in three dimensions, $R_{\max} = (\sigma_{\min})^{-1}$ is ~30 kOhm!

The fact that two is a marginal dimension for which the transport behavior changes from extended to localized can be understood easily from the simple classical-like expression for the conditional probability that an electron is found at the position r at time t, given that it was at the origin 0 at time $t = 0$. The stay-put probability is obtained by integrating this probability over all times, thus

$$P_0\,(0, r) \sim 1/(4\,\pi\,D)^{d/2} \int dt\,(\exp\,(-\,|r|^2/4Dt)\,/(t)^{d/2}. \tag{16.69}$$

Although a classical-like solution was employed above, the solution also emerges from a quantum treatment of diffusion of an electron as in the CTRW scheme in Chapter 15. It is clear that the integral diverges for $d < 2$, as it indeed would for localized states. We have thus linked the profound phenomenon of localization to the absence (or presence) of quantum diffusion.

References

Aharony, A., S. Gurvitz, O. Entin-Wohlman, and S. Dattagupta. 2010. Retrieving qubit information despite decoherence. *Phys. Rev.* B82, 245417.

Anderson, P. W. 1958. Absence of diffusion in certain random lattices. *Phys. Rev.* 109, 1492.

Ashcroft, N. W. and N. D. Mermim. 1976. *Solid State Physics*. New York: Holt, Rinehart, & Winston.

Costi, T. A. and C. Kieffer. 1996. Equilibrium dynamics of the dissipative two-state system. *Phys. Rev. Lett.* 76.

Dattagupta, S. 1987. *Relaxation Phenomena in Condensed Matter Physics*. New York: Academic Press.

Dattagupta, S. and S. Puri. 2004. *Dissipative Phenomena in Condensed Matter: Some Applications*. Heidelberg: Springer.

Egger, R., H. Grabert, and U. Weiss. 1997. Crossover from coherent to incoherent dynamics in damped quantum systems. *Phys. Rev.* E55, R 3809.

Gangopadhyay, G., M. S. Kumar, and S. Dattagupta. 2001. On dissipationless decoherence. *J. Phys.* A34, 5485.

Gorini, V., A. Kossakowski, and E. C. G Sudarshan. 1976. Completely positive dynamical semigroups of N-level systems. *J. Math. Phys.* 17, 821.

Huang, K. 1987. *Statistical Mechanics*, 2nd ed. New York: John Wiley & Sons.

Itakura, T. and Y. Tokura, 2003. Decoherence due to background charge fluctuations. *Phys. Rev.* B67, 195320.

Landau, L. D. 1930. Diamagnetism of metals. *Z. Phys.* 64, 629.

Leggett, A. J., S. Chakravarty, A. T. Dorsey, M. P. A. Fisher, A. Garg, and W. Zwerger. 1987. Dynamics of the dissipative two-level system. *Rev. Mod. Phys.* 59, 1.

Mishra, B. and E. C. G. Sudarshan. 1977. The Zeno's paradox in quantum theory. *J. Math Phys.* 18, 756.

Mott, N. F. and E. A. Davis. 1979. *Electronic Processes in Noncystalline Materials*. Oxford: Clarendon.

Mozyrsky, D. and V. Privman. 1998. Adiabatic decoherence. *J. Stat. Phys.* 91, 787.

Nagaoka, Y. and H. Fukuyama, Eds. 1982. *Anderson Localization*. Heidelberg: Springer

Ramakrishnan, T. V. 1983. In Agarwal, G. S. and S. Dattagupta, Eds., *Stochastic Processes: Formalism and Applications*, Vol. 184. Heidelberg: Springer.

Simonius, M. 1978. Spontaneous symmetry breaking and blocking of metastable states. *Phys. Rev. Lett.* 40, 980.

Slichter, C. P. 1963. *Principles of Magnetic Resonance*. New York: Harper & Row.

Van Vleck, J. H. 1932. *Theory of Electric and Magnetic Susceptibilities*. London: Oxford University.

Index

O

P

Q

T

Printed and bound by CPI Group (UK) Ltd, Croydon, CR0 4YY
23/10/2024
01778242-0014